Space Nuclear Propulsion and Power

Book 3

Space Nuclear Fission Electric Power Systems

David Buden

Space Nuclear Propulsion and Power Book 3: Space Nuclear Fission Electric Power Systems

Polaris Books
11111 W. 8th Ave., Unit A
Lakewood, CO 80215

www.polarisbooks.net

Manufactured in the United States of America

First Polaris Books Edition 2011

ISBN 978-0-9741443-4-4
Library of Congress Control Number: 2011933139

Cover image of Jupiter Icy Moons Orbiter provided by NASA. © NASA
Book and cover design by Alyssa Piccinni

CONTENTS

Foreword

Propulsion and power are defining technologies that enable man to perform more demanding space missions. In 1985, Orbit Book Company, Inc. published *Space Nuclear Power* by Joseph A. Angelo, Jr. and David Buden. Since that time, much has happened in the field of space nuclear systems. This series of three books on space nuclear power and propulsion:

> Book 1: *Space Nuclear Radioisotope Systems*
> Book 2: *Nuclear Thermal Propulsion Systems*
> Book 3: *Space Nuclear Fission Electric Power Systems*

brings together a summary of all of the developments in space nuclear systems.

Dr. Mohamed S. El-Genk organized the Symposium on Space Nuclear Power and Propulsion in Albuquerque, New Mexico from 1983 thru 2008. Dr. El-Genk is Regents' Professor, Chemical, Nuclear, and Mechanical Engineering, and Director of the Institute of Space Nuclear Power Systems at the University of New Mexico. A debt of gratitude is owed to Dr. El-Genk for his tireless efforts in bringing together the principle investigators in the space nuclear field and publishing symposium proceedings for each of the yearly meetings. The proceedings were extremely helpful in preparing this book.

Dr. Gary L. Bennett offered invaluable insight and assistance in preparing the manuscript of this book. He worked for NASA and the US Department of Energy (DoE) on advanced space power and propulsion systems. Dr. Bennett started his space nuclear career on the nuclear rocket program (NERVA) at NASA Lewis Research Center. Subsequently, he has held key positions in DoE's space radioisotope power program, including serving as the flight safety manager and Director of Safety and Nuclear Operations for radioisotope power sources used on Voyager 1 and 2 spacecraft to Jupiter, Saturn, Uranus, Neptune and beyond; Lincoln Laboratory's LES 8 and 9 communications satellites; and the Galileo mission to Jupiter. At NASA, he was the Manager of Advanced Space Power Systems in the transportation division of the Office of Advanced Concepts and Technology. He managed, among other things, fission nuclear propulsion activities.

Overview

Key elements in the exploration and exploitation of space are reliable and compact propulsion and power systems. Without adequate power and propulsion, space missions are severely limited. Nuclear powered systems are a key technology in meeting past and future propulsion and power needs. The current generation of nuclear systems can be organized in terms of radioisotope and fission nuclear reactor heat sources. Fission nuclear systems can be subdivided into power and propulsion applications. These are the bases for this series of three books:

> Book 1: *Space Nuclear Radioisotope Systems*
> Book 2: *Nuclear Thermal Propulsion Systems*
> Book 3: *Space Nuclear Fission Electric Power Systems*

It is vital to understand past activities in developing future programs to meet the challenges that lie ahead. In order to meet the needs and interest of individual readers, each book in this series is designed to stand-alone.

Solar energy has been the cornerstone of power systems for near-Earth missions. However, the solar flux drops as the inverse square of the distance from the Sun (See Fig. 1).[1] At Jupiter, for instance, the solar flux is only 4% of that at Earth. Solar arrays become too large and weigh too much at such distances. Also, surface operations in sun starved or shadowed areas at nearer distance cannot use solar technology. Near the Sun high temperatures also limits the use of solar technology.

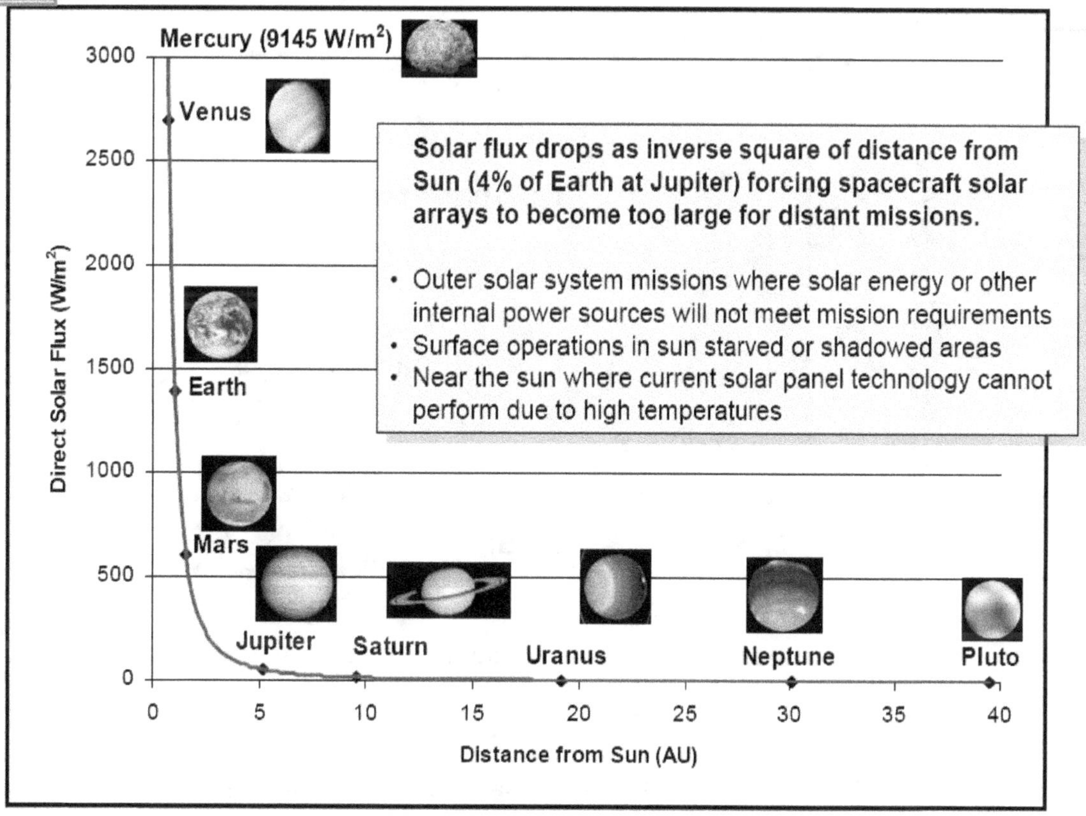

Fig. 1 Limited solar flux as distances increase from the Sun enables the need for nuclear space power systems.

To date, space propulsion has relied mainly on chemical fueled rockets. As missions with increasingly larger payloads are contemplated, the size of chemical rockets becomes unwieldy. Other more efficient propulsion systems, such as nuclear fission rockets, will be required.

Nuclear power has been used on many space missions by the United States in support of both civilian and military programs. These have taken the form of using the thermal energy from the decay of radioisotopes and converting this energy to electric power. Radioisotope power systems have proven to be highly reliable, operating for many years and in severe environments (e.g., trapped radiation belts, surface of Mars, moons of the outer planets) that make solar alternatives of limited use or unusable. Also, radioactive decay heat has been used to maintain temperatures in spacecraft at acceptable conditions for other components.

Radioisotope power systems are limited in power levels to a few kilowatts by the cost and availability of suitable radioisotope thermal heat sources. Nuclear reactors can provide tens-to-hundreds and even megawatts of power in future power systems. Extensive development works has already been performed on reactor-powered systems with the United States having flown one system in space. To meet the requirements of the more ambitious missions of the future, nuclear fission power will be a necessity.

Interest in nuclear rockets has centered on manned flights to Mars. The demands of such missions requires rockets that are several times more powerful than the chemical rockets in use today. Nuclear fission rockets have been extensively developed for this purpose. However, none of these developments have reached flight status.

The advantages of space nuclear systems can be summarized as: compact size; low to moderate mass; long operating lifetimes; operation in extremely hostile environments; operation independent of the distance from the Sun or of the orientation to the Sun; and high system reliability and autonomy.[2, 3, 4] In fact, as power requirements approach hundreds of kilowatts and megawatts, nuclear energy appears to be the only realistic power option (see Fig. 2).[5]

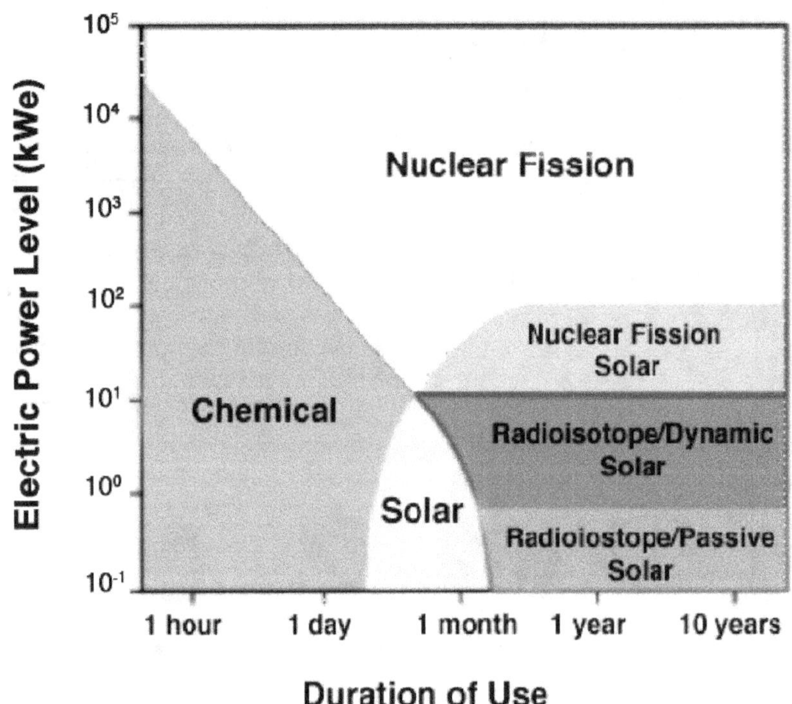

Fig. 2. Regimes of possible space power applicability.

The building blocks for space nuclear electric power system are depicted in Fig. 3. Radioisotope decay heat or the thermal energy released in nuclear fission can be converted to electrical using power generation equipment. These can take the form of either static electrical conversion elements that have no moving parts (e.g., thermoelectric or thermionic) or dynamic conversion elements (e.g., the Rankine, Brayton or Stirling cycle).   The options for nuclear energy heat sources and companion power generation subsystems are summarized in Fig. 4. Radioisotope and reactor power systems are further classified in Fig. 5.[6]

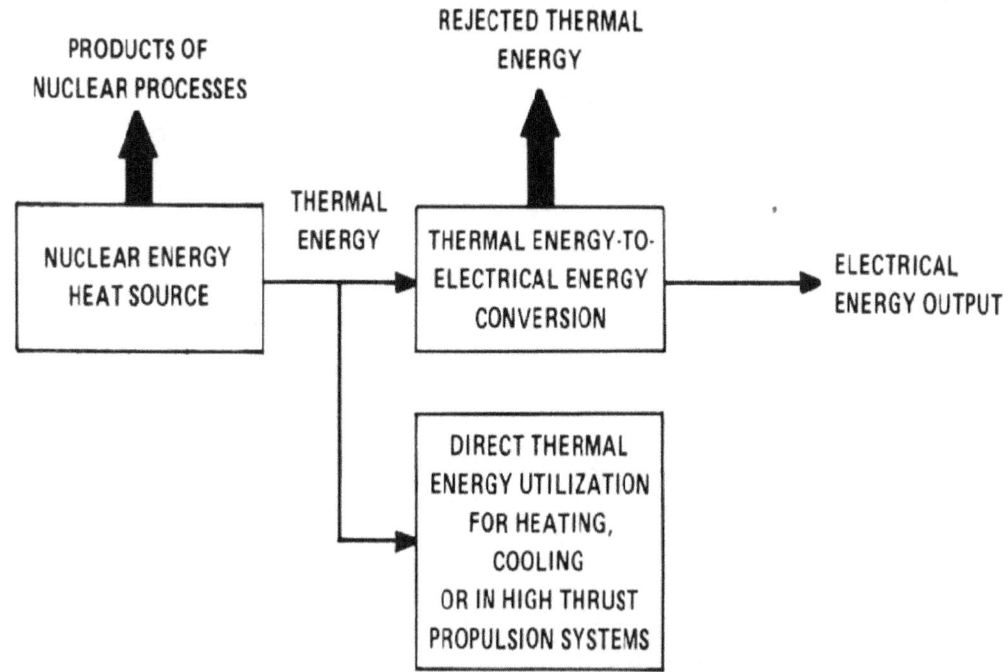

Fig. 3. Generic space nuclear power systems. *Courtesy of Los Alamos National Laboratory.*

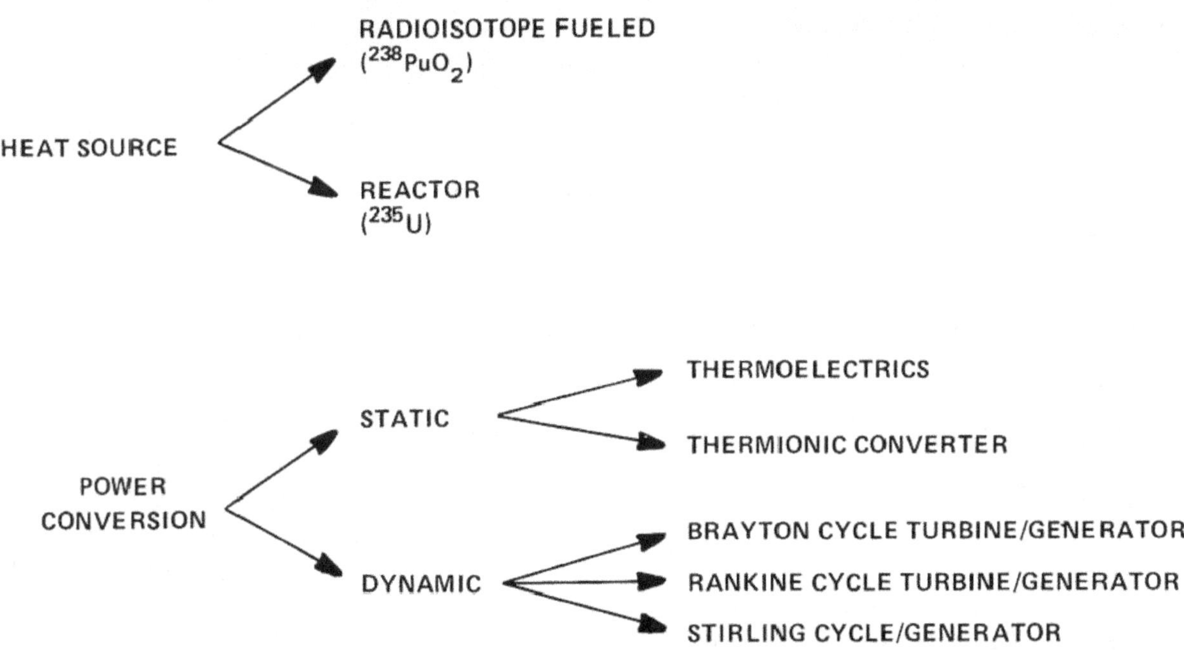

Fig. 4. Options covered in space nuclear programs. *Courtesy of Los Alamos National Laboratory*

Nuclear Power System	Electric Power Range (Module Size)	Power Conversion
Radioisotope Thermoelectric Generator (RTG)	Up to 500 We	Static: Thermoelectric

Radioisotope Dynamic Conversion Generator	0.5 to 10 kWe	Dynamics:
		  Brayton
		  Rankine
		  Stirling

Reactor Systems
  Heat Pipe
  Solid Core
  Thermionics
	10 kWe to 1,000 kWe	Static:
		  Thermoelectrics
		  Thermionics
		Dynamics:
		  Brayton
		  Rankine
		  Stirling

Reactor Systems
  Heat Pipe
  Solid Core
	1 to 10 MWe	Dynamics:
		  Brayton
		  Rankine
		  Stirling

Reactor Systems
  Solid Core
  Pellet Bed
  Fluidized Bed
  Gaseous Core
	10 to 100 MWe	Brayton Cycle (Open Loop)
		Stirling
		MHD

Fig. 5. Classification of nuclear power system types being considered for space applications.
Courtesy of Los Alamos National Laboratory.

Since 1961, the United States has launched twenty-seven National Aeronautics and Space Administration (NASA) and military space systems that derived all or part of their power requirements from nuclear energy sources. These systems are summarized in Table 1; some illustrations of nuclear powered spacecraft are shown in Fig. 6. As can be seen in this table, all but one of the previous missions used plutonium-238 as the fuel in various radioisotope thermoelectric generator (RTG) systems. The SNAP-10A was a compact nuclear fission reactor that used fully enriched uranium-235 as the fuel. The acronym SNAP stands for Systems for Nuclear Auxiliary Power, with the odd-numbered units representing radioisotope heat sources and the even-numbered units nuclear reactors.

Table 1. Summary of space nuclear power systems launched by the United States.[7, 8, 9]

Power Source	Spacecraft	Mission Type	Launch Date	Status
SNAP-3B7	Transit 4A	Navigational	6/29/1961	RTG operated for 15 years. Satellite now shutdown but operational.
SNAP-3B8	Transit 4B	Navigational	11/15/1961	RTG operated for 9 years. Satellite operated periodically after 1962 high altitude test. Last reported signal in 1971.
SNAP-9A	Transit 5-BN-1	Navigational	9/28/1963	RTG operated as planned. Non-RTG electrical problems on satellite caused satellite to fail after 9 months.

Power Source	Spacecraft	Mission Type	Launch Date	Status
SNAP-9A	Transit 5-BN-2	Navigational	12/5/1963	RTG operated for over 6 years. Satellite lost ability to navigate after 1.5 years.
SNAP-9A	Transit 5-BN-3	Navigational	4/21/1964	Mission was aborted because of launch vehicle failure. RTG burned up on re-entry as designed.
SNAP-10A REACTOR	SNAPSHOT	Experimental	4/3/1965	Successfully achieved orbit. Operated 43 days above 785 K.
SNAP-19B2	Nimbus-B-1	Meteorological	5/18/1968	Mission was aborted because of range safety destruct. RTG heat sources recovered and recycled.
SNAP-19B3	Nimbus III	Meteorological	4/14/1969	RTGs operated for over 2.5 years.
ALRH	Apollo 11	Lunar Surface	7/14/1969	Radioisotope heater units for seismic experimental package. Station was shut down 8/3/1969.
SNAP-27	Apollo 12	Lunar Surface	11/14/1969	RTG operated for about 8 years until station was shut down.
SNAP-27	Apollo 13	Lunar Surface	4/11/1970	Mission aborted on the way to the moon. RTG re-entered earth's atmosphere and landed in South Pacific Ocean. No radiation was released.
SNAP-27	Apollo 14	Lunar Surface	1/31/1971	RTG operated for over 6.5 years until station was shut down.
SNAP-27	Apollo 15	Lunar Surface	7/26/1971	RTG operated for over 6 years until station was shut down.
SNAP-19	Pioneer 10	Planetary	3/2/1972	RTGs still operating. Spacecraft successfully operated to Jupiter and is now beyond orbit of Pluto.
SNAP-27	Apollo 16	Lunar Surface	4/16/1972	RTG operated for about 5.5 years until station was shut down.
Transit-RTG	"Transit" (Triad-01-1x)	Navigational	9/2/1972	RTG still operating.
SNAP-27	Apollo 17	Lunar Surface	12/7/1972	RTG operated for almost 5 years until station was shutdown.
SNAP-19	Pioneer 11	Planetary	4/5/1973	RTGs operating. Spacecraft successfully operated to Jupiter, Saturn, and beyond.
SNAP-19	Viking I	Mars Surface	8/20/1975	RTGs operated for over 6 years until lander was shut down.
SNAP-19	Viking 2	Mars Surface	9/9/1975	RTGs operated for over 4 years until relay link was lost.
MHW-RTG	LES 8	Communications	3/14/1976	RTGs still operating.
MHW-RTG	LES 9	Communications	3/14/1976	RTGs still operating.
MHW-RTG	Voyager 2	Planetary	8/20/1977	RTGs still operating. Spacecraft successfully operated to Jupiter, Saturn, Uranus, Neptune, and beyond.
MHW-RTG	Voyager 1	Planetary	9/5/1977	RTGs still operating in 2010. Spacecraft successfully operated to Jupiter, Saturn, and in the heliosphere.

Power Source	Spacecraft	Mission Type	Launch Date	Status
GPHS-RTG	Galileo	Planetary	10/8/1989	Completed 34 orbits of Jupiter. Mission ended 21 Sept. 2003 when spacecraft was plunged into Jupiter's atmosphere.
GPHS-RTG	Ulysses	Planetary/Solar	10/6/1990	Spacecraft completed nearly three complete orbits of Sun. Operated more than 18 y, shutdown 6/30/2009.
GPHS-RTG	Cassini	Planetary	10/15/1997	RTGs still operating. Spacecraft orbiting Saturn.
GPHS-RTG	New Horizons	Planetary	1/19/2006	RTG still operating. Spacecraft en route to Pluto
MMRTG	Mars Science Laboratory	Mars Surface	Pending 2011	

The Russians have also have made extensive use of nuclear power systems in space. They launched two radioisotope systems in 1965. In the time period between 1971-1988, they launched some 35 nuclear reactor systems (see Table 2).

Table 2. USSR space power flight experience.[10]

TYPE	NAME	MISSION	NUMBER OF MISSIONS	LAUNCH DATES	STATUS	FAILURES
RTG		NAVIGATION SATELLITES	2	9/65	IN ORBIT	NONE KNOWN
RADIO-ISOTOPE HEATER UNIT		LUNAR ROVERS	4	9/69 TO 1/73	TWO SHUTDOWN ON MOON	TWO REENTRIES AFTER UPPER STAGE MALFUNCTIONS (1969)
REACTOR	RORSAT	OCEAN SURVEILLANCE	35	12/67 TO 3/88	31 SHUTDOWN AND BOOSTED TO HIGH ORBITS	- TWO LAUNCH ABORTS (1969, 1973) - TWO REENTRIES AFTER BOOST FAILURE (1977, 1982)
REACTOR	TOPAZ	OCEAN SURVEILLANCE	2	1/87 TO 10/87	SHUTDOWN AND BOOSTED TO HIGH ORBITS	NONE KNOWN

Safety has been and continues to be a key element in the development and deployment of space nuclear systems. The prevention of significant radiological risk to the Earth's population or to the terrestrial environment are guiding policies in all phases associated with space nuclear systems.[11] For radioisotope heat sources, this aerospace nuclear safety policy essentially consists of providing containment that is not prejudiced under any circumstance including launch accidents, reentry, or impact on land or water. For nuclear reactors, the safety mechanism consists of maintaining sub criticality under all conditions, normal and otherwise, in the Earth's atmosphere or on the Earth's surface. After the reactor has experienced power operation in space, the reactor will be prevented from reentering the terrestrial biosphere until the fission products and other radioactive materials no longer represent a radiological risk. Reactor operation is generally limited to orbits that have a lifetime in excess of three hundred years to support this safety approach.

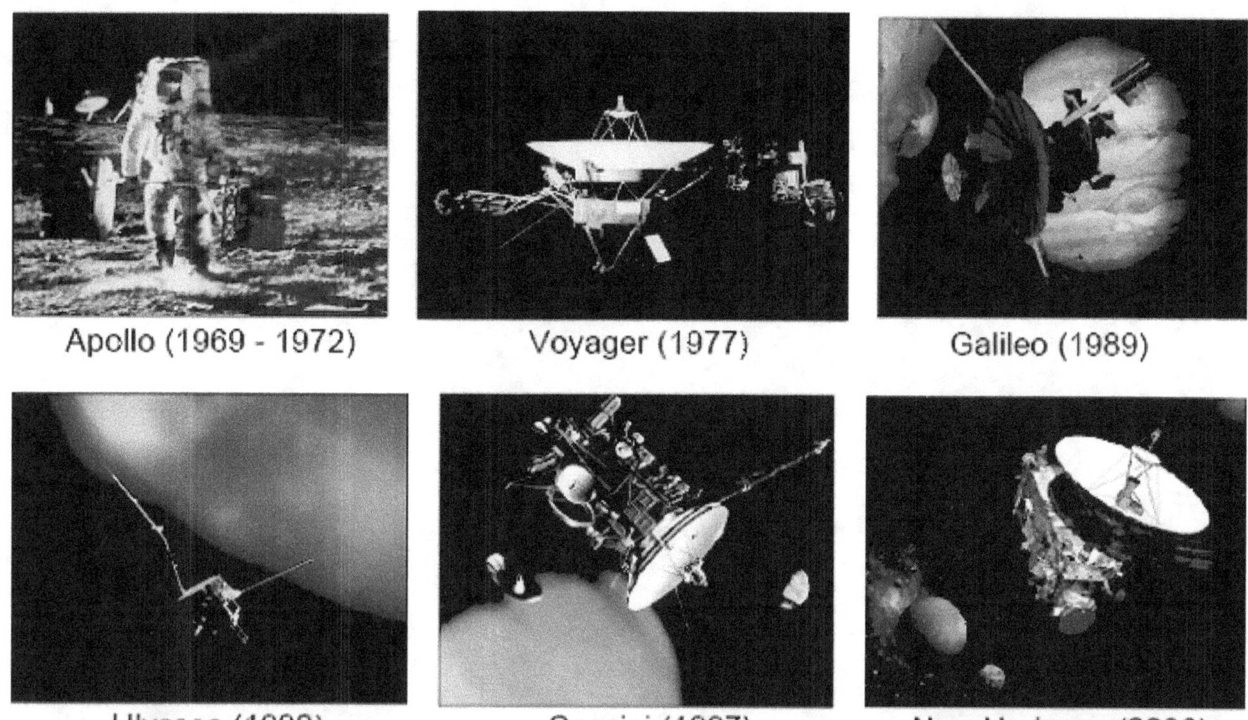

Fig. 6. Illustrations of missions using nuclear power. All missions operated far beyond their design life times. *Courtesy of Dennis Miotia.*

Chapter 1

Missions, History and Fundamentals of Space Nuclear Fission Systems

Potential Mission Power Needs

Table 1 summarizes major potential applications of space nuclear power. The table divides applications into near-Earth, solar system exploration,[1] and Lunar-Mars exploration. Near-Earth applications are further subdivided into non-proliferation and treaty verification, environmental monitoring, aviation, commercial, near-Earth defense, and near-Earth resources. The requirements presented are representative values; there are many possible optimizations for each specific application.

Table .1. Representative potential space nuclear power missions.

Mission	Key Requirements
Near-Earth	
Non-Proliferation and Treaty Verification	
Under ground measurements	To 40 kWe, lifetime 7 to 10 y
Moveable surface sensors	To 40 kWe, lifetime 7 to 10 y
Chemical, biological, nuclear (CBN) effluent monitoring	To 40 kWe, lifetime 7 to 10 y
Environmental Monitoring	
Earth observations	> 10 kWe, lifetime 7 to 10 y
Upper air turbulence	> 10 kWe, lifetime 7 to 10 y
Aviation	
Anti-collision aircraft radar	20 to 40 kWe, lifetime 7 to 10 y, high elliptical Earth orbit (HEEO)/medium Earth orbit (MEO)
Commercial	
Electronic information highway	25 to 100 kWe
Direct broadcast television	25 to 100 kWe
Near-Earth Defense	
Wide area surveillance	20 to 40 kWe, lifetime 7 to 10 y, rapid deployment, high elliptical Earth orbit (HEEO)/medium Earth orbit (MEO)
Battlefield communications	10 to 20 kWe, lifetime 7 to 10 y, rapid deployment, geosynchronous Earth orbit (GEO)
Electronic jammers	> 10 kWe, lifetime 7 to 10 y, rapid deployment, HEEO/MEO orbits
Submarine communications	10 to 40 kWe, lifetime 7 to 10 y, rapid deployment, HEEO orbit
Near-Earth Resources	
In situ probes	30 kWe, payload 1.5 Mg, lifetime 3 y

Solar System Exploration	
Neptune orbiter/probe	Payload 1.8 Mg, 100 kWe, power system mass 3.7 Mg
Pluto orbiter/probe	Payload 1.4 Mg, 56 kWe, power system mass 2.8 Mg
Uranus orbiter	Payload 1.4 Mg, 100 kWe, power system mass 3.7 Mg
Jupiter grand tour	Payload 1.4 Mg, 58 kWe, power system mass 2.9 Mg
Rendezvous	Payload 1.4 Mg, 40 kWe, power system mass 2.35 Mg
Comet Sample/Return	Payload 1.8 Mg, 100 kWe, power system mass 3.7 Mg
Lunar-Mars Exploration	
First lunar outpost	> 12 kWe
Enhanced outpost (ISRU)	> 200 kWe
Mars transportation	Flight time < 180 d, payload 52 MT
Mars stationary (600 d)	75 - 150 kWe
Mars in situ resources	> 200 kWe
Mars comsats	20 kWe

Possible missions in Earth orbit that can might need a nuclear reactor power source include performing underground measurements using ground penetrating radar or microwaves and chemical, nuclear and biological effluent monitoring using laser spectroscopy for non-proliferation and treaty verification. The Federal Aeronautics Administration has a need for oceanic anti-collision aircraft radar.

Environmental monitoring needs world-wide measurements of ozone and pollution and also upper air turbulence. Commercial uses could include a satellite component of the data information superhighway for remote and mobile sites and high definition television satellites.

Unique capabilities for defense missions are the need for rapidly deployable satellites from launch site storage in times of conflict and for redeployment of satellites already in space for better coverage of key geographic areas. Improved surveillance, communications, battlefield illumination, and electronic jammers are some of the unique systems that would be enabled by nuclear power. These require power levels of 10 to 40 kW_e with lifetimes of 7 to 10 y. The orbits-are sufficiently high to satisfy concerns about safety and possible interference with gamma ray observatories.

Near-Earth resource recovery from comets and asteroids is another possible application that might become a driver. Precursor missions are needed to establish the viability of this approach. Robotic exploration of the solar system and piloted exploration of the Moon and Mars using nuclear systems enable low cost orbiters to the outer planets.

Surface activities on the Moon will eventually require significant amounts of power estimated to be in the tens to thousands of kilowatts. Base power will be needed for habitats, mining, in-situ operations, fabrication, and power for regeneration of fuel cells used for transportation. Nuclear power systems offer lower mass and compactness compared to non-nuclear power sources. For a Moon base, the power source must operate reliably and continuously during the $\geq$14 Earth-day lunar night and over a range of temperature extremes of approximately 100 to 400 K. Fig. 1 compares solar and nuclear power systems for Lunar base applications. The nuclear systems in the range of 20 to 600 kW_e have about one-tenth the mass of the solar systems.[2]

Whereas transportation to the Moon is about a 7 day round trip, Mars transportation times involve a year plus orbital operations of up to 5 years. The spacecraft might require up to 30 kW$_e$ and thus nuclear power becomes a candidate. Habitats on Mars and in-situ operations that could require several hundred kilowatts-electrical power makes nuclear power the prime technology choice. The power source must operate reliably and continuously in dust storms and under insolation which at best is about 40% of that on Earth and through temperatures ranging from about 150 K to 310 K. Limited solar radiation as well as weight drives the need for nuclear power systems.

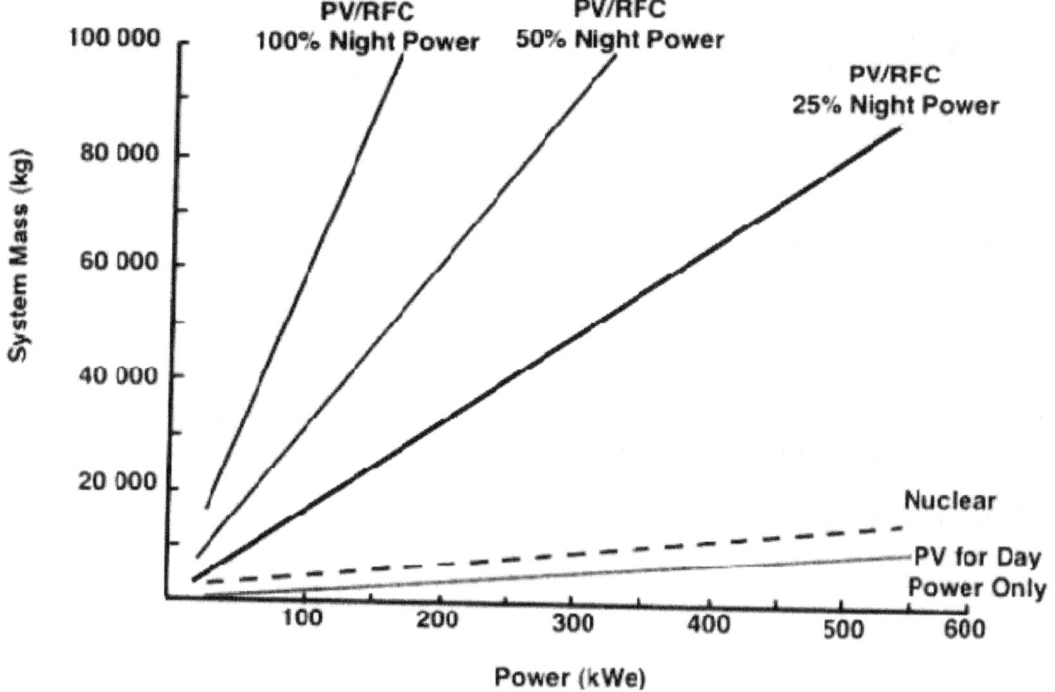

Fig. 1. Comparison of solar and nuclear power systems for lunar base applications. *From Gary L. Bennett, 2008.*

To meet these power needs, a number of technologies might be considered. Fig. 2 shows broad sectors defining these regions. Fission space nuclear power systems are an attractive option for power levels above a few kilowatts-electric and lifetimes measured in months and years. In addition, shorter lifetimes with very high power requirements are also attractive uses for space nuclear power.

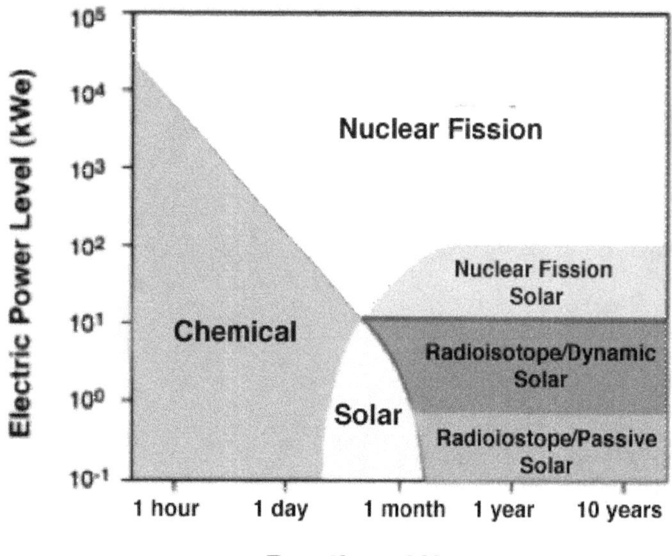

Fig. 2 Best technologies for maximizing specific power.
BOOK 3
SPACE NUCLEAR FISSION ELECTRIC POWER SYSTEMS

Summary of Past Programs[3]

Table 2 presents a list of programs both in the U.S. and Russia to develop fission space nuclear power plants. Active programs have existed intermittently from 1957 to the present time. A great variety of designs have been pursued at different times and to meet varying requirements.

The first design of space nuclear power reactors was initiated in 1955 to produce a 3 to 5 kW_e power system using a uranium-zirconium-hydride fuel and mercury-Rankine power conversion system. This system, was designated SNAP-2 where SNAP is an abbreviation for Systems for Nuclear Auxiliary Power. Two reactors were tested under the SNAP-2 program--the SNAP-2 Experimental Reactor (SER) and the SNAP-2 Developmental Reactor (S2DR). The SER achieved criticality in September 1959 and operated until December 1961. Testing included 1,800 hours at 920 K and 5,300 hours at temperatures greater than 750 K.

The S2DR was tested between April 1961 and December 1962. with 2,800 hours above 920 K and 7,700 hours above 750 K. The SNAP-2 program ended in 1967.

The SNAP-10A program, initiated in 1958, used a modified SNAP-2 reactor with thermoelectric power conversion subsystem. It resulted in the SNAPSHOT launch of the first reactor power system into space on April 3, 1965. The reactor was launched on a modified Agena vehicle and achieved a nominal orbit of 705/695 nautical miles with a lifetime of over 3,500 years. The operating power level was between 530-660 watts. After 43 days of successful performance, a spacecraft malfunction, probably a spurious command signal, resulted in the reactor being shutdown. SNAPSHOT demonstrated for the first time that nuclear reactor power systems could be safely launched and operated.

Table 2. Parameters for some of the key space nuclear fission reactor power plants.

Program	COUNTRY	TIME PERIOD	POWER LEVEL	DESIGN LIFETIME	THERMAL POWER	REACTOR FUEL	PRIMARY COOLANT FLUID	REACTOR OUTLET TEMPERATURE	ELECTRICAL CONVERTER	DEVELOPMENT STATUS
SNAP-2	USA	1957-1967	3 KWe	1 y	55 KWt	U-ZrH	NaK-78	920 K	Rankine, Hg	Experimental Reactor (SER) 1,800 hr at 920 K; Developmental Reactor (S2DR) 2,800 at 920 K, 7,700 hr above 750K.
SNAP-10A	USA	1958-1965	0.5 KWe	1 y	30 KWt	U-ZrH	NaK-78	810 K	Thermoelectrics, SiGe	Flight System FS-3 417 days at approximately 800 K; SNAPSHOT space flight 43 days above 785 K.
SNAP-8	USA	1960 - 1973	35 KWe	1 y	600 KWt	U-ZrH	NaK-78	975 K	Rankine, Hg	Experimental Reactor (S8ER) 1 y above 400 KWt at 975 K; Developmental Reactor (S8DR) 429 hr at 1,000 KWt.
SNAP50/SPUR/ ADVANCED LIQUID METAL COOLED REACTOR	USA	1958-1958-	300-1200 KWe	10,000 hr	2.2 MWt	UC or UN	Li	1425 K	Rankine, K	UN fuel tested 6,000 hr. UC for 2,500 hr with higher swelling rates. Non-nuclear components
710 POWER	USA	1965-1968	200 KWe	10,000 hr	872 KWt	UO$_2$	Inert gas (He-Xe, neon, or argon)	1444 K	Brayton	Fuel element tested 7,000 hr
ROMASKA (RORSAT)	USSR	1959-1988	3 KWe		100 KWt	U-Mo alloy	NaK	around 970 K	Thermoelectrics, SiGe	33 Radar Ocean Reconnaissance Satellites (RORSAT) flown, Cosmos 954 and Cosmos 1402 reentry accidents.
TOPAZ I	USSR	1961 - 1988	5 to 10 KWe	3 y	130 - 150 KWt	UO2	NaK	1,773 K emitter	Thermionics	Topaz I demonstrated 1 y operation in space on Cosmos 1818 and 1867.
TOPAZ II	USSR	1972 - 1992	6 KWe	3 y	115/135 KWt	UO2	Na (79%)K (21%)	1,800 - 2,100 K	Thermionics	Topaz II demonstrated 1.5 y nuclear ground testing. Non-nuclear testing also performed up to 14,000 h in USA.
SPAR/SP-100	USA	1979 - 1983	10 - 100 KWe	7 y	1,600 KWt	UO2	Li heat pipes	1,500 K	Thermoelectrics, SiGe/GaP	Heat pipes tested to >20,000 h at 1,500 K.
SP-100	USA	1983 - 1993	10 - 100 KWe	7 y	2,500 KWt	UN	Li	1,375 K	Thermoelectrics, SiGe/GaP	In-pile fuel tests for swelling and fission gas release verified for 10 y
TFE DEVELOPMENT	USA	1986 - 1993	2 MWe	7 y	22.5 MWt	UO2	Li	1,020 K	Thermionics	TFEs tested to 13,500 h.
MULTI-MEGAWATT	USA	1985 - 1990	10s - 100s MWe	100s + 1y		UO2, UC	Hydrogen,	around 1,200 K	Brayton, Rankine, Thermionics	Limited fuels scoping tests.
PROMETHEUS	USA	2003-2006	200 KWe	20 y	1 MWt	UO2	HeXe mixture	1,150 K	Brayton	Pre-conceptual design studies underway.
FISSION SURFACE	USA	2007-Present	40 kWe	8 y		UO2	Liquid Metal (Na or NaK)	900 K	Stirling or Brayton	Develop system options by 2013

The SNAP-8 program was initiated in 1960 to develop a 30-60 kW$_e$ system for space power. (SNAP-4 and SNAP-6 were terrestrial reactor concepts). It used a modified SNAP-2 core with 211 fuel-moderated elements and a mercury Rankine power conversion system. Two reactors were ground tested, the SNAP 8 Experimental Reactor (S8ER) and the SNAP-8 Development Reactor (S8DR). The S8ER was used as a proof-of-concept test reactor and tested from May 1963 to April 1965. It had an outlet temperature of over 974 K, but post-test examination showed 80 percent of the fuel elements had cracked cladding. The S8DR, a prototype flight system, was tested from January 1969 until December 1969. After approximately 7,000 hours of operation it was found that many fuel elements experienced cracked fuel cladding which caused an increase in the coolant activity and decrease in reactivity. Efforts continued until program cancellation in 1973 to correct these problems.

The SNAP-50/SPUR (Space Power Unit Reactor) program was initiated in 1958 as part of the Aircraft Nuclear propulsion Program (ANP). After the ANP program termination, work was reoriented to meet power needs from 300 to 1,000 kW$_e$ for electric propulsion and power. The design was for a 1,363 K fast, lithium cooled, refractory alloy reactor coupled to a potassium Rankine power conversion system. Work included reactor and system conceptual design, nuclear criticality mockups, small scale testing, and fuel materials testing. The program was shut down in 1968.

The Medium Power Reactor Experiment (MPRE) was begun in 1959 as a small but intensive design and development program to meet power needs up to 150 kW$_e$. Potassium was boiled in the reactor core and passed directly to a Rankine cycle turbine. Both the reactor and test facility were designed and a series of component and system test rigs were designed, built and operated to check the design work. Funding ended in 1966 to support other programs.

In 1965, the 710-Gas Cooled Reactor Program was initiated to make use of technologies developed during the Aircraft Nuclear Propulsion (ANP) program. The goal was to produce a design for 200 kW$_e$ with up to 10,000 hours of operation using a fast reactor of refractory metal-fuel cermets. The design included a reactor cooled by inert gas coupled to a direct-cycle, Brayton power conversion system. In 1968 the program was reduced to a fuel element development program, and then discontinued.

A thermionic reactor program was started in 1959 and lasted until 1973. Thermionic reactors potentially offer minimal moving parts, high redundancy, high efficiency, small radiator area, low specific mass, and a broad power range capability. Improvements had been achieved in high temperature metallurgy, emitter and collector surfaces, and their spacings. Problems continue in diode lifetimes at high temperature and have involved emitter cracking and shorting, fission gas release, fuel/clad interactions, and fuel swelling.

A series of reactor safety tests were conducted--SNAPTRAN 1, 2, and 3. These included small transients and destructive excursion tests of full scale reactors to simulate water flooding and impact. Environmental restrictions would make it difficult to get approval to reproduce these type of tests today. They answered questions concerning reentry disassembly and the burnup of fuel elements.

In 1979, a modest space nuclear reactor component technical program called Space Power Advance Reactor {SPAR) was started. After evaluating several advanced concepts, the Heat Pipe Reactor was selected and component work performed on development of high temperature heat pipes. This continued until 1983 when the National Aeronautics and Space Administration (NASA), the Department of Energy (DOE) and the Department of Defense (DoD) entered into an agreement to fund a joint space nuclear reactor power program. This program, called SP-100, superseded the heat pipe reactor development program (whose name had been changed to SP-100) that had started in 1979. The goals of the new program were to develop the technology to provide tens-to-hundreds of kW$_e$ of electric power with the specific initial design concentrating on 100 kW$_e$ at an operational time of 7 y and lifetime of 10 y. Starting with a broad range of candidates [4, 5, 6, 7, 8] including liquid-metal, gas-cooled, thermionic, and heat pipe reactors with various combinations of thermoelectric, thermionic, Brayton, Rankine, and Stirling energy conversion systems, three concepts were selected for further evaluation. These were: (1) a high temperature, pin-fuel element reactor with thermoelectric conversion, (2) an in-core thermionic power system, and (3) a low-temperature pin-fuel element reactor with Stirling cycle conversion. In 1985, the high-temperature pin-fuel element reactor with thermoelectric conversion was selected for development to flight readiness. The thermionic power plant offered a more compact heat rejection subsystem and lower temperature structural materials, but issues of lifetime excluded its selection. The Stirling system offered higher energy conversion efficiency and lower temperature materials, but the technology risk was considered greater at that time because of the preliminary status of its development. Because sufficient merit was recognized in other options, technology programs were started on in-core thermionic fuel elements, called the Thermionic Fuel Element Verification Program, and another program on developing a high-temperature Stirling engine that can be mated with the high temperature pin-fuel reactor. The SP-I00 power system progressed from a concept, through a generic flight system design, to the design, development, and testing of specific components. until the program was discontinued in 1993

During 1985-1990, a number of activities occurred in developing multi-megawatt-level power systems to support power needs for directed energy weapons and electric propulsion. The Multimegawatt Program evaluated systems for: (1) open loop, power levels of tens-of-megawatts for hundreds of seconds, (2) closed loop, power levels of tens-of-megawatts for one year, and (3) open loop, power levels of hundreds-of-megawatts for hundreds of seconds. Six concepts were considered during the Phase I pre-conceptual activities. The purpose of Phase I was to identify key technology feasibility issues. Phase II was to resolve the issues prior to technology selections. This program was terminated in 1990, because of funding constraints, before Phase II design contracts were awarded.

Project Prometheus was established in 2003 with a goal of developing the first nuclear reactor-powered electrical propulsion system for long-duration civilian deep-space exploration missions. The initial application of space fission power was to be the Jupiter Icy Moons Orbiter (JIMO), a nuclear electric propelled spaceship. The selected concept was a gas-cooled reactor with a directly coupled Brayton energy conversion system for further development. The program was cancelled in 2006 as being too complex and expensive to fit NASA's budget.

Under the NASA Exploration Technology Development Program, a project to develop Fission Surface Power (FSP) was initiated in 2007. A primary objective is to develop viable options by 2013 to support an expected flight power system for lunar outposts and later missions to Mars. Current plans presume the use of a fast-spectrum, uranium disoxide (UO_2) fuel pins with stainless steel cladding and structure. Both Stirling and Brayton cycle heat engines are being considered for their compatibility with the lunar and Martian environments.[9]

The Russian also had a very active space nuclear power program starting around 1959. Since 1967, the U.S.S.R. has orbited approximately 33 thermoelectric reactor power systems as a power source for ocean surveillance radars in satellites called RORSAT. The last one was on March 14, 1988. Power levels ranged from several hundred watts to a few kilowatts. Limited information is available on the details of the RORSAT power system. We know that the RORSAT power systems are fast reactors using SiGe thermoelectric conversion system.

The reactor powered RORSAT satellites have experienced three known malfunctions or accidents. The most serious, Cosmos 954, reentered the Earth's atmosphere over Canada on 24 January 1978. The reactor burned up on reentry, as designed. This was the same safety philosophy used by the U.S. during the 1960's and 1970's, namely, burnup to reduce doses and to eliminate the possibility of re-criticality in the event of a reentry accident. Following the Cosmos 954 accident, a satellite re-boost capability was added to increase safety. Cosmos 1402, on 7 February 1983 failed to re-boost, but burned up in the upper atmospheres after the core was ejected from the spacecraft. Cosmos 1900, on September 30, 1988, was boosted to a lower then planned disposal orbit. However, the orbit lifetime is about 500 years with almost no radiological risk at the expected time of re-entry into the Earth's atmosphere.

In 1987-1988, the U.S,S,R, tested a different type of reactor power system using thermionic power conversion. Two space tests were performed, with one operating six months (Cosmos 1818) and the other operating 346 days (Cosmos 1867). These power plants are designated in the U.S. as Topaz I. Topaz I design output is 10 kW_e. The flight-tested units used a multi-cell thermionic fuel element with an output power of approximately 5 kW_e, one with a molybdenum emitter and the other with a tungsten emitter. The power system with the tungsten emitter operated for the longer period of time; degradation of performance occurred, with the thermal power increased to compensate for this degradation.

Fundamentals of Space Nuclear Fission Systems

Space nuclear fission electric power systems building blocks are depicted in Fig. 3. The nuclear fission reactor acts as a source of thermal energy. Heat from the reactor is transported to the electric power conversion generator. Electrical power conversion can take the form of either static electrical conversion elements that have no moving parts (e.g., thermoelectric or thermionic) or dynamic conversion (e.g., the Rankine, Brayton or Stirling cycle). Waste heat is rejected to space. In addition, the reactor also emits radiation that can be damaging to the spacecraft and, if a manned mission, to the crew. Radiation protective shielding is used to mitigate this potential damage.

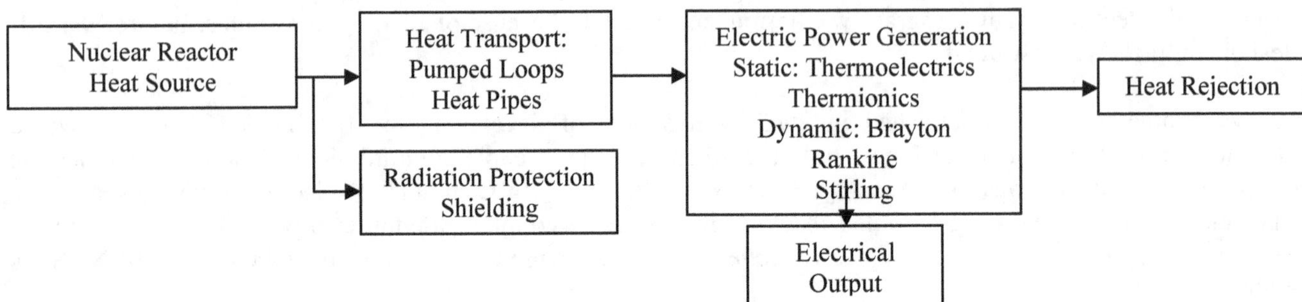

Fig. 3. Power system elements.

Fundamentals of Nuclear Reactors[10]

Reactor Elements

The heart of a fission nuclear power system is the nuclear reactor. A nuclear reactor is device in which the controlled fissioning or splitting of certain atoms occurs. A product of the fissioning process is the release of significant amounts of energy. A fissile nuclide is a nuclide (e.g., $^{233}_{92}U$, $^{235}_{92}U$. or $^{239}_{94}Pu$) that is capable of being split or fissioned upon absorbing neutrons of any energy. Thus, a nuclear reactor is configured to contain fissile nuclear material, such that a chain reaction of fission events can be maintained and controlled.

A chain reaction is a reaction that stimulates its own repetition. For a fission chain reaction, a fissile nucleus absorbs a neutron, splits, and releases additional neutrons (see Fig. 4). A fission chain reaction is self-sustaining when at least one neutron per fission event survives to create another fission reaction. The two lighter elements produced in the splitting of the heavy nucleus are called fission products (FP).

The multiplication factor, k, is used to describe the fission chain reaction. The definition of k is: [11, 12, 13]

$$k \equiv \frac{\text{number of nuclear fissions (or neutrons) in one generation}}{\text{number of nuclear fissions (or neutrons) in the immediately preceding generation}} \tag{8}$$

Fig. 5 illustrates the behavior of the fission chain based on the value of the multiplication factor k. When

$k = 1$, the fission reaction is critical or self-sustaining and power production occurs at a steady rate.

$k < 1$, the chain reaction is called subcritical and the number of fissions occurring per generation (or the neutron population) will eventually reach zero.

$k > 1$, the chain reaction is supercritical and the number of fission reactions (or the neutron population) increases each generation (see Fig. 5).

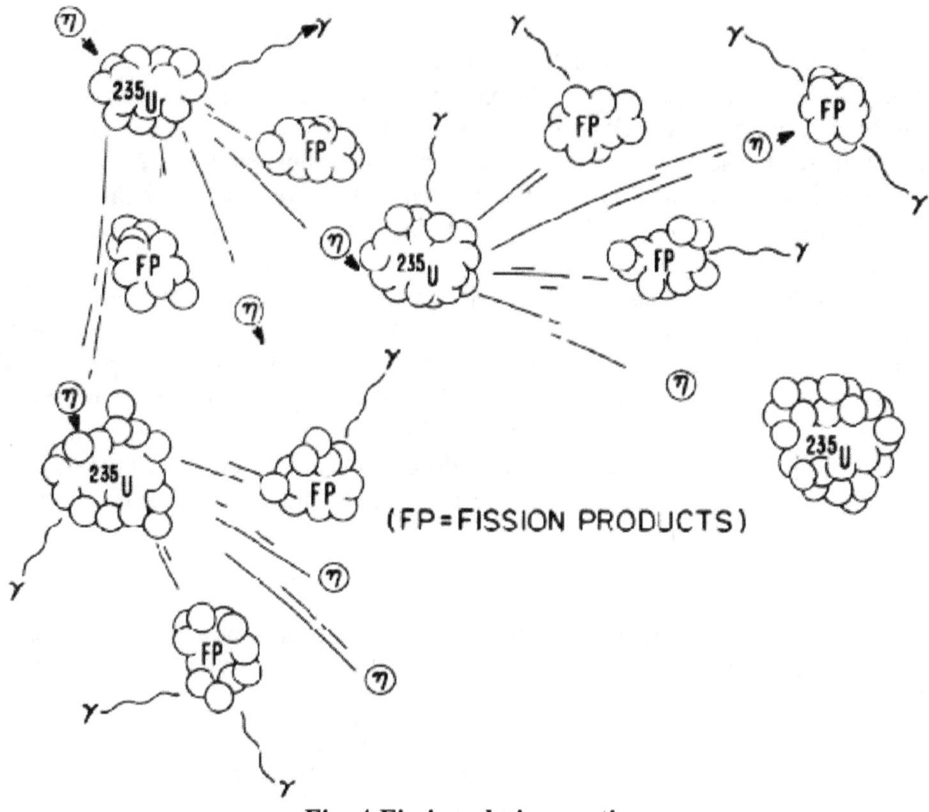

Fig. 4 Fission chain reaction

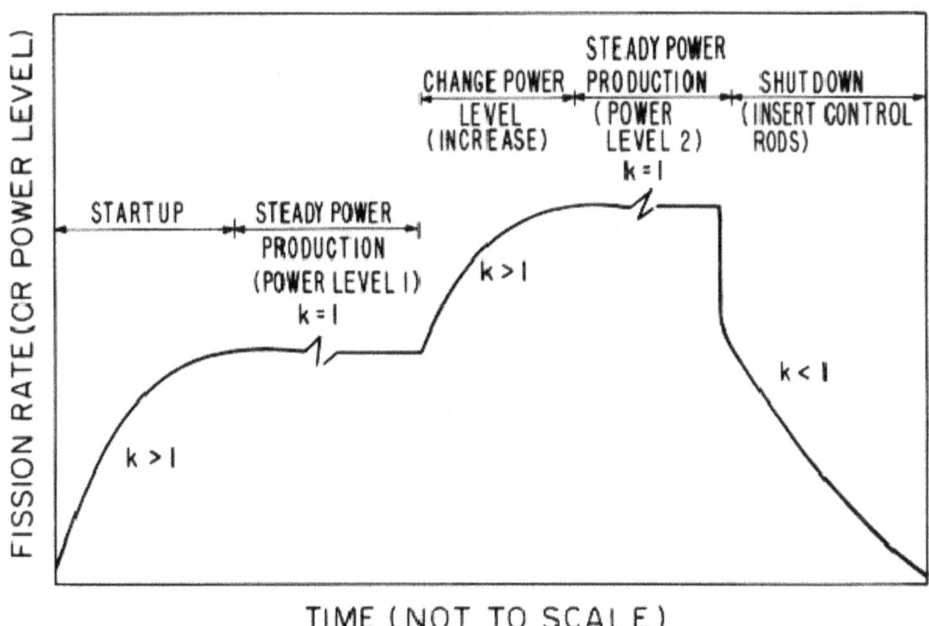

Fig. 5. Multiplication factor as a function of reactor conditions.

Space nuclear reactor components include: a core with the nuclear fuel of fissile material, coolant to remove energy generated in the core, a reflector to minimize the lost of neutrons from the core, control drums or rods to regulate the multiplication factor, and a radiation shield to protect components outside the reactor from destructive radiation [14, 15, 16, 17] (see Fig. 6). In addition, a moderator component maybe present to regulate the neutron spectrum in the region of nuclear fissioning. The predominant fissioning energy spectrum depends on the reactor design. A thermal reactor is one in which thermal neutrons are the predominant cause of fission reactions. A fast reactor is one designed so that the majority of fissions occur at fast neutron energies, for instance, > 100 keV. A moderator is found only in thermal reactors.

The arrangement of the nuclear reactor has the core as the central reactor region. It contains the nuclear fuel, a moderator (thermal reactors only), suitable structural materials, and coolant passages for heat removal. The nuclear fuel fissile material in a space reactor is usually highly enriched (typically 93.5 percent) uranium-235. It sustains the fission chain reaction, is responsible for the criticality of the reactor, and provides for the release of large quantities of energy in the nuclear fission reaction. Fuel may be in solid, liquid, or gaseous form. If a moderator is present, it is a low-mass material, such as hydrogen, that slows down or moderates neutrons from a fission energy spectrum to a thermal energy spectrum, or perhaps an epithermal spectrum. (Epithermal neutrons are neutrons with energies above thermal values.) This moderating material is found only in thermal or epithermal reactors and is not used in fast reactors.

Coolant is used to remove the thermal energy from the reactor core. This heat removal function is accomplished using a pumped-working fluid or heat pipes.

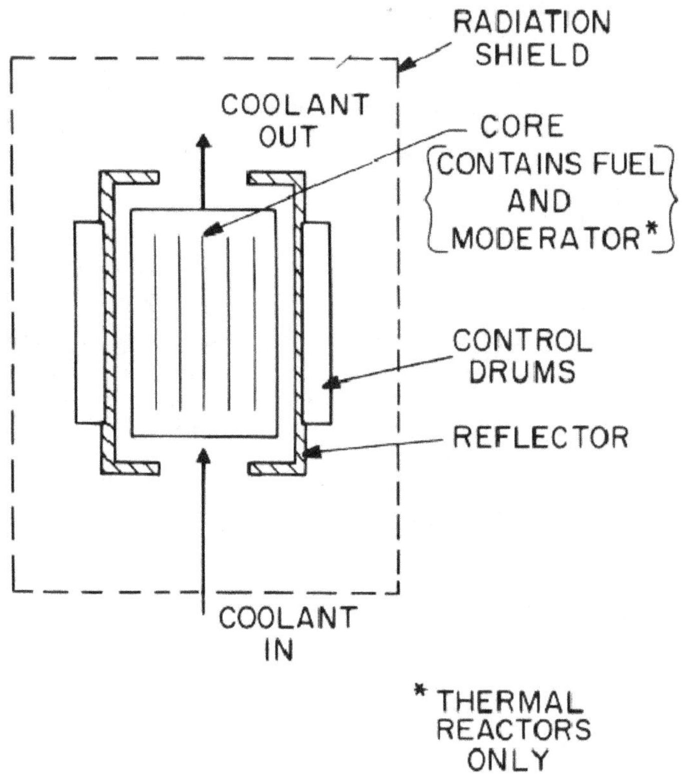

Fig. 6. Generic space nuclear reactor system.

The reflector is a material that scatters neutrons back into the core. It is located adjacent to the reactor core. Beryllium and graphite are typical reflector material. These materials have high scattering cross section and low neutron absorption characteristics. The reflector reduces neutron leakage and permits a more uniform power production within the core through flux flattening. Neutrons leaking out of the core scatter in the reflector material and some of these return to the core. The use of a reflector helps reduce the critical mass of the system and supports flux flattening and a more uniform power generation in the core.

The reactor is controlled by positioning various types of neutron absorber materials. These can be arranged in several forms, such as rods in the core or drums in the reflector. Boron carbide (B_4C) is a representative material used to regulate the nuclear reactor. Movement of these rods or drums adjusts the multiplication factor, thereby controlling the reactor's power level (see Fig. 5).

Astronauts and equipment outside the reactor must be protected from neutrons and gamma rays escaping the reactor core. A protective radiation shield is used for this purpose. In fission space power systems, partial shields maybe used with the reactor located at the end of the spacecraft or the reactor can be completed enclosed in a radiation shield. Typical space reactor shield materials include lithium hydride (LiH) for neutron attenuation and tungsten (W) for gamma ray absorption.

The Fission Process
A nucleus must be in an excited state for the nuclear fission process. Fission may occur if the excitation energy exceeds a certain critical energy. Or, emission of gamma radiation to return the compound nucleus to its ground state may occur. Fission threshold energy are shown in Table 3 calculated for a wide range of atomic masses.[18, 19] Only the very heavy nuclides (mass number (A) > 230) have reasonably low threshold energies. Experimental values of the fission thresholds for selected heavy nuclides are presented in Table 4. Negative threshold energy (i.e., $^{233}_{92}U$, $^{235}_{92}U$, or $^{239}_{94}Pu$) mean that neutrons with essentially zero kinetic energy (thermal neutrons) can cause these nuclides to undergo "thermal" fission. $^{233}_{92}U$, $^{235}_{92}U$. and $^{239}_{94}Pu$ are called "fissile" nuclides while the other heavy nuclides shown (i.e., $^{232}_{90}Th$, $^{234}_{92}U$, $^{236}_{92}U$, and $^{237}_{93}Np$) can only undergo fast fission.

Table 3. Neutron fission thresholds as a function of nuclear mass (calculated).

Mass Number (A)	Fission Threshold (MeV)
16	18.5
60	48
100	47
140	62
200	40
236	~ 5

Table 4. Neutron fission thresholds of heavy nuclides (experimental)

Target Nucleus	Compound Nucleus	Fission Threshold (MeV)
^{232}Th	^{233}Th	1.3
^{233}U	^{234}U	< 0
^{234}U	^{235}U	0.4
^{235}U	^{236}U	< 0
^{236}U	^{237}U	0.8
^{238}U	^{239}U	1.2
^{237}Np	^{238}Np	0.4
^{239}Pu	^{240}Pu	< 0

Fission causes the highly unstable compound nucleus to almost always splits into two fission fragments. The nucleons are more tightly bound within the fission fragments than they were in the original heavy nucleus. Large amounts of energy (typically 200 MeV per fission) is released in this nuclear reaction. This energy is the difference in binding energy between the original heavy nucleus and its fission products.   Table 5 presents a typical energy distribution for uranium-235 fission. Nuclear fission results in the prompt emission of several neutrons. The average number of neutrons released per fission (v) is a function of the neutron energy and the fissile nuclide. Thermal fission of uranium-235 releases about 2.5 neutrons per fission. In addition, some of the fission products themselves are nuclides which possess more neutrons than necessary for nuclear stability. These unstable fission fragments (with half-lives of up to approximately one minute) emit delayed neutrons. For thermal fission of uranium-235, the delayed neutron fractions (β) are about 0.7 percent (see Table 6).[20] Because the fission fragments are neutron-rich, they undergo radioactive decay, emitting mainly beta and gamma radiations.

Table 5. Typical energy distribution for Uranium-235 fission.

Energy Form	Energy Released (MeV)	Energy Potentially Recoverable (MeV)
Kinetic Energy of Fission Fragments	168	168
Decay of Fission Products		
- Beta Radiation	8	8
- Gamma Radiation	7	7
- Neutrinos	12	--
Prompt (Fission) Gamma Radiation	7	7
Kinetic Energy of Fission Neutrons	5	5
Capture Gamma Radiation	--	3-12
Totals	207	198-207

Table 6. Delayed neutron fraction (β) for fast and thermal fission.

Nuclide	Fast Fission	Thermal Fission
^{233}U	0.0027 ± 0.0002	0.00264 ± 0.0002
^{235}U	0.0065 ± 0.0003	0.0065 ± 0.0003
^{238}U	0.0157 ± 0.0012	—
^{239}Pu	0.0021 ± 0.0002	0.0021 ± 0.0002
^{240}Pu	0.0026 ± 0.0003	—

Fig. 7. illustrates the fission product mass numbers as a function of fission yield for the thermal and fast (14 MeV) fission of uranium-235. The fission yield is defined as the percent of the total number of fissions that produces fission products of a given mass number. As the energy of the incident neutron increases, the probability of producing two fission products of nearly the same size increases and the characteristic dip that is present in the thermal fission-yield plots at approximately mass number 120 disappears

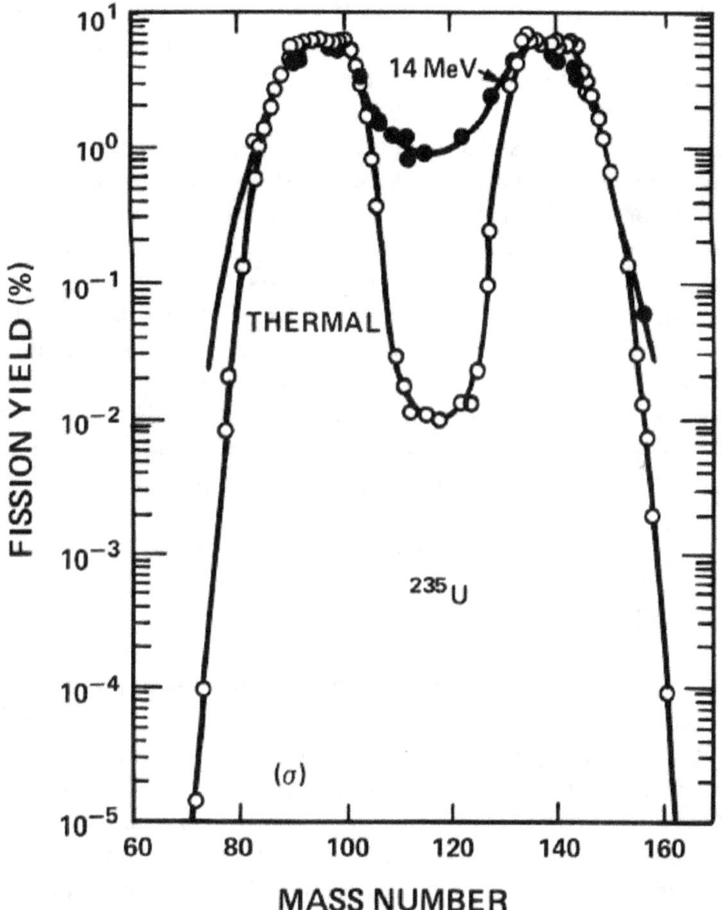

Fig. 7. Fission product mass distribution for uranium-235 fission.

Electric Power Generation

Static Power Generators

Static electric conversion devices include thermoelectrics and thermionic electric power generators. Both of these have been extensively developed for space applications. The only U.S. reactor flown in space used thermoelectric power conversion; the Russian have flown 33 mission with thermoelectric converters and two with thermionic converters.

Thermoelectric Power Conversion

Most U.S. power conversion generators used in space nuclear systems has been in the form of thermoelectric converters. All U.S. radioisotope power systems have used thermoelectric power conversion subsystems for converting the heat from the radioisotope heat source to electricity. The same technology can be used in space nuclear fission systems. These devices are considered to be passive power conversion generators in that they have the advantage of having no mechanical moving parts. However, the devices tend to have relatively low power conversion efficiencies--less than 10%.

Thermal energy is directly converter into electricity in a thermoelectric (TE) device based upon the Seebeck effect. Thomas Seebeck (1770-1831), in 1821, observed that an electromotive force (emf) is generated when the junctions of two dissimilar metals are maintained at two different temperatures. Basic to the thermoelectric effect is the fact that a temperature gradient in a conducting material results in heat flow; this results in the diffusion of charge carriers. The flow of charge carriers between the hot and cold regions in turn creates a voltage difference.

Ideal thermoelectric materials have a high Seebeck coefficient, high electrical conductivity, and low thermal conductivity. Low thermal conductivity is necessary to maintain a high thermal gradient at the junction. The semiconductors are connected electrically in series and thermally in parallel.

Practical thermoelectric converters have resulted with the development of special semiconductor materials that combine a high Seebeck coefficient (α), relatively low electrical resistivity (ρ), and low thermal conductance (k). N-type and p-type semiconductor materials are used in modern TE devices in order to create larger voltage outputs per degree of temperature difference. The basic operating principle of a thermoelectric conversion device is illustrated in Fig. 8. An individual TE converter consists of two semiconductor legs that are bonded to two heat transfer surfaces called the hot and cold shoes or junctions. One of the semiconductor materials is a p-type material in which the cold shoe region becomes positively charged by the migration of holes under the influence of a thermal gradient. The other semiconductor material is an n-type material in which a temperature difference causes electrons to diffuse to the cold shoe. One region of this TE couple is maintained by the bonded hot shoe at the high temperature, while thermal energy is removed from the other end of the couple at the cold shoe. The thermally driven flow of electrons and holes creates a voltage across both cold shoe plates. By connecting an external load across the two cold shoe plates, a current is made to flow through this external circuit. The power flowing through this external circuit is maximized when the load resistance is matched to the internal TE converter resistance. Such TE unicouples can then be connected in an external series-parallel circuit to provide redundant against failures that might create open circuits.[21, 22]

The output of a TE converter cell depends on the operating temperatures, properties of the n-type and p-type semiconductor materials, and specific design details. The induced voltage in the cell is equal to the product of the overall Seebeck coefficient (n and p type material) and the temperature difference. The current flow is equal to the voltage divided by the sum of internal and external resistances. The net output power is equal to the current squared divided by the external load resistance. Again, maximum power occurs when this load resistance is equal to the internal resistance. Thermal energy transferred from the hot shoe to the cold shoe through either semiconductor material leg represents a loss. This conduction process is dependent upon the temperature gradient, the thermal conductivity of the TE materials, and the device design.[23]

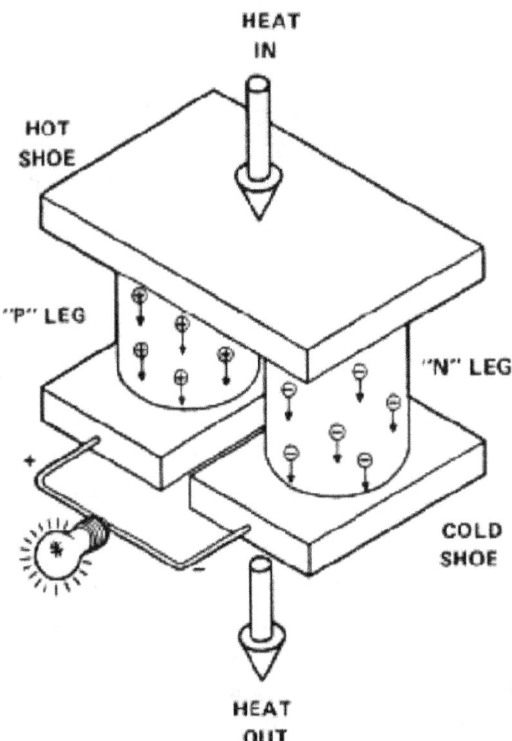

Fig. 8. Operating principle of the thermoelectric converter *Courtesy R. V. Anderson, et al.*

If the n type and p type materials have equal thermal conductivities $(k)_{n,p}$ and equal electrical resistivities $(\rho)_{n,\,p}$, then the figure of merit for our idealized TE converter becomes

$$ Z = \alpha^2 / k\rho. $$

(9)

where α is the material Seebeck coefficient (μV / K).
        k is the thermal conductivity (W / K-cm).
        ρ is the electrical resistivity of semiconductor material (mΩ-cm).

The figure-of-merit is usually considered the most significant parameter in the selection of materials for TE power generators. Table 7 and Fig. 9 present figure-of-merit data for selected TE materials.[24] The larger the value of Z, the higher the overall efficiency of the TE converter.

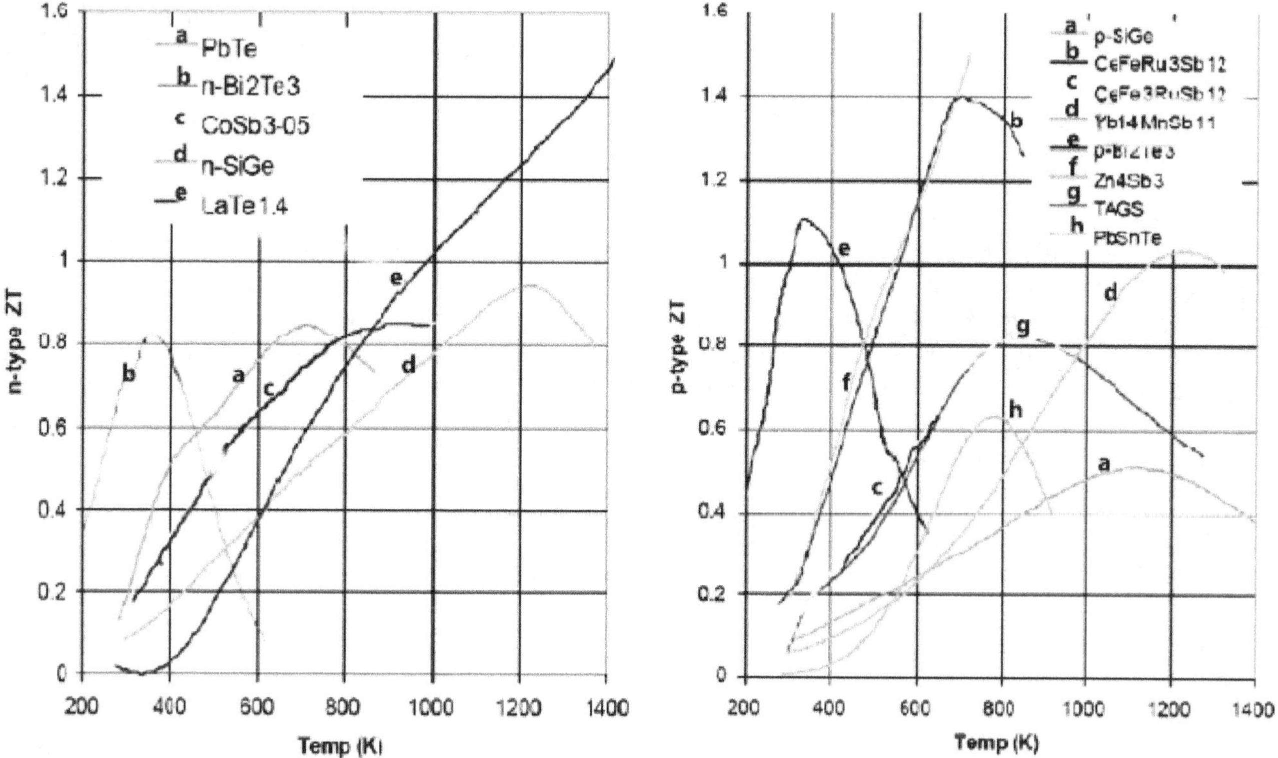

Fig. 9 Figure-of-merit of selected thermoelectric materials. *From T. Caillat, et al, "Advance Thermoelectric Power Generation Technology Development at JPL," 3rd European conference on Thermoelectric, Nancy, France, Sept. 2005.*

Table 7. Typical calculated figure-of-merit values.

MATERIAL	TEMPERATURE	SEEBECK COEFFICIENT (α)	FIGURE OF MERIT (Z)
METALS	300 K	5 μV/K	3×10^{-6} K^{-1}
SEMICONDUCTORS	300 K	200 μV/K	2×10^{-3} K^{-1}
INSULATORS	300 K	1 mV/K	5×10^{-17} K^{-1}

In many configurations the arrays of thermocouples maybe mechanically assembled between the heat source and the heat sink. Springs have been utilized in most telluride TE systems assemblies. These provide good thermal and electrical contacts at the junctions, mechanical forces for withstanding operational or launch payloads and accommodate thermal expansion changes that might occur during thermal cycling of the converter system. Two designs have been created without springs. These are the light-weight, unsealed, bonded panels (launched on the Transit spacecraft) and the hermetically sealed, close-packet tubular module. In SiGe converter systems, such as the MHW unit, the thermocouples are cantilevered from the cold side, and the heat input to silicon molybdenum hot shoes is accomplished by radiative transfer from the heat source.

Thermal insulation is used to reduce heat losses from the ends of the generator and between the thermocouples. Insulation materials depends on the operating temperature, the geometry of the system, and the chemical compatibility of its various components. Good thermal insulation ensures that the thermal energy from the heat source passes through the power conversion thermocouples with losses of less than 10 to 15 percent.

Most nuclear power space experience has been with spacecraft incorporating thermoelectric converters. All of the U.S. launched nuclear power systems, that includes 27 NASA and military spacecraft with radioisotope generators and one reactor (SNAP 10A), have used thermoelectric converters. They have demonstrated a high degree of redundancy and durability. For instance, the Voyager spacecraft continued to operate after 30 years. Thermoelectric converters have been used in several space reactor programs. These include Snap 10A, The Russian Rorsat spacecraft, SPAR/SP-100 and SP-100.

Thermionic Converter Principles[25]

Another method of directly transforming heat into electricity is by use of thermionic converters. The thermionic (TI) converter can be described as a static energy conversion device that essentially "boils electrons" from a hot emitter surface (typically 1800 K) across a small interelectrode gap ($\leqslant 0.5$ mm) to a cooler (~ 1000 K) collector surface. The operating principles with projected operating parameters and potential component materials are illustrated in Fig. 10.[26] The TI converter can be thought of as a heat engine that uses electrons as its working fluid. In a thermionic converter, the voltage potential between the emitter (cathode) and collector (anode) drives a current through an externally connected electric load.

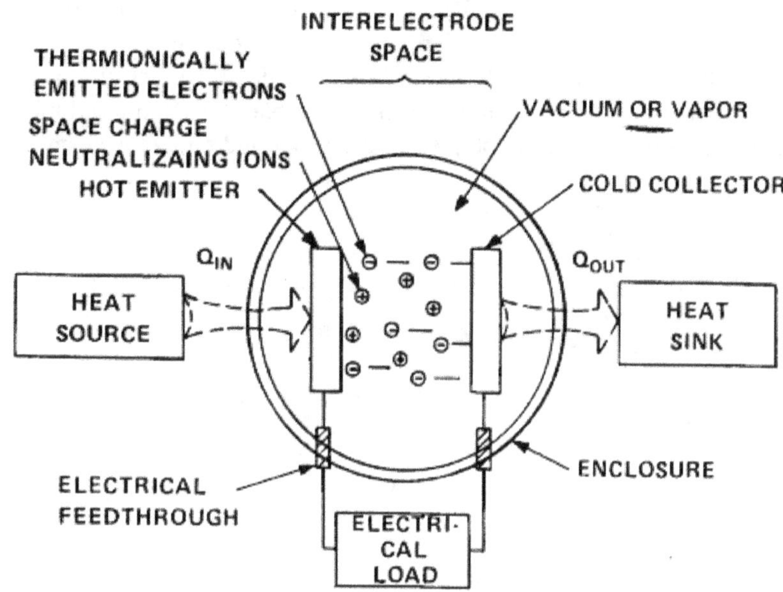

TYPICAL OPERATING REGIME

EMITTER TEMPERATURE	1600 - 2000 K (2420 - 3140° F)
COLLECTOR TEMPERATURE:	800 -1100 K (980 - 520° F)
ELECTRODE EFFICIENCY:	UP TO 20%
POWER DENSITY:	1 - 10 w/cm ^{2}

MATERIALS

EMITTER TEMPERATURE	W, Re, Mo
COLLECTOR TEMPERATURE:	Nb, Mo
ELECTRODE EFFICIENCY:	Al_2O_3, AL_2O_3/Nb CERMET
POWER DENSITY:	Cs AT 1 Torr

Fig. 10. Schematic of thermionic converter

The upper performance limit can be considered in terms of a Carnot Cycle efficiency. Achieving the Carnot efficiency in thermionic converters is limited by: (1) radiation heat transfer between the cathode and the anode; (2) the space charge effect which impedes or prevents the flow of electrons from the emitter to the collector; and (3) thermal energy losses to the surroundings. The space charge effect can be reduced in a thermionic converter by either moving the electrodes closer or by injecting positively charged ions into the interelectrode gap. This neutralizing the electron produced negative space charge. A vacuum diode thermionic converter has closely spaced electrodes and no vapor in the interelectrode gap. Whereas, a plasma diode has a vapor, like cesium, in the gap to reduce the space charge effect

A simple analytic model (refer to Fig. 10) of thermionic converter performance can be formulated using an ideal thermionic diode. Thermal energy is supplied to the emitter such that a sufficiently high temperature is obtained for electrons to be emitted or boiled-off. These electrons cross the interelectrode gap and are collected at a cooler electrode, called the collector or anode. The collector temperature is maintained by removing thermal energy to the radiator; this keeps the temperature sufficiently low to prevent electron back emission. The electrons flow from the anode through an external load and return to the cathode.

An ideal diode motive diagrams and converter current voltage characteristics are shown in Fig. 11. In the motive diagram ϕ_e and ϕ_c represent the emitter and collector work functions, respectively with the other equations defined in equations (11) and (12) below. The work function is the potential barrier that must be overcome by electrons leaving either electrode. The Fermi-Dirac statistical model for electrons can be used to predict the number of electrons that will break through the (surface) potential barrier found in all metals. The output voltage (V) of the thermionic converter is the difference between the emitter and collector Fermi levels.

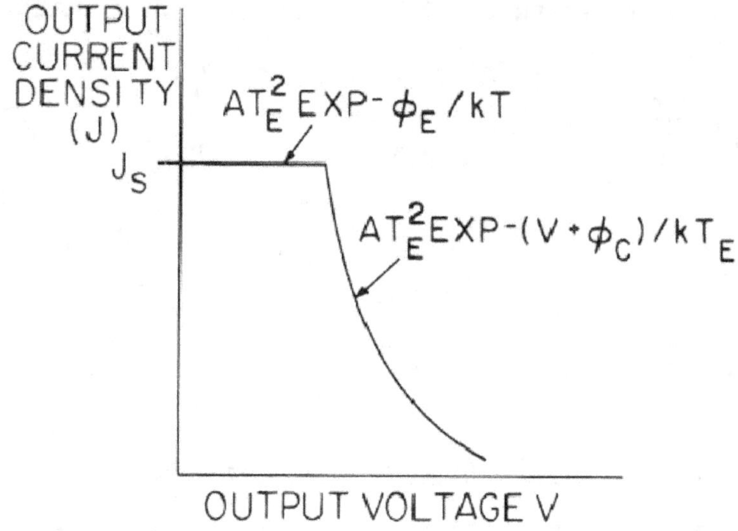

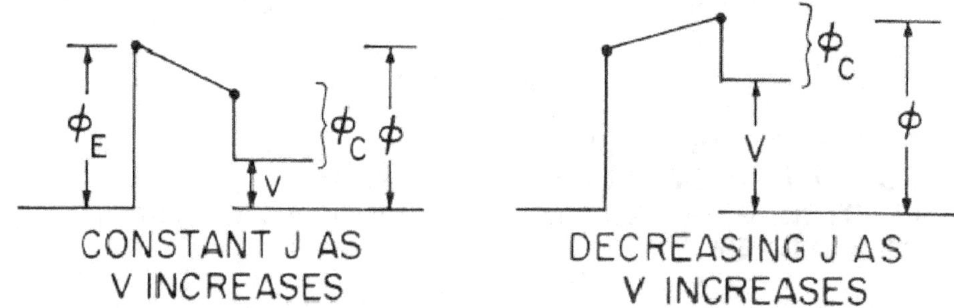

Fig. 11. Ideal thermionic diode characteristics

The Fermi level (E_F) is the electron energy at which the Fermi-Dirac distribution function is 0.5. In solid-state physics, this distribution function represents the probability that an electron will occupy a particular quantum state at thermal equilibrium (see Fig. 11). The Fermi level may also be viewed as the maximum electron energy at $T = 0$ K. At temperatures above absolute zero, electrons acquire thermal energy and their distribution in particular energy levels is described by Fermi-Dirac statistics.

In the absence of space charge and other such phenomena between the electrodes, the current density (J) from the emitter to the collector is determined by the emitter temperature (T_e) and the potential barrier (ϕ) the electrons must overcome, In an ideal diode, the current density is given by the Richardson-Dushman equation

$$J \,(\text{amps}/\text{cm}^2) = AT_e^2 e^{-\phi/(kT_e)}$$

(10)

where A is a constant ($A = 120$ amp / cm^2 - K^2)

     ϕ is the work function (eV)

     k is the Boltzmann constant (8,6168 X 10^{-5} eV / molecule-K),

There are basically two distinct regions of operation of the TI diode. As long as the sum of the collector work function (ϕ_c) and the converter output voltage (V) is less than the emitter work function (ϕ_e), the barrier to electron flow is $\phi = \phi_e$. The current density (J) is independent of the output voltage and this region of the current-voltage (I-V) curve is called the saturation region. The saturation current density is expressed as

$$J_s = A T_e^2 e^{-\phi_e/(kT_e)}$$

(11)

where ϕ_e is the emitter work function (eV) ,

As soon as the output voltage is increased to the point where $\phi_c + V = \phi_e$, the potential barrier becomes $\phi = \phi_c + V$, and the converter current density falls exponentially with further increases in output voltage. The current density in this region of the I-V curve, called the retarding region, is given by

$$J = A T_e^2 e^{-[(\phi_c + V)/(kT_e)]}.$$

(12)

The ideal diode converter is not achieved in real operational converters. To begin with, the work functions of refractory metals are too high to permit their application at reasonable operating temperatures. For example, polycrystalline tungsten has a work function of approximately 4.6 eV. To obtain a current density of 1 amp / cm^2 with this material requires an operating temperature of 2,600 K. However, tungsten has a significant evaporation rate at such temperatures. In addition, a vacuum diode configuration electron space charge effects in a practical interelectrode gap configuration limits current density to less than one milliampere per centimeter squared. However, interelectrode gaps on the order of 10^{-2} mm are actually needed to achieve useful current densities in a vacuum diode.

These engineering difficulties are mitigated by adding a low pressure cesium vapor (typically 13 Pa to 1.33 kPa) in the interelectrode gap. This type of low pressure vapor can be established by including a reservoir of liquid cesium in the TI converter enclosure at a suitable temperature, generally between 500 K and 600 K. In the presence of this cesium vapor, a partial monolayer of cesium will form on both the emitter and collector, reducing their work functions to near optimum values. By varying the cesium temperature (and therefore its pressure), a range of work functions can be obtained at given emitter temperatures. To maximize the output voltage, the collector is operated at a low work function. The electron space charge effect to TI current flow can be neutralized by ionizing the cesium vapor, creating a plasma. Cesium is a particularly good choice because it has the lowest ionization potential of any element. In the unignited mode, the ions are generated using high emitter temperatures so that a sufficient number of ions are created thermally by means of contact ionization with the emitter. The ignited mode is used at lower temperatures. Here, ions are created electrically in a low voltage arc between the electrodes. In this situation, a portion of the electric power generated by the TI converter is dissipated internally to maintain the plasma. The plasma-sustaining voltage drop across the interelectrode gap is called the arc drop (V_d). Including the influence of electron scattering, its effective value is usually about 0.5 volt.

Since losses such as arc drop and electron scattering are incurred when cesium is added to the converter, its performance is no longer accurately represented by the ideal diode model. It has been observed, however, that a thermionic converter can be well-characterized by simply adding another voltage loss term in the ideal diode equation. As illustrated in Fig. 12, the sum of these losses, including the collector work function, is called the barrier index (V_B). The actual converter performance is then given by :

$$J = A T_e^2 e^{-[(V + V_B)/(kT_e)]}.$$

(13)

Typical converter values for V_B are -1.9 eV to 1.5 eV

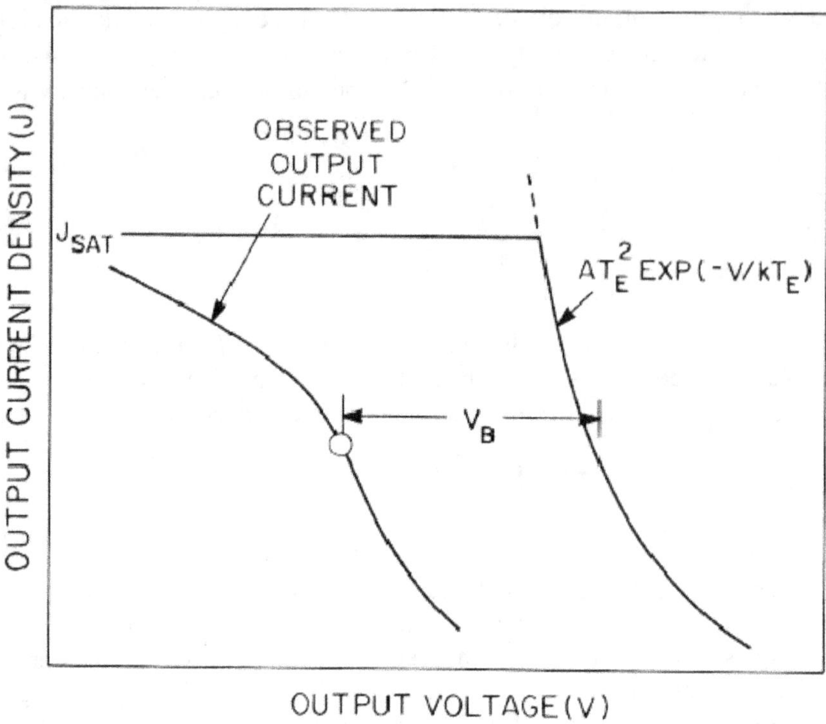

Fig. 12. Relationship between ideal and observed thermionic diode behavior.

While the actual power output of the converter is the product of current and voltage, only about 90 percent of this power is available at the external load due to voltage drops in the converter's electrical leads. Consequently, the useful output power density of a practical TI converter is given by:

$$P_L(\mathrm{W/cm^2}) = 0.9J\left[kT_e\ln\left(AT_e^2/J\right) - V_B\right].$$

$$(14)$$

The collector temperature (T_c) is of crucial importance to the power conversion system. The highest possible collector temperature (and therefore heat rejection temperature) is desired in space power applications, since all waste heat must be radiated away to space and the size of the radiator is inversely proportional to $T_c{}^4$. However, if the collector has too high a temperature, electrons will be emitted back into the flow of electrons from the emitter. This collector back emission must be compensated for either by increasing the collector work function (thereby reducing back emission) or by increasing the number of electrons leaving the emitter. As a general consequence, increasing the collector temperature decreases the thermionic converter performance. So a design trade-off between heat rejection optimization and converter performance has to be made. In most TI converters the back emission is limited to 10 percent or less of the emitter current density.

The emitter input power ($\dot{Q}_{in}$) is determined by emitter radiative heat transport, cesium gas thermal transport, conductive losses through the electric leads, and electron cooling of the emitter. Of these, electron cooling and radiative heat transfer are the dominant mechanisms. A good approximation to many converters in the region of interest is provided by the expression

$$\dot{Q}_{in}(\mathrm{W/cm^2}) = 1.8 \times 10^{-3}JT_e + 1.2 \times 10^{-12}\left(T_e^4 - T_c^4 \right).$$

$$(15)$$

Fig. 13 shows the input power as a function of both emitter temperature and current density. If we combine eq. 14 and eq. 15, the efficiency of a thermionic converter (η_{TI}) at its leads is approximated by

$$\eta_{TI} = \frac{0.9J\left[kT_e\ln(AT_e^2/J) - V_B\right]}{1.8 \times 10^{-3}JT_e + 1.2 \times 10^{-12}(T_e^4 - T_c^4)} \, .$$

(16)

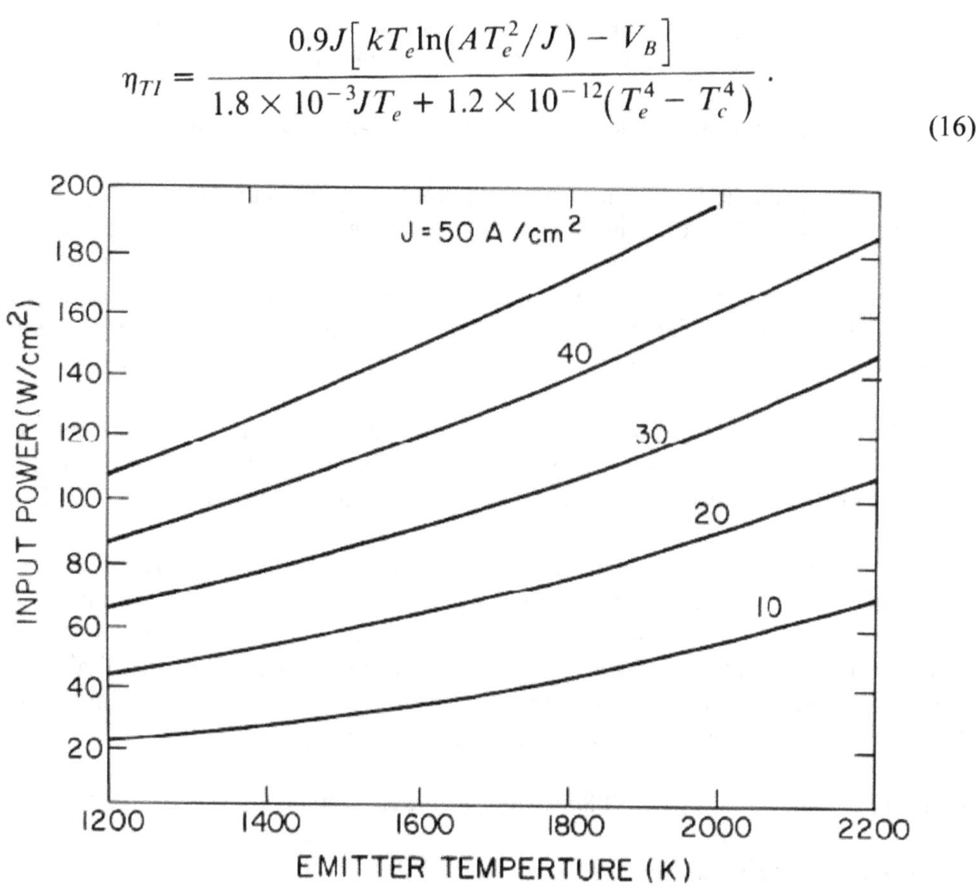

Fig. 13. Input power requirements for thermionic converter

As illustrated in Fig. 14, η_{TI} is relatively insensitive to the current density. Consequently, converter performance remains near its design level under partial load conditions. The output power density, however, is almost directly proportional to current density. Thus, the design current density must be carefully selected to optimally match the heat flux from the nuclear heat source, the electric load requirement, and space system size and mass constraints.

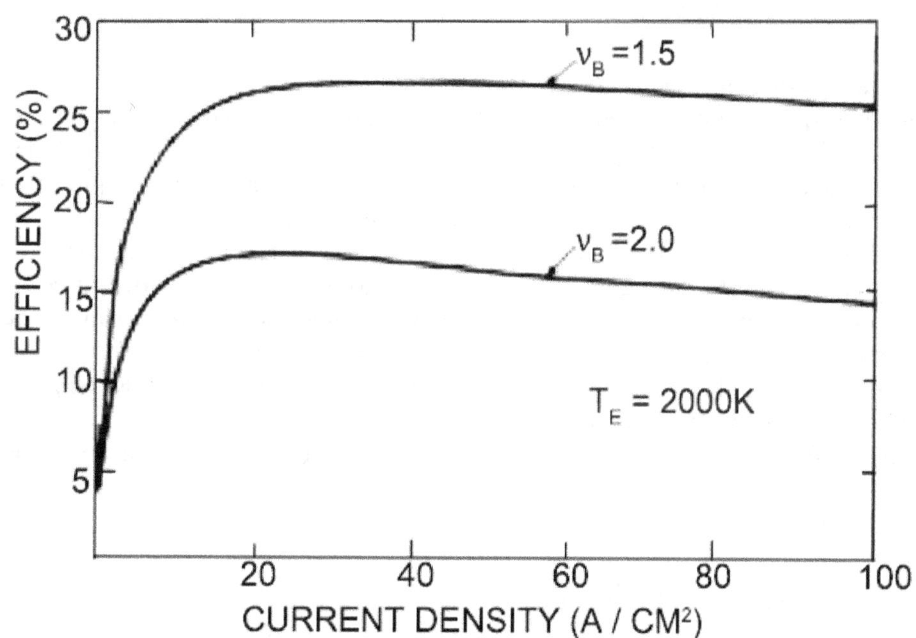

Fig. 14. Thermionic converter efficiency as a function of current density and barrier index [I].

Thermionic converter technology has been under development in the United States since the 1960s at varying levels of support. First generation converters tested under the in-core (i.e., in the reactor core) thermionic program in the early 1970s produced power densities of 6 W / cm^2 at 1850 K. This development effort was cancelled in 1972, when the majority of the U.S. space nuclear power effort was terminated.

Continuing thermionic research at lower emittance temperatures permits the converter system to become an out-of-core design--that is, the reactor core and the power conversion system are physically separated, resulting in a much more effective optimization of both the nuclear reactor and the thermionic power conversion system. Out-of-core thermionic systems separate the radiological and fission products problem from the thermionic element development. Now, however, other components in the reactor become subjected to high temperatures and very high temperature heat transport is needed.

During the SP-100 program, significant progress had been achieved in the development of thermionic space reactors. For in-core thermionic systems, the good design feature is that all of the components outside of the thermionic fuel element are at temperatures that are straight forward for design purposes and experience.

Collector temperatures and other components outside the thermionic convertors are 1000 K or less. Difficulties with the TFE design result from very high temperature operation with emitter temperatures of around 1,800 K, the integral nature of the fuel and thermionic converter resulting in radiological and fission product interactions design problems, very tight tolerances needed in the converter designs, and the limitations on performing accelerated testing because materials are operating near their limits. The latter limits the rate that in-core thermionic fuel elements can be developed and demonstrated for long life systems.

Two Russian reactors launched in 1987-1988 incorporated thermionic converters.

Dynamic Power Conversion Generators

Dynamic power conversion units can also be coupled with fission reactor heat sources to provide much higher energy conversion efficiency than static converters. However, such dynamic power conversion systems have moving parts that require long-lived bearings for years of maintenance-free operation in space. The most reliable bearings are those that support the moving parts on a film of the working fluid during operation. In addition, dynamic systems have inherent rotational torques and vibrational forces that must be considered in spacecraft integration.

Principles of Dynamic Thermal-to-Electric Conversion Cycles[27]

A heat engine can be defined as a controlled-mass, thermodynamic device that receives thermal energy from a source, converts a portion of this energy into mechanical work (which is transferred across the system boundaries to the surrounding application), and then rejects the unused amount of input thermal energy to a lower temperature sink. The output work of interest here is in the form of electrical power. The second law of thermodynamics establishes the maximum amount of input thermal energy that can actually be converted into work. Real heat engines fall short of this performance optimum as a result of irreversibilities caused by the heat transfer process with large temperature differentials and working fluid-induced frictional effects. The working fluid within the heat engine undergoes cyclic operation; after some period of time all the working fluid within the device returns to its initial state. A generic heat engine that operates between a high temperature heat source (T_H) and a low temperature heat sink (T_L) is shown in Fig 15.[28, 29] This type of engine is called a $2T$ heat engine, since thermal energy is transferred into the system at one temperature and rejected from the system at another (lower) temperature.

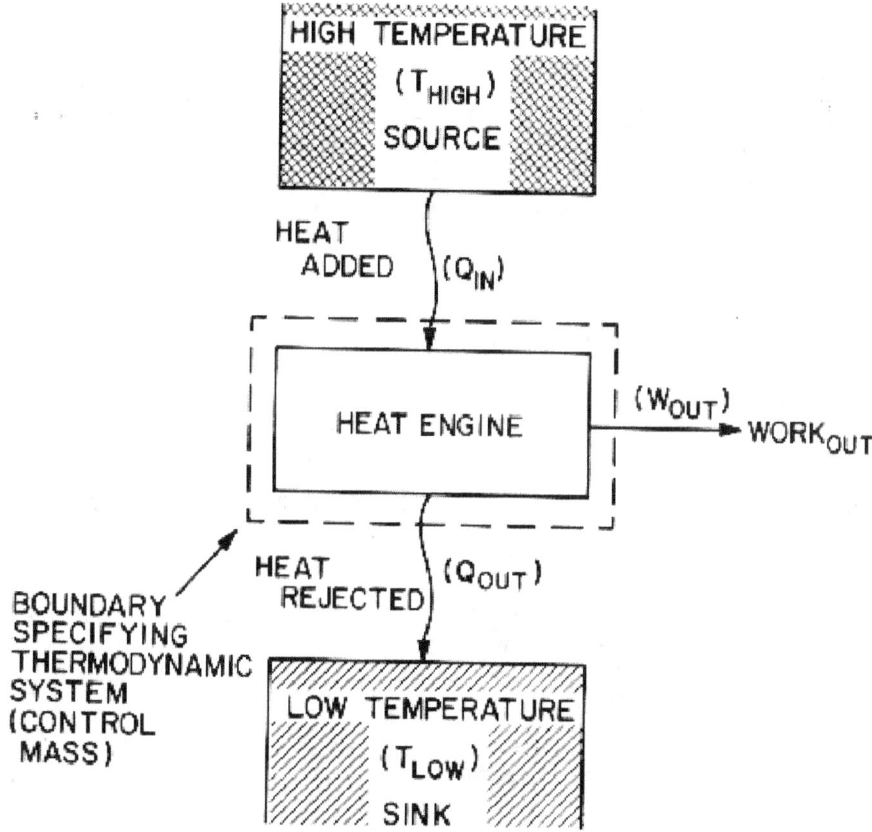

Fig. 15. The reversible two temperature heat engine.

Efficiency expresses the effectiveness of using the heat available. The efficiency of a system can be defined as:

$$\eta \equiv \frac{\text{some useful effect}}{\begin{array}{c}\text{energy that must be expended}\\ \text{to achieve that effect}\end{array}} .$$

$$(17)$$

The maximum thermal efficiency of a heat engine is called its Carnot efficiency (η_{Carnot} or (η_{th}). The Carnot cycle is a fundamental theoretical concept in dynamic energy conversion, because the thermal efficiency of this cycle is the maximum possible value for any heat engine operating between the same two temperature limits.[30, 31] Carnot (1796-1832) was a French military engineer who performed pioneering examinations of heat engine efficiencies in the early 19th century. Fig. 16 shows the pressure-volume and temperature-entropy diagrams of the Carnot cycle. In this idealized cycle, a working fluid experiences: reversible, isothermal heat addition from state I to state 2; reversible, adiabatic (i.e., isentropic) expansion from 2 to 3 during which work is transferred from the system to its environment; reversible, isothermal heat rejection from state 3 to state 4; and finally, a reversible, adiabatic (i.e., isentropic) compression process from 4 to I during which work is transferred to the system from the surroundings. By definition, an isentropic or constant entropy process is a reversible, adiabatic process

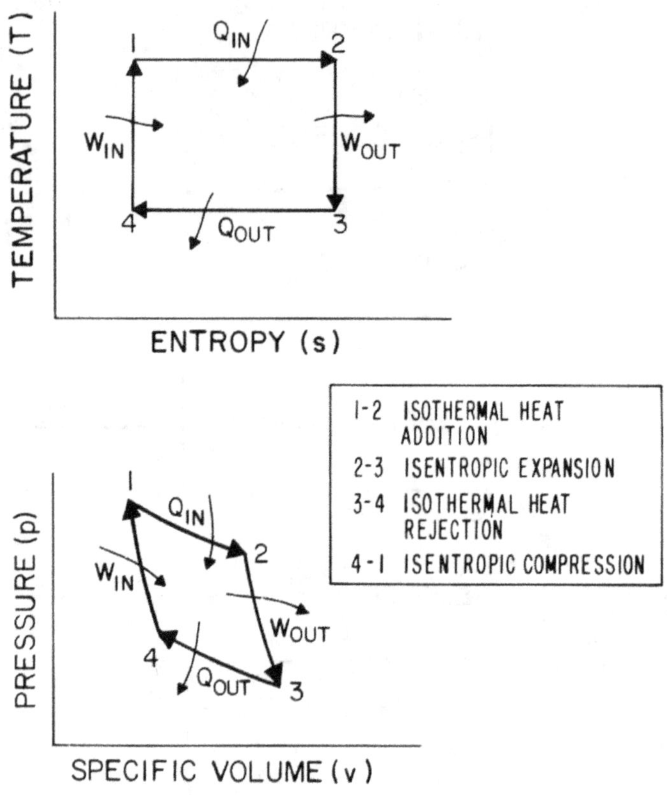

Fig. 16. Carnot cycle.

In theory, isothermal heat addition and rejection can be achieved by using suitable heat exchangers, while isentropic turbines and compressors or pumps may be used to transfer energy as work to or from the working fluid. The theoretical efficiency of this cycle is given by

$$\eta_{CARNOT} = \frac{\text{WORK}_{out}}{\text{HEAT}_{IN}} = \frac{T_H - T_L}{T_H} = 1 - T_L/T_H \tag{18}$$

where T_H is the higher absolute temperature (K) at which heat is added
$\quad T_L$ is the lower absolute temperature (K) at which heat is rejected.

Eq. 18, shows that the thermal efficiency of the system is greater for the higher the heat addition temperature (T_H) and the lower the heat rejection temperature (T_L). The Carnot efficiency represents the maximum thermal efficiency (η_{th}) of any heat engine operating between the same source (T_H) and the sink (T_L) temperatures.

Sometimes in treating a spacecraft's energy conversion system, thermal efficiencies (i.e., Carnot limitations) are separated from other system losses. These other losses are both electrical or mechanical. In that case, a power conversion system (PCS) efficiency can be defined as

$$\eta_{PCS} = \eta_{th} \cdot \eta_D \tag{19}$$

where $\quad \eta_{th}$ is the thermal efficiency
$\quad \eta_D$ is the device efficiency.

Brayton Cycle Principles

The Brayton cycle is probably the more straight-forward of the dynamic conversion cycles. The working fluid is always in the gaseous state. A closed cycle gas turbine power system is illustrated in Fig. 17. An idealized Brayton cycle has heat addition and rejection occurring at constant pressure, while expansion and compression processes are assumed to be isentropic. The working fluid process is: constant pressure heat addition from state 1 to 2; isentropic expansion in the turbine from state 2 to 3 (a portion of the output work goes to compressing the working fluid); constant pressure heat rejection from state 3 to 4; an isentropic compression from state 4 to 1.

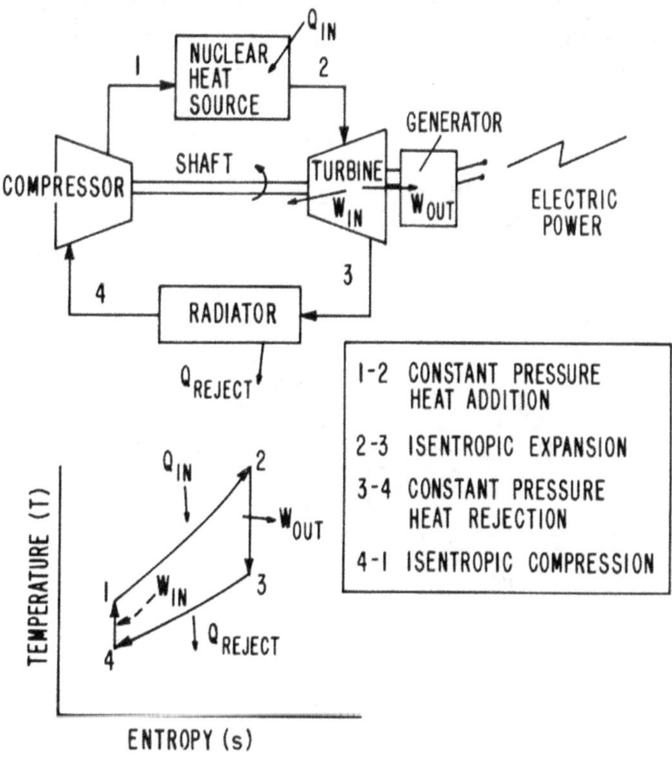

Fig. 17. Closed Brayton cycle (ideal).

The thermal efficiency of the ideal closed Brayton cycle is given by: (neglecting working fluid potential and kinetic energy changes)

$$\eta_{th} = \frac{(W_{out})_{turbine} - (W_{in})_{compressor}}{Q_{in}}$$

(20)

or,

$$\eta_{th} = \frac{(h_2 - h_3) - (h_1 - h_4)}{(h_2 - h_1)}.$$

(21)

If the working fluid is assumed to be an ideal gas, the thermal efficiency is given by

$$\eta_{th} = 1 - (p_1/p_4)^{(1-k)/k}$$

(22)

where $k = c_p / c_v$, the ratio of the specific heats of the gas at constant pressure and volume respectively.

To improve the thermal efficiency of the basic closed Brayton cycle, a regenerator (or recuperator) can be incorporated into the cycle, see Fig. 18. In the regenerator, the hot gases, which are exhausted from the turbine (i.e., state 3), preheat the working fluid as it exits the compressor (state 6) and returns to the nuclear power supply (state 1).

Another way to improve Brayton cycle efficiency is through the use of multistage compression with intercooling and multistage expansion with reheat. In an ideal intercooler (see Fig. 19) the gas is cooled to the initial inlet temperature T_1 before entering the next state; that is, $T_1 = T_3$ and $p_2 = p_3$.

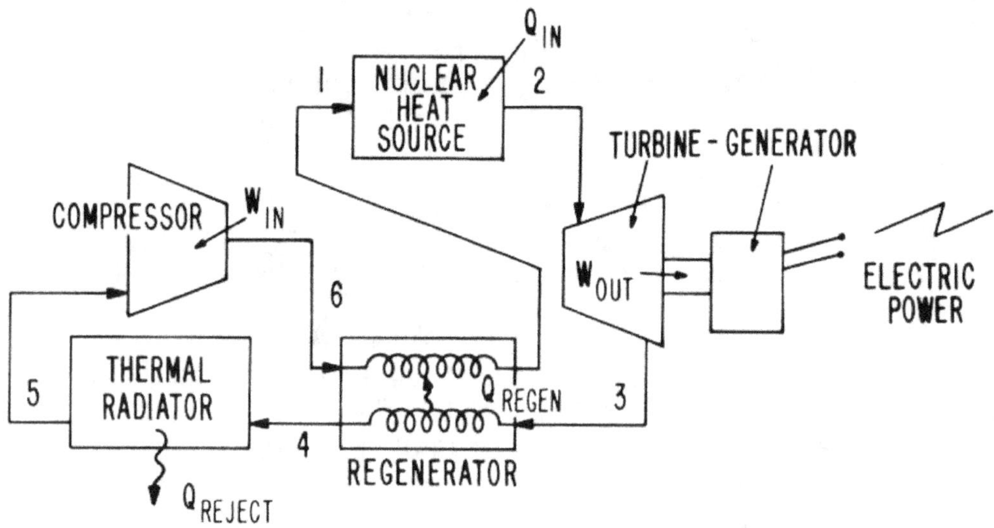

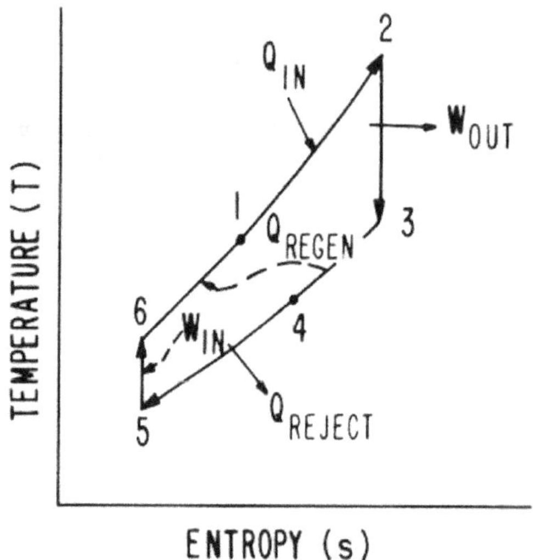

Fig. 18. Closed Brayton cycle with regeneration (ideal).

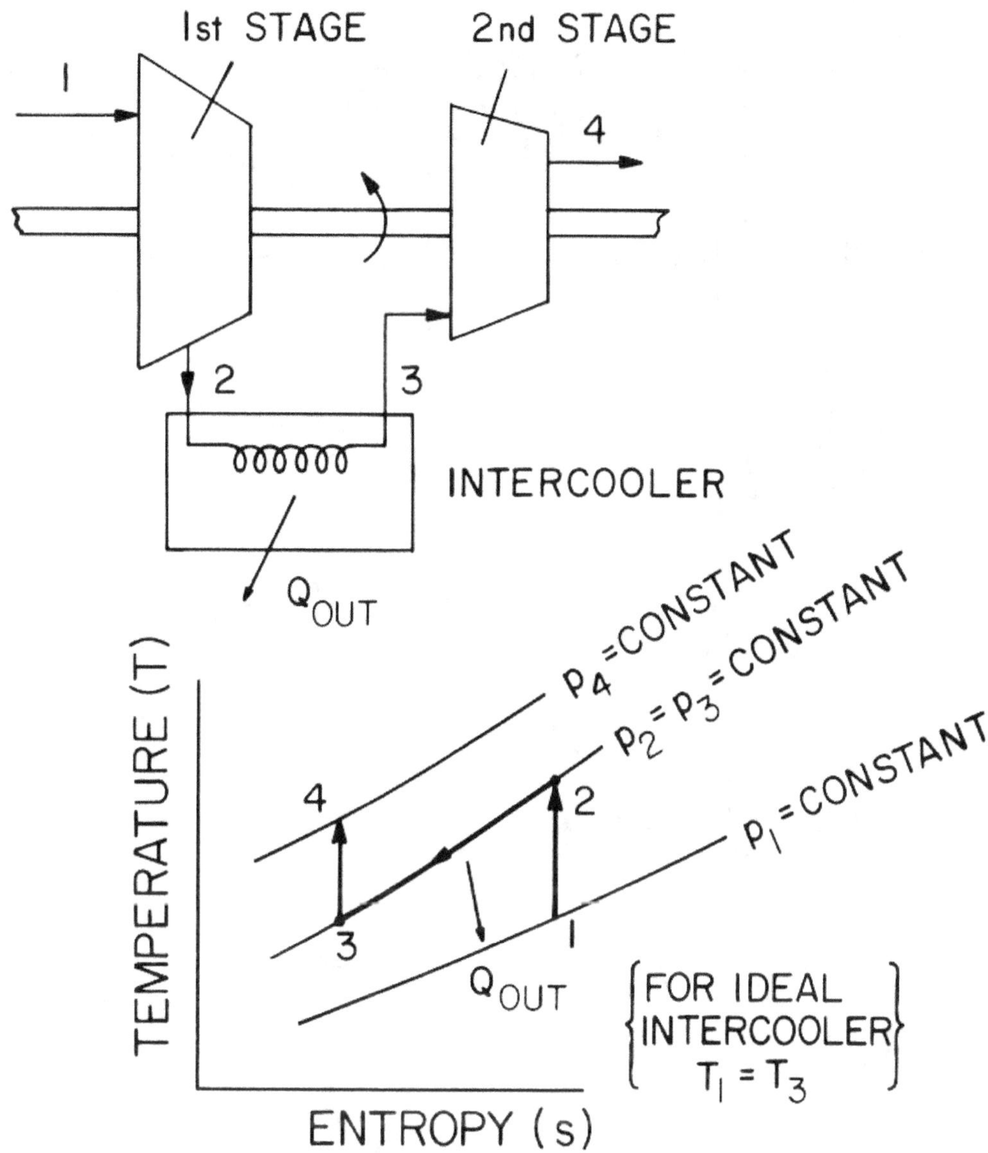

Fig. 19. Intercooling: ideal two-stage compressor.

The combination of reheating, intercooling and regeneration provides the best practical Brayton cycle efficiency. If regeneration is used along with a very large number of reheating stages and intercooling states, then, in principle, all thermal energy input eventually appears in the reheat-heat exchangers (when the gas is at its maximum temperature) and all heat rejection occurs in the intercoolers (when the gas is at its lowest temperature). This condition can approach the Carnot efficiency of a reversible $2T$ heat engine (see Fig. 20).

The Brayton cycle converter was selected by the Prometheus Program as the generator of choice for that system. Also, it is a candidate for the Fission Surface power system.

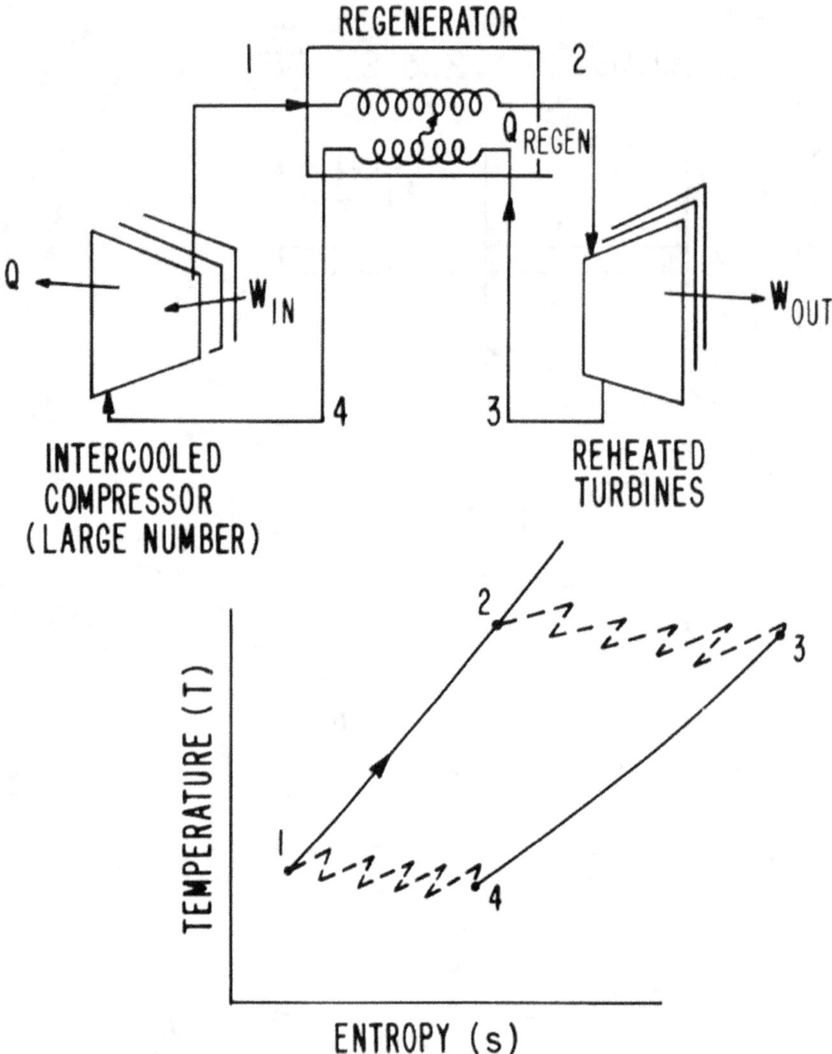

Fig. 20. Regenerative Brayton cycle (ideal) with a large number of reheating and intercooling stages.

Rankine Cycle Principles

The Rankine thermodynamic cycle is named after the Scottish engineer William Rankine (1820-1872). He made fundamental contributions to engineering thermodynamics, especially with respect to the steam engine. The basic Rankine cycle is a form of heat engine that includes a working fluid having two-phases as part of the cycle.

In simple terms, the Rankine cycle heat engine is composed of a heat source, in this case the reactor fuel; a turbine-generator where thermal energy is converted to electric power; a heat rejection radiator where the fluid undergoes a phase change and waste heat is rejected to space; and a pump to increase working fluid pressure. In thermodynamic terms, the Rankine cycle (Fig. 21) working fluid undergoes an isentropic expansion in the turbine from state 1 to state 2. The "wet" liquid-vapor mixture then experiences a phase change to a saturated liquid as it undergoes isothermal heat rejection in the condensing radiator (state 2 to state 3). This is followed by isentropic compression in the pump from state 3 to state 4. Here, the liquid is assumed to be incompressible and actually goes from the saturated liquid state (3) to the sub-cooled liquid state (4). Finally, the working fluid experiences constant pressure heat addition from state 4 to state I. At state 4' sensible heat addition gives way to phase change--so that, as more thermal energy is added, the working fluid gradually changes from a saturated liquid (point 4') to a saturated vapor (point 1) of 100 percent quality. This phase change process occurs at both constant temperature and constant pressure.

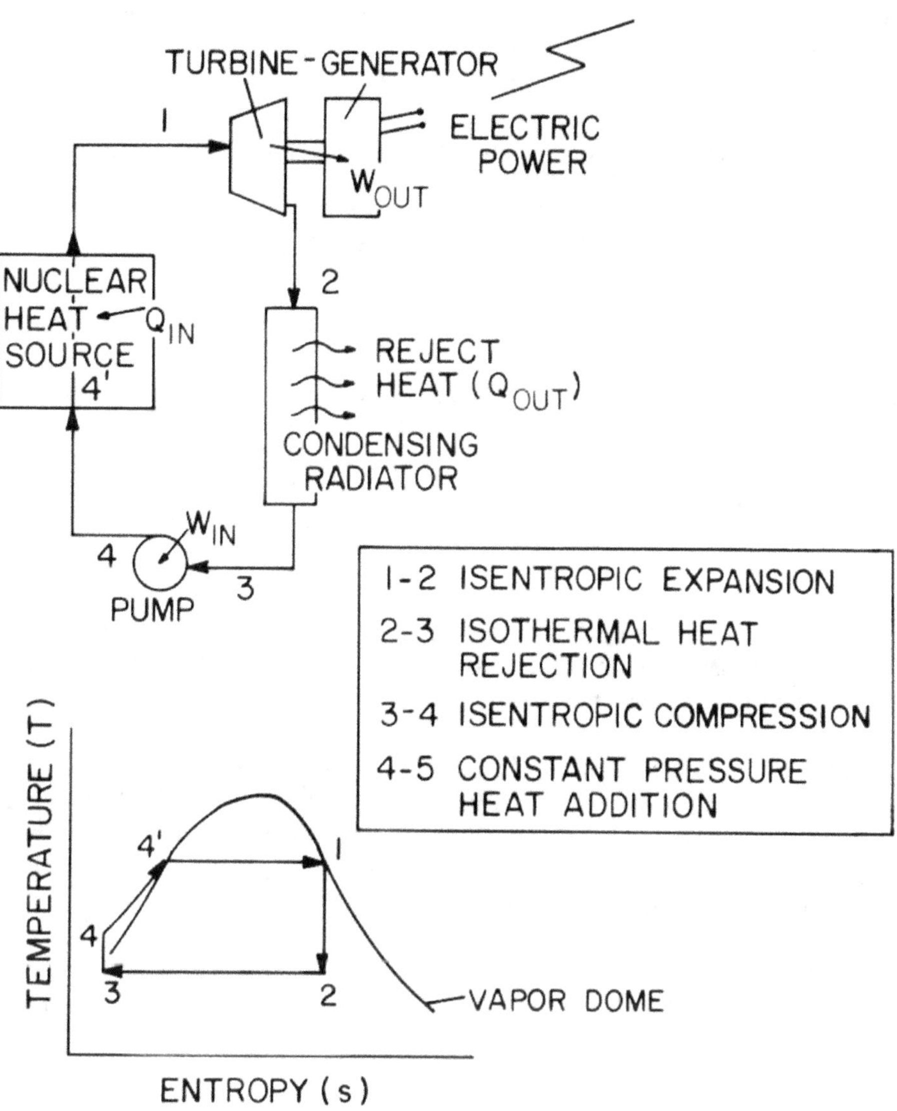

Fig.21. Basic Rankine cycle (ideal).

The quality (x) of a liquid-vapor mixture is defined in thermodynamics as the fraction of mass of a working fluid in the vapor phase. Similarly, the wetness of a mixture ($1 - x$) describes a liquid-vapor mixture with a quality less than 100 percent. The thermodynamic properties of a two-phase mixture are designated in terms of the specific enthalpies:

$$h_{\mathrm{mix}} = (1 - x)h_f + xh_g$$

(23)

where h_{mix} is the enthalpy of the mixture (J / kg)

      h_f is the enthalphy of the saturated liquid (subscript "f" for liquid phase) (J / kg)

      h_g is the enthalphy of the saturated vapor (subscript "g" for vapor phase) (J / kg)

      x is the quality of the mixture.

A subcooled or compressed liquid is one that exists at a temperature lower than the saturation temperature, corresponding to its pressure; a superheated vapor is a vapor that exists at a temperature greater than the saturation temperature, corresponding to its pressure.

Neglecting changes in the potential and kinetic energy of the working fluid in the basic Rankine cycle, the thermal efficiency is:

$$\eta_{th} = \frac{(\text{work out})_{turbine} - (\text{work in})_{pump}}{\text{heat added}}.$$

(24)

Assuming ideal components throughout the system yields

$$\eta_{th} = \frac{(h_1 - h_2) - (h_4 - h_3)}{(h_1 - h_4)}.$$

(25)

The thermal efficiency of the Rankine cycle can be improved by superheating the working fluid, as shown in Fig. 22. Here, the constant pressure heating of the working fluid is continued past the saturated vapor state (state I) and into the superheated vapor region (state 1').

The thermal efficiency of the basic Rankine cycle can also be improved using the reheating process shown in Fig. 23. The working fluid is expanded in the first turbine stage (from point l' to 2) until it reaches pressure p_2 (on the saturated vapor line). Then, the working fluid is returned to the nuclear power source where it is reheated at constant pressure to temperature T_2 in the superheated vapor region (state 2^1). Next, it undergoes expansion to state 3 in the second stage of the turbine.

Superheat and reheat reduce the amount of moisture in the working fluid entering the turbine. This prevents liquid droplet erosion of turbine blades and raises the average temperature at which thermal energy is added to the working fluid. Both techniques are used to improve overall Rankine cycle efficiency.

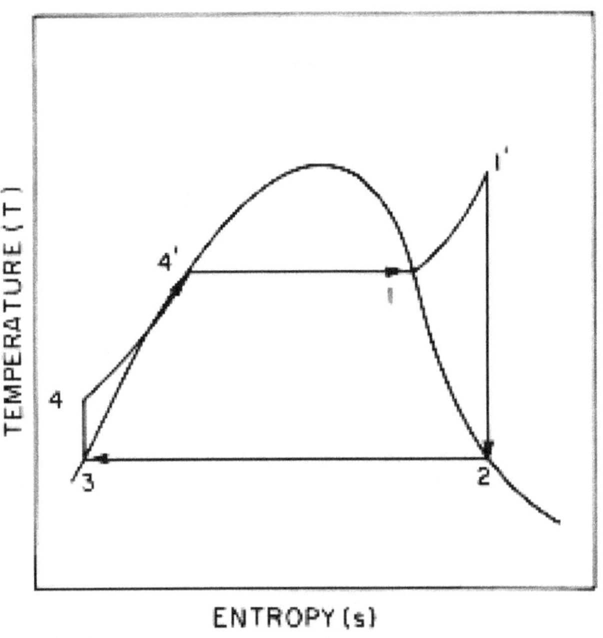

Fig.22. Rankine cycle with superheat.

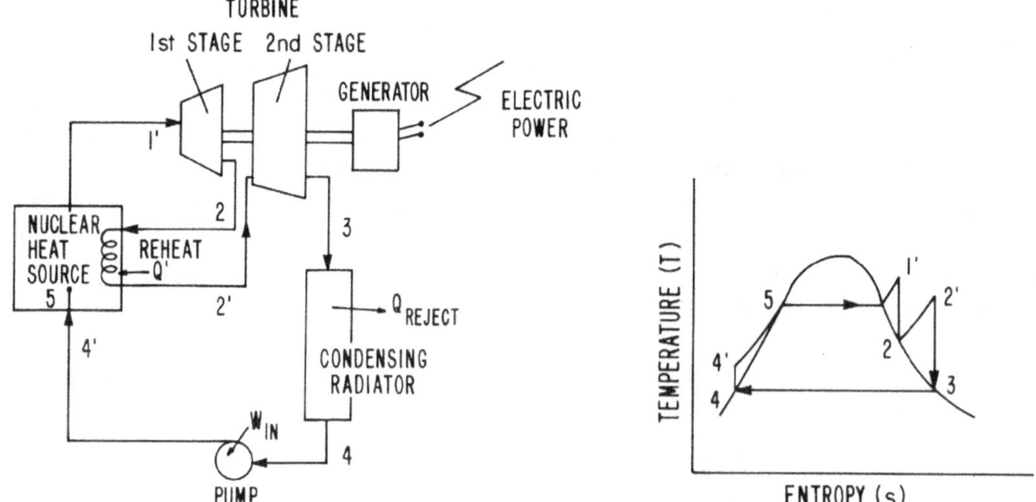

Fig. 23. Rankine cycle with reheat.

Regeneration also can be used to improve the overall efficiency of the Rankine cycle. As shown in Fig 24, when the working fluid expands in the turbine some of its thermal energy content is used to preheat the liquid-phase working fluid before it enters the nuclear heat source. This technique allows a close approach to the ideal Carnot cycle efficiency, since thermal energy transfer in the overall cycle now occurs isothermally. An ideal turbine as a power producing device is not an ideal heat exchanger. Therefore, many engineering trade-offs are made in practical Rankine cycle applications, including the use of superheat, reheat, and regeneration.

Extensive development work was performed on mercury Rankine cycle converters for SNAP 2 and 8 and potassium Rankine cycle for SNAP 50.

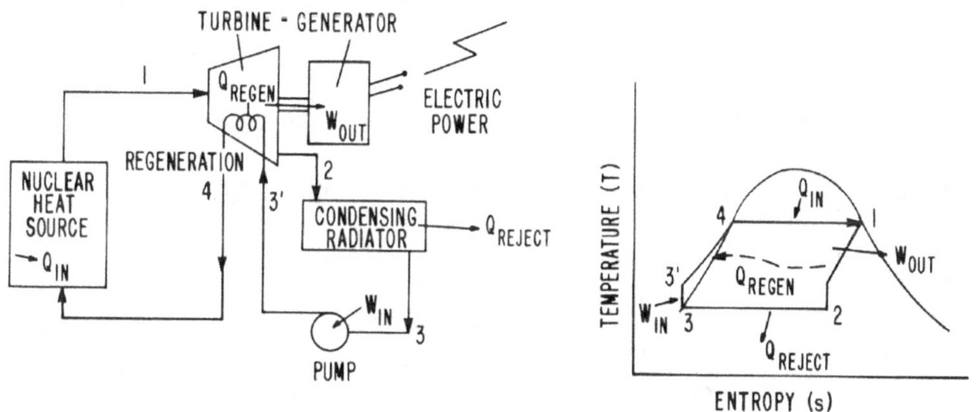

Fig. 24. Ideal Rankine cycle with regeneration.

Principles of Stirling Cycles[32]
To achieve a totally reversible heat engine involves devising a means by which all thermal energy transfer to and from the system takes place both isothermally and reversibly. Robert Stirling (1790-1878) designed such a cycle in the early 19th century. In fact, the Stirling engine is the earliest example of a reversible heat engine. Robert Stirling, along with his brother James, used regeneration in developing this engine. Stirling engines enjoyed considerable success as quiet pumping engines. However, with the development of more compact internal combustion engines, Stirling engines fell into disuse. Space power applications have revised interest in the Stirling cycle.[33]

A heat engine may incorporate a component called an ideal regenerator. Through the regenerator, heat is alternately stored and then recovered, reversibly. To accomplish this, the working fluid transfers heat on a temporary basis, while dropping in temperature from some upper value (T_{high}) to some lower value (T_{low}). Inherent in this ideal regenerator concept is the assumption that heat transfer occurs reversibly -- that is, there is no temperature difference between the heat absorbing regenerator material (such as wire mesh or tiny thin-walled tubes) and the working fluid at any point where they are in the thermal contact (see Fig. 25). Then, when the working fluid passes back through the regenerator, entering T_{low}, it recovers the heat which was originally stored and leaves at T_{high}. Since the regenerator is in exactly the same state after completion of a cycle as it was before, it is considered as a component of the heat engine and not as a part of some external heat source or sink. In the Stirling engine this reversible heat exchange in the regenerator occurs at constant volume.

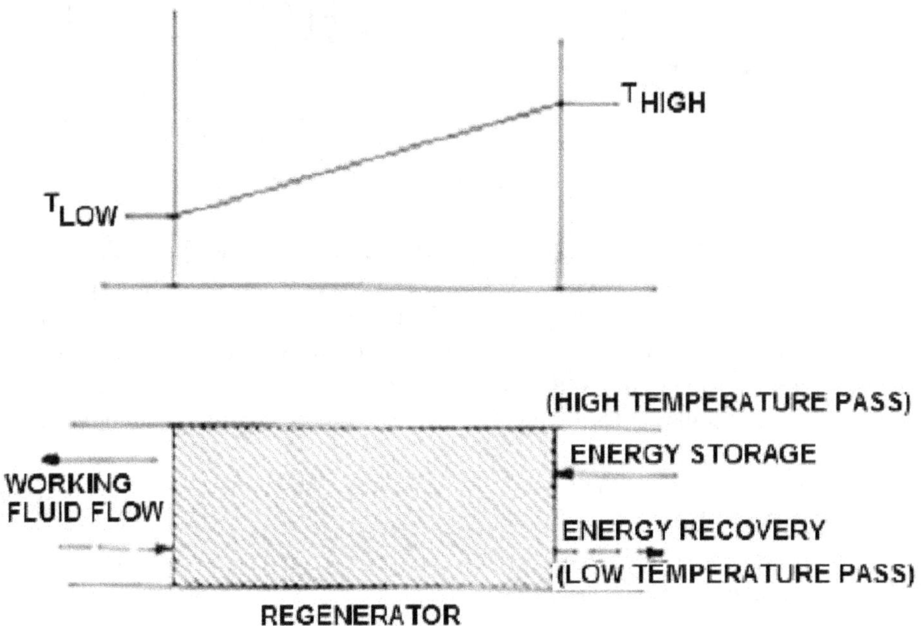

Fig. 25. Flow of a working fluid through an ideal regenerator.

Fig. 26 illustrates the basic Stirling cycle with an ideal gas working fluid. From state 1 the gas, initially at the lower temperature limit, recovers the stored thermal energy from the regenerator in a reversible, constant volume process. This continues until the working fluid reaches the upper temperature limit, state 2. Then, from state 2 to state 3 the working fluid experiences reversible heat addition from the external heat source (also at T_{high}) and expands isothermally to state 3. From state 3 to state 4, the working fluid again interacts with the regenerator in a constant volume process. This time, however, thermal energy is transferred from the working fluid to the regenerator. The ideal gas goes from T_{high} at state 3 to T_{low} at state 4. An isothermal compression takes place from state 4 to state 1. During this final portion of the Stirling cycle, the working fluid rejects heat to an external sink that is also at T_{low}. The thermal efficiency of the ideal Stirling cycle is

$$\eta_{Stirling} = 1 - T_{low}/T_{high}$$

(26)

where T_{low}, T_{high} are absolute values of temperature (K). Equation 12 is the identical mathematical expression for the thermal efficiency of the Carnot cycle operating between the same temperature limits.

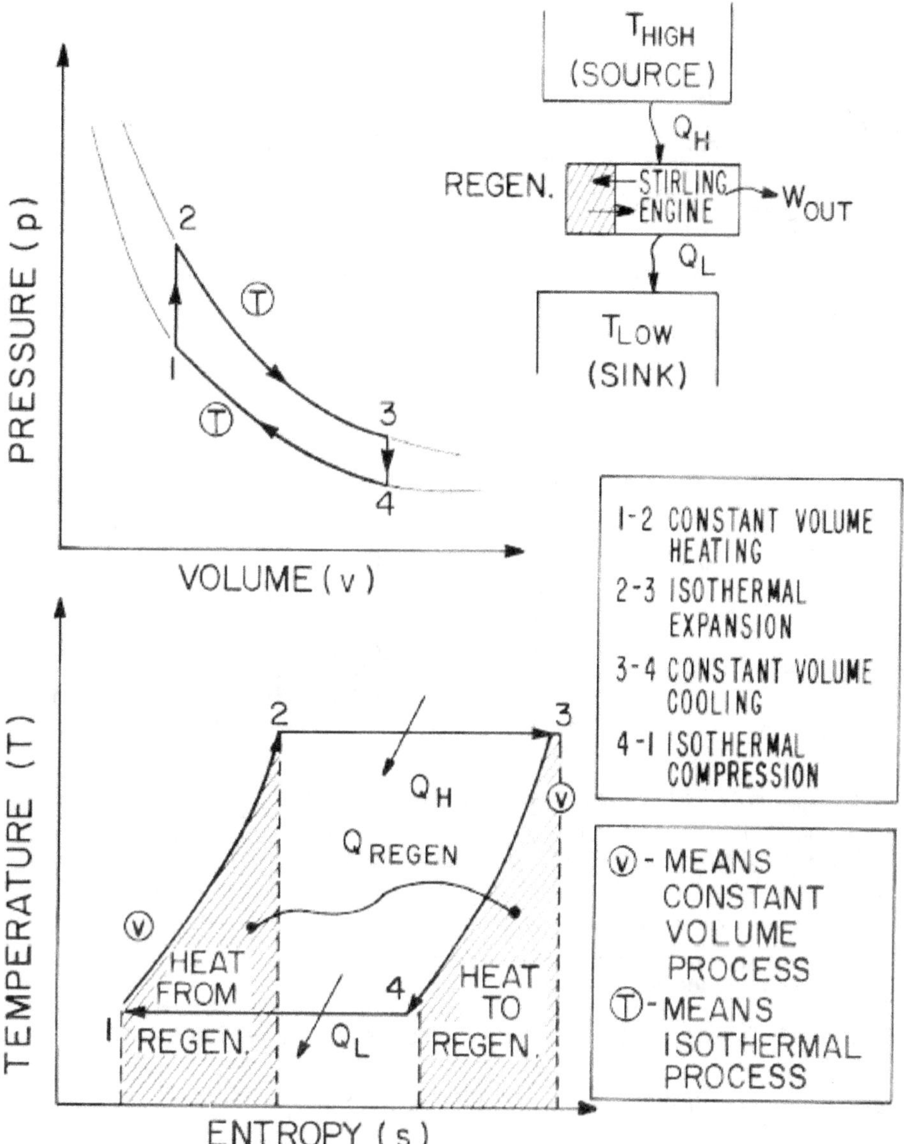

Fig. 26. Ideal Stirling cycle.

The free piston Stirling engine (FPSE) of interest in space applications is a thermally driven mechanical oscillator, using the Stirling cycle and deriving its output power from heat flow between a source and a sink.[34, 35] The FPSE displacer motion is produced by gas pressures rather than by mechanical linkages. This type of engine operates at the highest device efficiency of all known heat engines and is uniquely suited to drive direct-coupled reciprocating loads (such as linear generators) in a hermetically sealed configuration and without requiring high pressure shaft seals or contaminating lubricants.

The Free-PistonStirling Engine (FPSE) consists of three basic components (see Fig. 27): a power piston (labeled 'piston' on the figure), the displacer piston and a sealed cylinder. The displacer and the power piston assembly are the only two moving parts in the FPSE engine-alternator unit. The displacer rod passes through the power piston and is in communication with the bounce space. The bounce space acts as a constant pressure chamber. The power piston assembly is composed of the power piston directly connected to the alternator armature. This eliminates the need for converting reciprocating motion to rotary motion and importantly the need for lubricating such a Stirling device. The assembly is supported by two pressurized (hydrostatic) helium gas bearings. Long-term reliability is inherent in this design, because compressor inlet and discharge flow is controlled by means of fixed ports rather than valves.

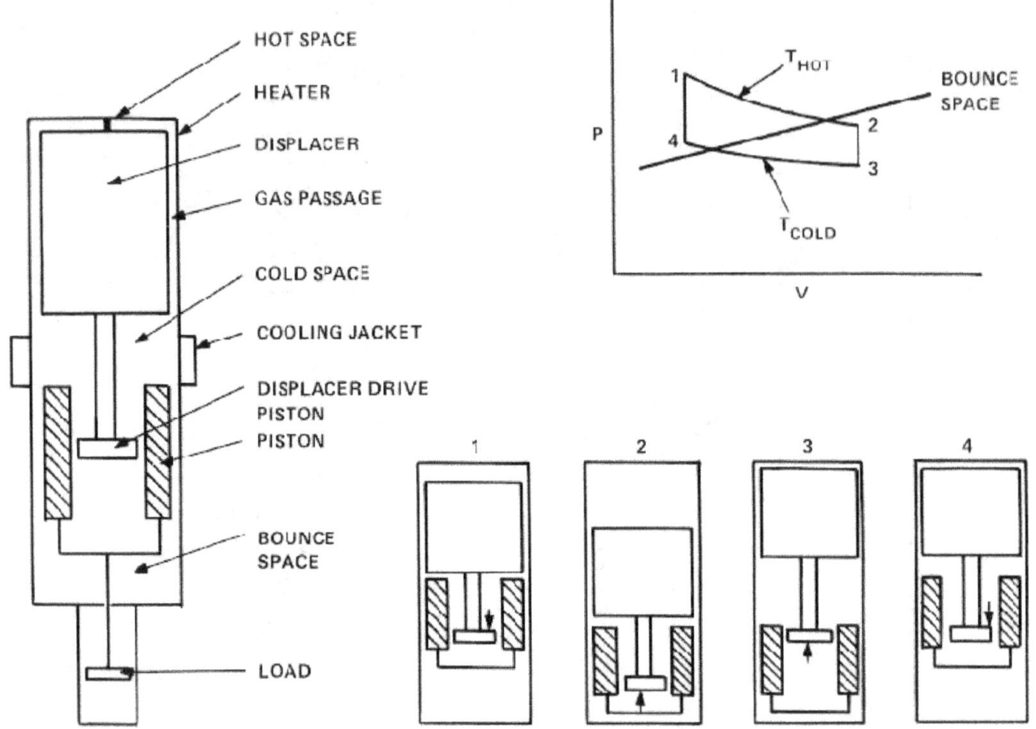

Fig. 27. Operational principles of free piston Stirling engine (FPSE).

The Stirling cycle is a candidate for development for the Fission Surface Power (FSP) system.

Heat Transport

Energy must be moved from the reactor to the power generating components and waste energy from the power generation components to the heat rejection subsystem. The mechanism usually takes the form of pumped loops, though sometimes heat pipes are used. The heat transport must: (1) be hermetically sealed to prevent fluid leakage; (2) have a minimum mass and be mass-compatible with the overall power system mass budget; (3) have a satisfactory interface with the other components of the power system, e.g., nuclear heat source, shield, power conversion system and radiator; (4) be capable of removing all the heat from the nuclear heat source and its components so that no component exceeds its design temperature; (5) be reliable; (6) accommodate all thermally induced volumetric and dimensional changes; (7) minimize thermal energy losses; (8) be capable of startup with proper functioning after launch and after a normal or scheduled shutdown; and (9) for pumped-loop design systems, the fluid pumping power requirements must be consistent with the overall system mass and power requirements. Primary heat transport arrangements are depicted in Fig. 28.

Pumped-Loops

Heat transport performed by pumped-loops at the temperature levels of interest in space nuclear power systems are generally metals in either the liquid or vapor phase or are gases. The working fluids are typically transported in thin-walled, metal alloy tubing.

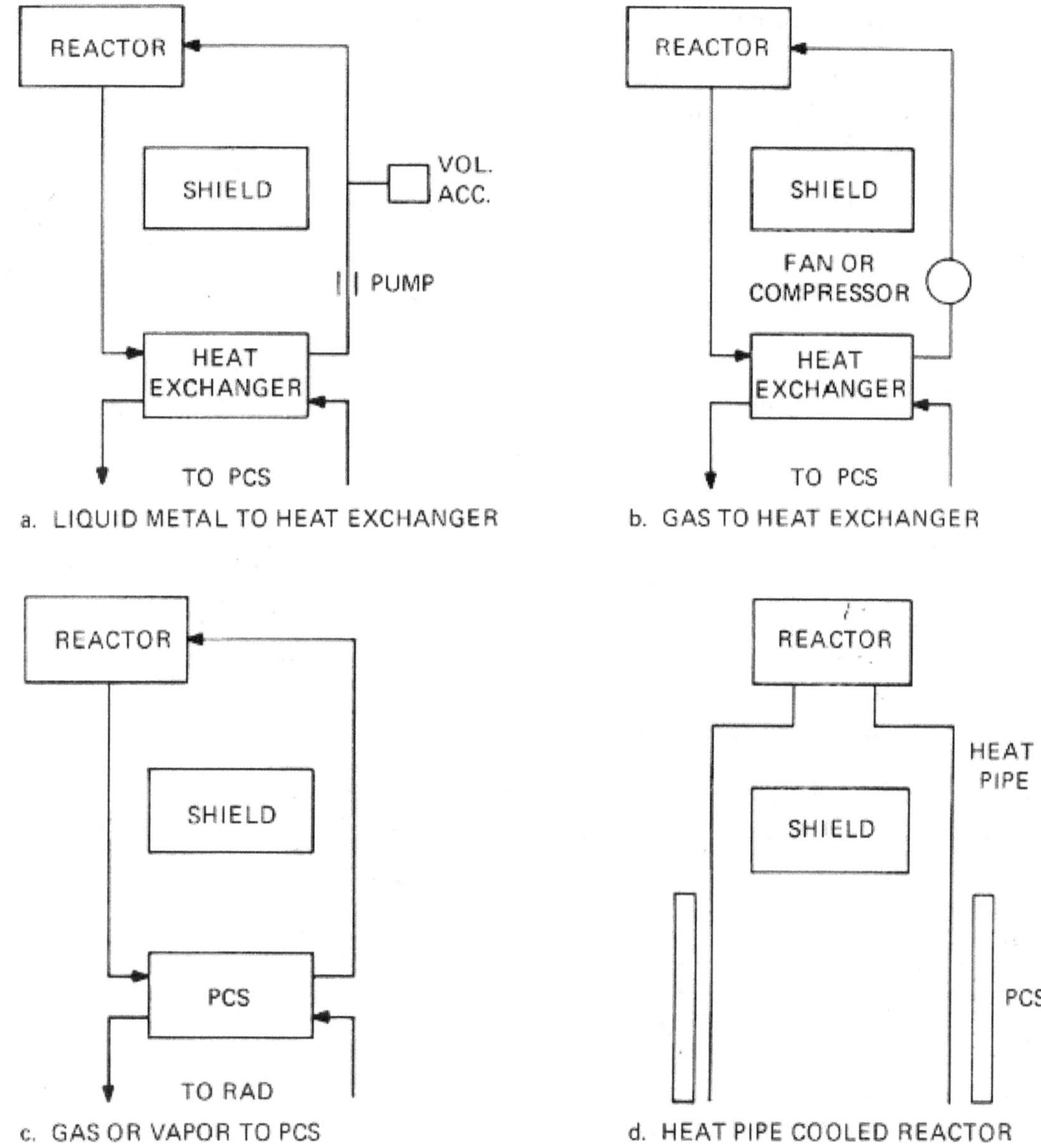

Fig. 28. Various primary heat transport loop configurations for a space reactor power plant. *Courtesy of Department of Energy and Rockwell International.*

High thermal fluxes and high operating temperature environments has resulted in the selection of liquid metals as working fluids in many systems. The liquid metals offer superior thermophysical properties such as high thermal conductivity and low vapor pressure. Also, they are stable at high temperatures and in intense radiation fields. Low atomic weight liquid metals, such as lithium and sodium, also have relatively high specific heats and volumetric heat capacities. However, liquid metals are: (1) generally extensively chemical reactivity at high temperatures; (2) require special containment materials to avoid corrosion; and (3) must be protected from oxidation. Physical properties of liquid metals most frequently considered as coolants for space nuclear power systems are summarized in Table 8. Sodium-potassium mixture (NaK) is often favored as the coolant because of its low melting point and, therefore, its simplified startup procedures. Sodium (Na) has limited application in very high temperature systems because of its low boiling point. For very high temperature systems, Lithium (Li) is favored, mainly because of its high boiling point.

Table 8. Physical properties of commonly used liquid metals [1].

General physical properties	Lithium (natural)	Mercury	Potassium	Sodium	Na (22%) K (78%)
Atomic number	3	80	19	11	
Atomic weight	6.94	200.61	39.100	22.997	
Density [kg/m³ (lb/ft³)]	479.0 (29.9)	13,231 (826)	714.4 (44.6)	823.3 (51.4)	741.7 (46.3)
Viscosity [MN · s/m² (lb/h · ft)]	339 (0.82)	1141 (2.76)	169 (0.41)	227 (0.55)	178 (0.43)
Surface tension [N/m (lb/ft)]	0.3472 (0.0238)	0.4482 (0.0307)	0.0834 (0.0057)	0.1471 (0.0101)	0.1138 (0.0078)
Electrical resistivity ($\mu\Omega$/cm)	36	108.3	48	29	71
Thermal properties					
Melting point [K (°F)]	454 (357)	234.5 ($-$37.97)	337 (146)	371.2 (208.1)	262 (12)
Boiling point [K (°F)]	1604 (2428)	631 (675)	1034 (1402)	1154 (1618)	1057 (1443)
Specific heat [J/g · K (Btu/lb · °F)]	1.286 (0.996)	0.0421 (0.0326)	0.235 (0.182)	0.389 (0.301)	0.270 (0.209)
Thermal conductivity [kJ/h · m · K (Btu/h · ft · °F)]	172 (27.6)	36.1 (5.8)	132 (21.2)	231 (37)	93.8 (15.05)
Heat of fusion [kJ/kg (Btu/lb)]	88.96 (185.9)	2.39 (5.0)	12.20 (25.5)	23.31 (48.7)	
Heat of vaporization [kJ/kg (Btu/lb)]	3995 (8349)	59.8 (125)	408.2 (853)	822.2 (1718)	
Prandtl number ($c_p\mu/k$)	0.0295	0.0154	0.0035	0.0044	0.0060
Change of volume on fusion (% solid volume)	1.5	3.6	2.41	2.5	2.5
Nuclear properties					
Fast activation cross section (mb)	0.03	89.0	0.59	0.87	0.64
Nonelastic scattering cross section at 2 MeV (barns)	< 1	2	4 (3.7 MeV)	< 1	NA
n, γ daughter half-life	None	48 days	12.4 h	15 h	See K, Na
n, γ daughter activity (MeV)	None	0.28	0.32, 1.5	2.775, 1.368	NA
ξ, average log energy decrement	0.2643	0.0100	0.0507	0.0852	NA
General properties					
Container material	Cb-Zr	Ferrous metals	304 SS	304 SS	304 SS
Reaction with uranium	Slight	Soluble	Slight (1)	Slight (1)	Slight
Reaction with plutonium	Slight	Soluble	Slight (1)	Slight (1)	Slight
Toxicity	Moderate	High	High	High	High
Fire or explosion hazard	High	None	High	High	High

These metals can usually be treated in their liquid phase as incompressible liquids. For an incompressible liquid the specific volume (v) is a constant and the specific internal energy change is given by:

$$u_2 - u_1 = \int_1^2 c(T)\,dT$$

$$(27)$$

where $c(T)$ is the temperature-dependent specific heat of the liquid (J / kg-K).

The specific entropy change for idealized incompressible liquid is given by:

$$s_2 - s_1 = \int_1^2 \frac{c(T)}{T}\,dT.$$

$$(28)$$

Equations 27 and 28 can be simplified by assuming that the specific heat is a constant ($\bar{c}$). Thus, the change in specific internal energy becomes:

$$u_2 - u_1 = \bar{c}\left[T_2 - T_1 \right]$$

$$(29)$$

and the change in specific entropy is:

$$s_2 - s_1 = \bar{c} \ln(T_2/T_1).$$

(30)

Finally, the specific enthalpy change for an idealized incompressible fluid with constant specific heat is given by

$$(h_2 - h_1) = \bar{c}(T_2 - T_1) + (p_2 - p_1)v.$$

(31)

Turning to gases, gases as a working fluid in a pumped-loop subsystem provides several distinct advantages: (1) thermal and radiation stability; (2) ease of handling; and (3) the absence of hazardous conditions. However, gases require a significant increase in pumping power. Possible coolant gases are Hydrogen (H_2) and helium (He). There characteristics are summarized in Table 9. Heat transfer and pumping power considerations favors hydrogen. However, its containment, especially under pressure, requires the use of materials which are not subject to attack or embrittlement. Helium, though not as favorable in terms of heat transport as hydrogen, has: (1) negligible neutron capture cross section; (2) thermal and radiation stability; and (3) the fact that helium is chemically inert and, therefore, neither hazardous nor corrosive to structural and containment materials. Space power systems using helium as the working fluid require high gas pressures to decrease reactor core size.

To remedy some of the limitations of helium, a mixture of helium and xenon has been used. The advantages of this mixture arise from a trade-off between the higher heat transport rates obtainable with low molecular weight gases versus the greater overall cycle efficiency (due to increased pumping efficiencies) obtained with high molecular weight gases, like xenon.

A gaseous working fluid can often be considered using the ideal or perfect gas approximation:

$$pv = \frac{\mathscr{R}}{M} T = RT$$

(32)

where p is the pressure (N / m^2)

     v is the specific volume (m^3 / kg)

     $\mathscr{R}$ is the universal gas constant (8314.5 J / kg-mole K)

     M is the molecular weight of the gas

     R is the gas constant for a particular gas (J / kg-K).

Table 9. Thermal and nuclear properties of coolant gases [1].

General Physical Properties	Hydrogen	Helium
Molecular weight	2.016	4.0
Density [kg/m^3 (lb/ft^3)]	0.030 (0.00185)	0.060 (0.00375)
Viscosity [N · s/m^2 (lb/h · ft)]	1.74×10^{-5}	3.8×10^{-5}
	(0.042)	(0.094)
Thermal properties		
Heat capacity [J/g · K (Btu/lb · °F)]	4.68 (3.625)	1.61 (1.245)
Thermal conductivity	1.40 (0.224)	0.99 (0.159)
[kJ/h · m · K (Btu/h · ft · °F)]		
Prandtl number	0.660	0.735
Nuclear properties		
σ_a , microscopic absorption cross section (barns)	0.332	0.0070
n, γ daughter half-life	None	None
n, γ daughter activity (MeV)	None	None
ξ, average log energy decrement	1.0	0.4281
General properties		
Container material	Carbon steel	Carbon steel
Reactions with U	Reacts	None
Reactions with Pu	Reacts	None
Toxicity	None	None
Fire or explosion hazard	High	None

The gas constant for a particular gas is given by:

$$R = \mathscr{R} / M \ \text{(J/kg-K)}.$$

(33)

Equation 33, called the ideal gas equation of state, has many useful forms and establishes relationships between the thermophysical properties needed to define the state of a thermodynamic system. A gas generally exhibits ideal behavior when the density is low, that is, at pressures low compared to the critical pressure p_{cr} or at temperatures high when compared to the critical temperature T_{cr}. The critical temperature and critical pressure are associated with the critical point, the thermodynamic state where the pure liquid phase has identical properties with the pure vapor phase at the same pressure and temperature. Helium, T_{cr} is 5.19 K and p_{cr} is 229.0 kilopascals (33.21 psi); hydrogen (H_2), T_{cr} is 33.24 K and p_{cr} is 1296.7 kilopascals (188.06 psi).

A microscopic interpretation of the ideal gas model includes the following physical interpretations: (I) the gas atoms or molecules behave as point masses and are widely separated; (2) because of wide particle spacing, intermolecular forces exert very little influence on collisional processes; and finally (3) the physical volume occupied by the gas molecules is very small compared to the total volume occupied by the gas.

Since the ideal gas model is just an approximation of real gas behavior, a compressibility factor Z has been developed for non-ideal gas behavior of such real gases. That is,

$$Z = pv / RT.$$

(34)

As Z approaches unity, the real gas approaches ideal gas behavior..

Heat Pipes

The heat pipe first emerged in the 1960s as a by-product of the early space nuclear program. Since, it has undergone an interesting transition from laboratory curiosity to a very useful thermal energy transfer device in a variety of applications.[36, 37] The heat pipe is a self-contained device that attains very high thermal conductances by means of two-phase fluid flow with capillary circulation. In ordinary conduction heat transfer, thermal energy is transported by the motion of atoms and electrons. Metals, because of their special atomic structure, have exceptionally high thermal conductivities. For example, pure silver has the highest known thermal conductivity of all metals. However, the effective thermal conductivity of a heat pipe can be thousands of times greater than that of solid silver.

Fig. 29 illustrates the elements of a heat pipe. An operating heat pipe is basically a container in which a working fluid circulates between heated and cooled regions. In the heat pipe, this fluid exists as two phases: liquid and vapor. The high thermal conductance property of the heat pipe is achieved using the latent heat of vaporization [L] (also called the enthalpy of evaporation) of a working fluid in the heated or evaporator region. Some typical values of L are shown in Table 10. The thermal energy transfer to the liquid causes it to experience phase change and become a vapor. The vaporized working fluid then flows from the evaporator region to the heat pipe's cooled or condenser region. There, the thermal energy transfer process is reversed. As the vapor cools, it becomes a liquid and gives up energy (i.e., it releases the enthalpy of evaporation). A wick structure within the heat pipe provides capillary action, permitting the return of the liquid to the evaporator region.

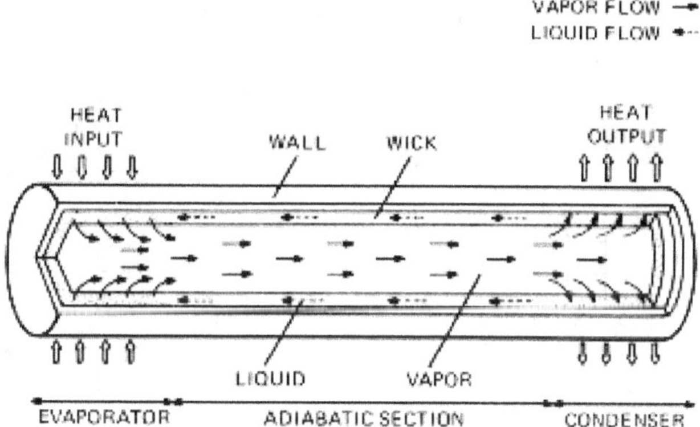

Fig. 29. Heat pipe operating principle. *Courtesy of Los Alamos 'National Laboratory.*

In a properly designed heat pipe, working fluid evaporation and condensation occur at approximately the same temperature and thermal energy transfer between a source and sink occurs isothermally. Although the working fluid circulates within the heat pipe, there are no moving mechanical parts and it appears externally as a passive device.

Table 10. Typical heat pipe working fluids. *Courtesy of Los Alamos National Laboratory.*

Working Fluid	Temperature (K)	Latent Heat of Vaporization (kJ / kg)
Ammonia	273	1263
Freon II	273	190
Acetone	273	564
Methanol	273	1178
Water	273	2492
Mercury	773	291
Cesium	773	512
Potassium	773	2040
Sodium	773	4370
Lithium	1273	20525

Heat Rejection Subsystem

Waste heat in space systems must be rejected by thermal radiation. Radiator design depends on both operating temperature and the amount of heat to be rejected. The amount of waste heat that can be radiated to space by a given surface area is determined by the Stefan-Boltzmann law and is proportional to the fourth power of the radiating surface temperature (see Fig. 30). To minimize system mass and size, this fourth power relationship drives the power generator design to higher temperatures.

In addition, the thermodynamic cycle efficiency is also sensitive to the heat rejection temperature. The lowest possible rejection temperature yields the highest thermal (Carnot) efficiency. Space power plants will usually operate at much lower thermodynamic efficiencies than experienced in terrestrial plants operating on similar cycles. This occurs because a high premium is often placed in space power system design on minimizing radiator area and mass, resulting in the selection of higher heat rejection temperatures.

Launch vehicle compatibility is another design constraint placed on the radiator system.

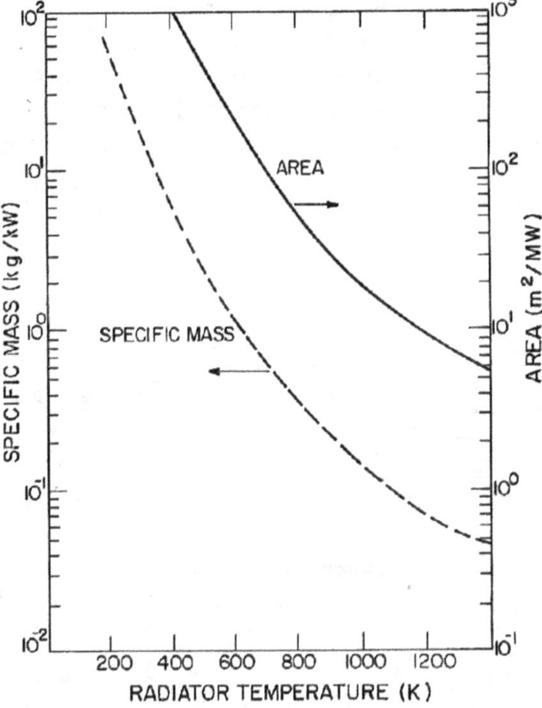

Fig. 30 Radiator area and mass as a function of temperature.

The thermal energy balance for a radiator must consider the following factors: (1) direct solar radiation; (2) Earth-emitted radiation; (3) Earth-reflected solar radiation (2 and 3 are frequently called "earthshine"); and (4) any internally generated heat load from the spacecraft and its nuclear power plant. Earthshine consists of reflected sunlight and thermal radiation from the Earth and its atmosphere. All of these factors represent thermal energy that must ultimately be rejected to space by the radiator. The thermal environment for a spacecraft operating in Earth orbit includes solar radiation (1371 ± 5 W / m²); Earth radiation (~ 240 W / m²); and the sink temperature for outer space (~ 0 K). [38, 39] From the first law of thermodynamics, the overall thermal energy balance equation for the radiator is:[40]

$$\alpha_s F_s G_s + \alpha_r F_r A_p G_s + \alpha_e F_e E_e + P_i / A = \epsilon \sigma T^4$$

$$(35)$$

where α_s is the solar absorptivity (0 to 1.0)

F_s is the cosine of the angle between the unit surface normal vector and the direction to the Sun   (0 to 1.0)

G_s is the solar radiation incident on a plane normal to the Sun [At one astronomical unit (AU) the solar constant = 1371 ± 5 W / m^2]

α_r is the absorptivity to solar radiation reflected from the Earth (0 to 1.0)

F_r is the view factor for solar radiation reflected by Earth (Typically, $F_r = 0.1$ for low-Earth-orbit and 0.02 for geosynchronous orbit)

A_p is the Earth's albedo; the fraction of incident solar radiation that is reflected by the Earth and   its atmosphere to space (Typical value ~0.3)

α_e is the absorptivity to radiation emitted by the Earth (0 to 1.0)

F_e is the view factor for radiation emitted by the Earth to the radiator surface

E_e is the earthshine radiation (typically ~ 240 W / m^2)

P_i is the internal waste heat load (W)

ε is the emissivity of the radiator surface

σ is the Stefan-Boltzmann constant (5.67×10^{-8} W / m^2 - K^4)

A is the area of the radiator (m^2)

T is the radiator operating temperature (K)

Potential radiator materials are a function of the operating temperatures. Fig. 31 shows some candidate materials.

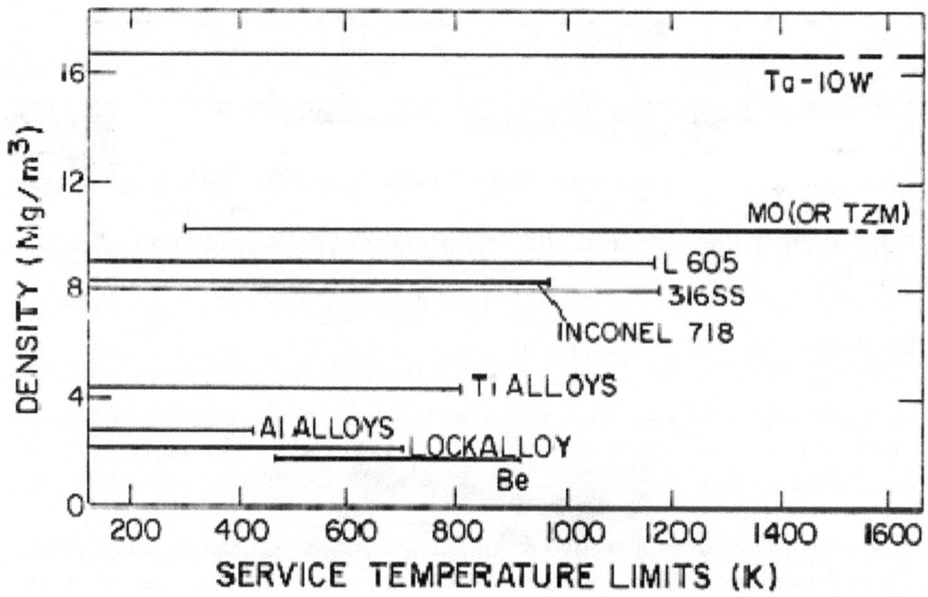

Fig. 31. Potential radiator materials for space power systems.

In order to improve radiator performance, the radiator components are covered with coating materials that have a high emissivity-to-absorptivity ratio at normal operating temperatures. Radiator temperatures are a significant factor in the optimization of the system size, power, and mass.

Shielding

Radiation attenuation shielding is used to protect the astronaut crew, payload and radiation-sensitive spacecraft equipment from the damaging effects of radiation. Fig. 32[41] presents typical radiation damage thresholds for certain materials, components, and systems that might be used in a nuclear-powered spacecraft. When such materials and components are exposed to nuclear radiation, their properties are changed in a variety of ways including rate effects and cumulative effects that reduce their overall established lifetimes.

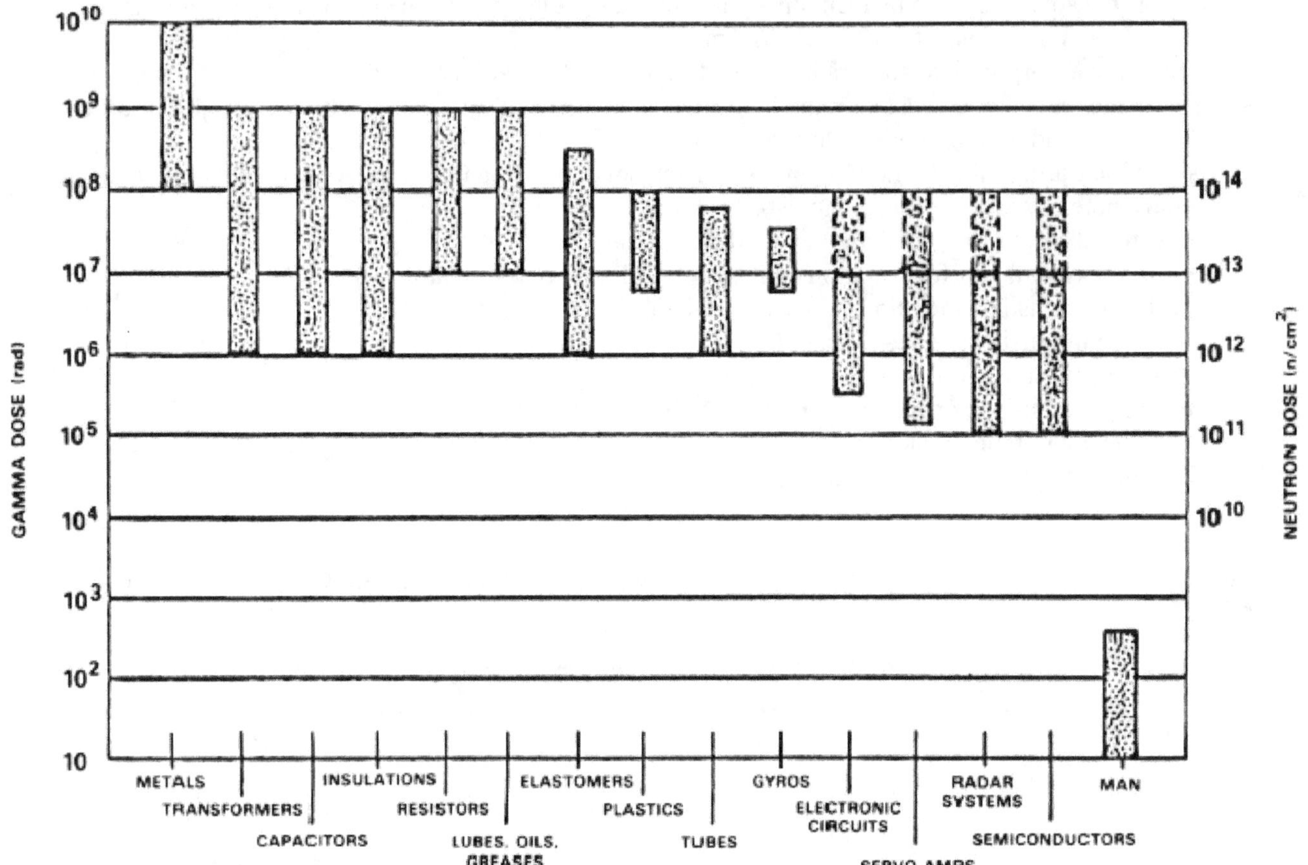

Fig. 32. Typical radiation damage thresholds. *Courtesy of U. S. Department of Energy and Rockwell International.*

Rate effects usually involves the intense radiation dose in a short period of time that leads to ionization-induced damage. For example, sensitive electronic circuits will experience noise, electronic up-set, or even burnout under influence of a rate effect. Cumulative radiation effect typically involves the long-term buildup of radiation-induced lattice defects in the material that then produce significant changes in their thermophysical properties.

As Fig. 32 illustrates, humans are more sensitive to radiation damage than material components. Table 11 list some of the effects of radiation on the human body. As a point of reference, the estimated annual average wholebody dose rates to people in the United States is ~ 180 mrem / yr or ~ 1.8 mSv / yr. To reduce the radiation levels to acceptable levels for the crew, radiation shielding is included in the spacecraft. The absence of a surrounding gaseous medium or atmosphere which would scatter neutrons back toward the system makes it feasible to design the crew radiation shield within a protected cone located between the reactor and other elements of the spacecraft.

Organ of Body	Acute Irradiation Level (rem)	(sievert)	Acute and Delayed Biological Effects
SKIN	300	3.0	Erythema or "sunburn" effect noticeable.
	1500	15.0	Raw, moist skin surface where irradiated.
	5000–7000	50–70	Ulceration, slow healing, possible skin cancer
GONADS	50	0.5	Brief functional sterility in males only.
	250	2.5	Sterility for 1 to 2 years in both male and female.
	600	6.0	Permanent sterility.
EYE	200	2.0	Change in optic lens opacity.
	600	6.0	Clinically significant cataract.
FETUS	10–20	0.1–0.2	Significant probability of malformation, if irradiation occurs in first 3 months of pregnancy.

Table 11. Effects to the body from localized exposure to X and gamma radiation.

The total radiation doses (neutron and gamma) in the vicinity of an unshielded 1.6 megawatt-thermal space reactor that has operated for over seven years is shown in Fig. 33.[42] Space reactors are operating in the vacuum of space so radiation levels decrease as the square of the distance from the core.

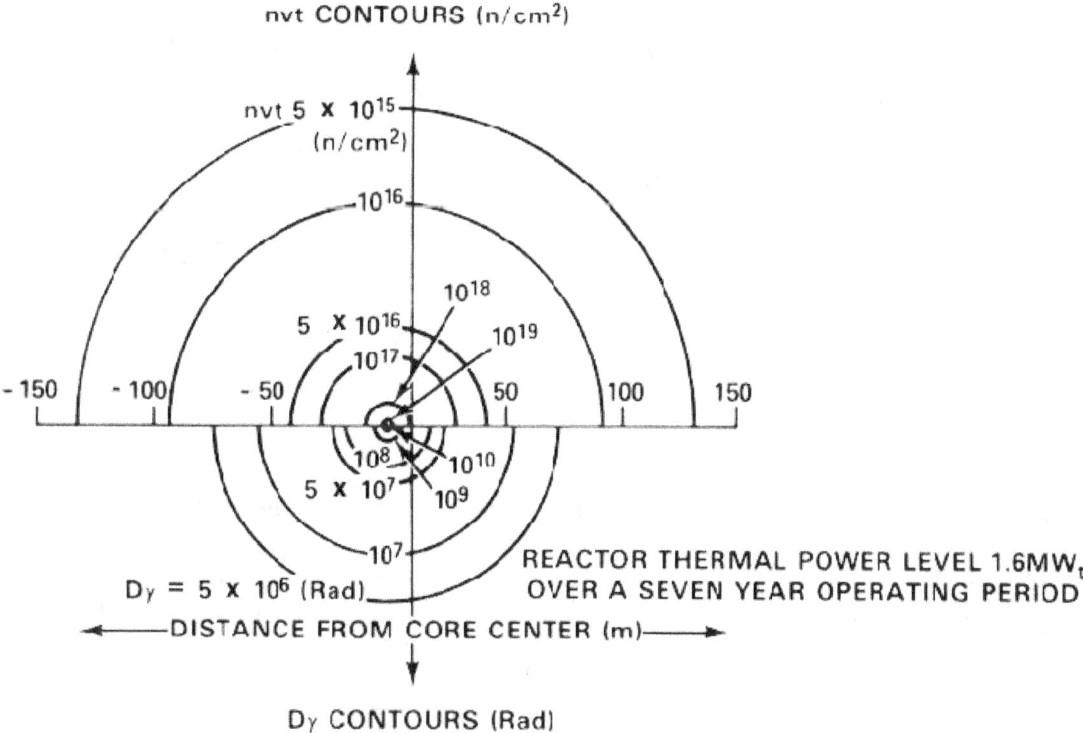

Fig. 33. Radiation levels around an unshielded reactor. *Courtesy of NASA.*

Shields for space reactors are sometimes independent of the application for which the reactor is intended, while other times the shielding issues are very much application-specific or mission-specific. Some conditions encountered in a space mission that influence reactor shielding design include:

1. The absence of a surrounding gaseous medium or atmosphere which would scatter neutrons back toward the system.
2. The need to reject waste thermal energy to space by radiation heat transfer.

3. Unmanned missions will usually have less restrictive radiation dose criteria than missions involving human crews.

4. The shield mass-efficient design practice of locating the reactor and its payload at opposite ends of the spacecraft or space platform in some elongated geometric configuration.

The vacuum of space makes it desirable to configure the spacecraft in a manner that components of the spacecraft are enclosed in a relatively small volume. A "shadow shield" configuration, illustrated in Fig. 34,[43] is used that leaves a major portion of the reactor completely unshielded. This configuration is usually considered for unmanned spacecraft and represents a major savings in shield mass. By locating the reactor at one end of the spacecraft, using an elongated spacecraft design geometry, and placing the payload and sensitive spacecraft components in the dose plane or shadow of the shield, designers take advantage of the inverse square law for radiation attenuation as well as using the minimum shield mass needed to achieve a desired level of radiation protection.[44] Notice that the end of the reflector has been rounded to reduce both shield mass and the possibility of secondary scattering. Neutron attenuation maybe accomplished using lithium hydride (LiH) fabricated in the shape of a frustum, while a heavy metal (typically tungsten) gamma shield is added at the reactor end of the shield configuration.

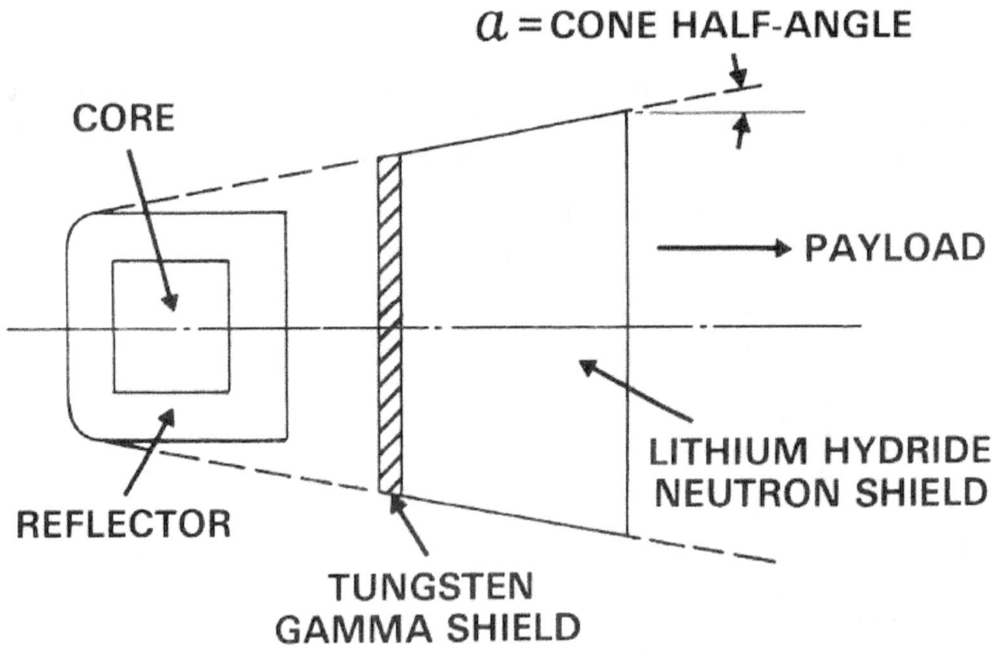

Fig. 34. Typical shadow shield configuration. *Courtesy of Los Alamos National Laboratory.*

Space reactor systems are often operated at fairly high temperatures to reduce the size and mass of thermal radiators. This design condition also can create an elevated (750-900 K) temperature environment for the radiation shield.

Shields can be configured to fit a variety of spacecraft or space station platform requirements. Fig. 35 shows a reactor-powered space base and the variety of shield configurations that might be selected. For example, a shadow shield will only partially surround the nuclear power source. The area within the shadow of the shield is protected, while radiation escapes to space from the reactor's non-shielded surfaces. If a shadow shield geometry is chosen for applications involving human beings or very sensitive equipment, then special consideration must be given to the regions of space around the vehicle that might be used for rendezvous and docking operations, on-orbit maintenance, or extravehicular activity (EVA) by the astronaut crew.[45] The upper limit of a shield is the four-pi (4π) shield--a shield design that provides approximately the same amount of shielding material in all directions around the radiation source. Numerous variations of these two basic shielding configurations can also be created, for example the 2π shield and the preferential 4π shield.

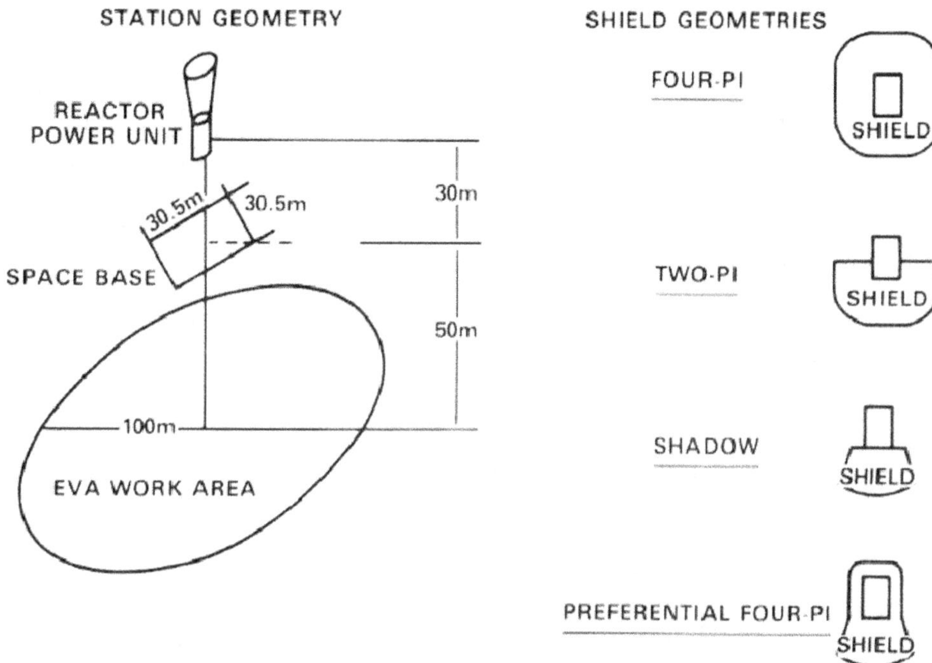

Fig. 35. Various shield design concepts. *Courtesy of NASA.*

For a nuclear reactor used on a manned spacecraft or space station, a 4π shield is generally favored to avoid exposure during docking and rendezvous operations and to prevent prohibitive constraints on crew activities in space around the facility. As shown in Fig. 36, a variety of design features, including booms and tethers, can be used to complement the 4π geometry shield mass in achieving acceptable radiation dose levels. Fig. 37 identifies typical 4π shield masses needed for a manned space system, as a function of separation distance from the reactor.

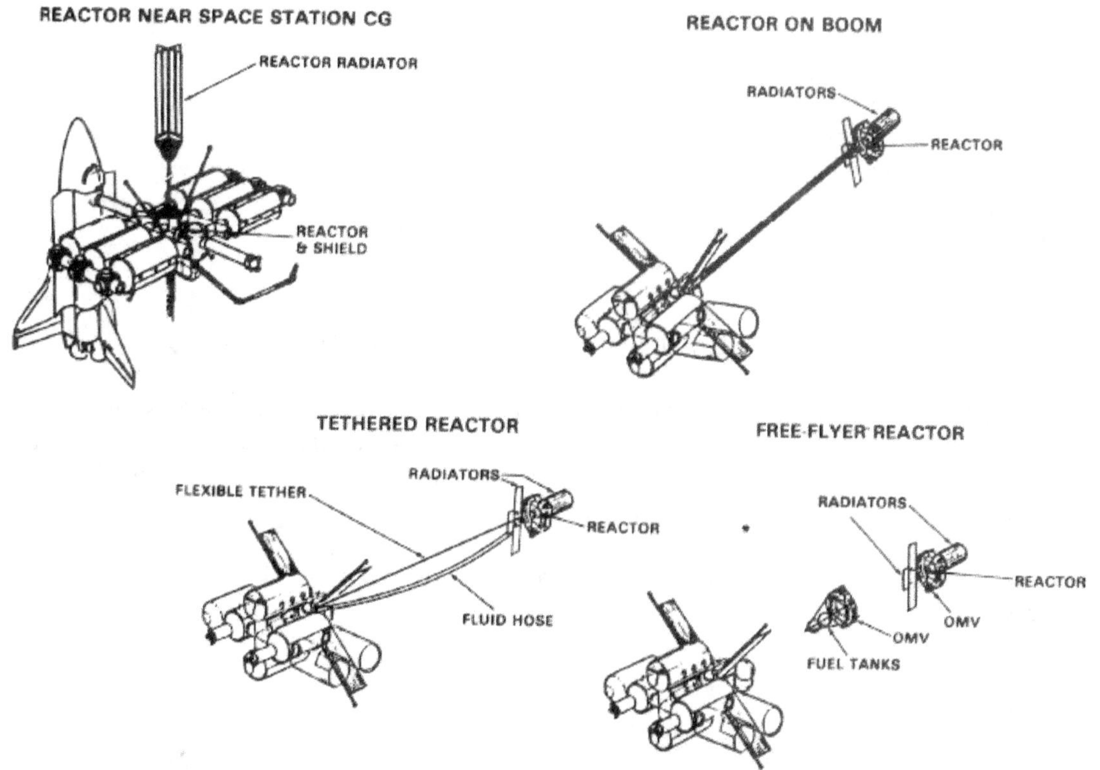

Fig. 36. Various reactor deployment options for a cluster-concept space station. *Courtesy of NASA*

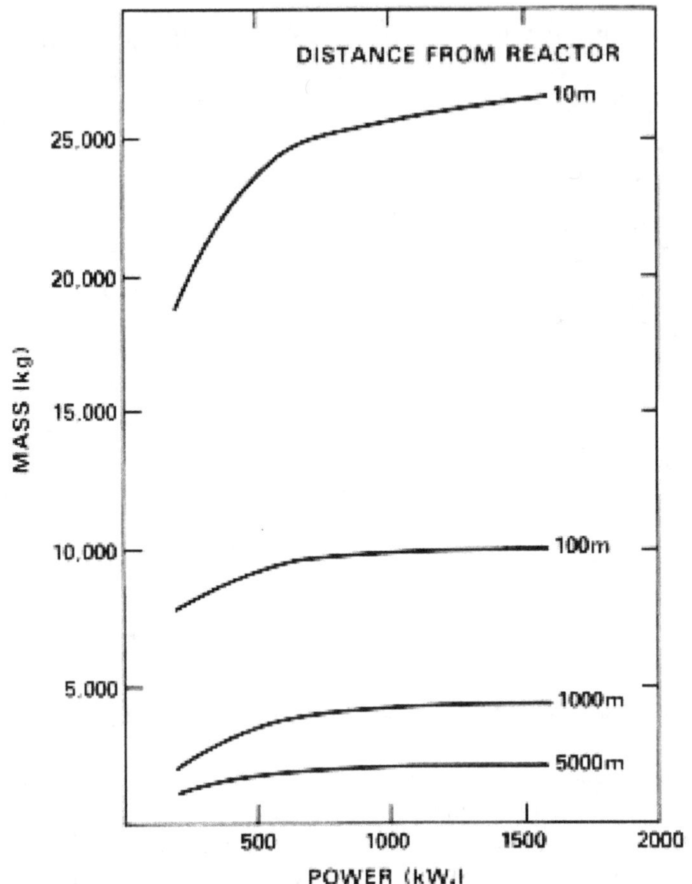

Fig. 37. Typical 4π shield mass needed for a manned space mission. *Courtesy of NASA.*

Shielding materials are often chosen on the basis of:

1. Ability to attenuate radiation
2. Minimum mass
3. Resistance to radiation-induced thermophysical damage
4. Stability at elevated operating temperatures, i.e., 700-900 K regime
5. Ease of fabrication
6. Availability
7. Cost

Space reactor shield design activities are an iterative process consisting of the following steps.

First, the overall guidelines for the shield must be established. In particular, requirements must be established defining what portions of the spacecraft or platform are to be protected; what is the overall mission, including rendezvous and docking requirements if appropriate; and what vehicle configurations can actually be considered in light of launch vehicle and on-orbit assembly constraints.

Second, requirements defining the acceptable radiation exposure levels for the payload, spacecraft, and if appropriate, crew must be established.

Third, the primary and secondary sources of radiation that can contribute to the particular shielding problem must be defined. With shadow shields, particular care must be taken to ensure all radiation scattering sources are accounted for. A scattering source is an object that projects beyond the conical shadow shield and reflects or scatters radiation into the protected zone (dose plane) beyond the shadow shield. A rendezvousing spacecraft or a derelict space object can become an unintentional and undesirable scattering source. Neutrons and gamma rays streaming through penetrations in shields also represent major shielding concerns that must be identified and quantified. These shield penetrations can include the primary coolant loop or core heat pipes, reactor control actuator drives, shield cooling pipes, and electrical cable ducts and conduits.

Fourth, the shield designer must select materials on the basis of their ability to attenuate the radiation sources, function over time in the powerplant environment, and to withstand the rigors of the space environment. Minimum mass, fabricability and reliability are key design parameters that are invoked in selecting shield materials.

Finally, the designer must perform detailed shielding analysis in an effort to arrive at an optimum shield design. A thorough shield design effort involves numerous iterations of these steps, with each iteration characterized by many design trade-offs and compromises. Iterations will involve updating the design as other components in the powerplant change and applying knowledge gained from fabrication and testing materials and components.

Materials containing large amounts of hydrogen make the best neutron attenuating materials. One of the most important parameters for rating shielding materials for space applications is the number of hydrogen atoms (N_H) the material contains per unit volume. Water, with an N_H of approximately 6.7×10^{22} atoms of hydrogen / cm^3, is commonly used to moderate and shield low temperature terrestrial reactors, but it is not suitable as a shield above its critical point. Because of their high temperature stabilities and hydrogen content values, metal hydrides are the prime candidates for neutron shielding applications in space. Of these materials, lithium hydride (LiH) is the favored choice because of its high hydrogen density ($N_H = 5.9 \times 10^{22}$ hydrogen atoms / cm^3), low mass density (0.775 g / cm^3), and moderately high melting point of ~ 960 K. Lithium hydride also produces a minimum amount of secondary radiation. Some LiH thermophysical properties are presented in Table 12, while Fig. 38 depicts typical LiH neutron attenuation. The thermal environment of a space reactor shield is a critical design consideration. For example, when LiH melts, it expands some 25 percent. Consequently, the shield must be kept below the LiH melting point temperature to avoid undesirable thermomechanical stresses.

Table 12. Representative properties of lithium hydride (LiH).

Density (g / cm^3)	0.775
Molecular weight	5.95
Hydrogen content N_H (atoms H / cm^3)	5.85×10^{22}
Hydrogen content (weight %)	12.68
Melting point (K)	960
Volume change on melting(%)	$+25 \pm 2$
Crystal Structure	Face centered cube (fcc)
Heat of fusion (kJ / mole)	21.77 ± 1.3
Compressive strength, cold pressed (MPa)	96.5 - 165.5
Molar volume (cm^3 / mole)	10.254

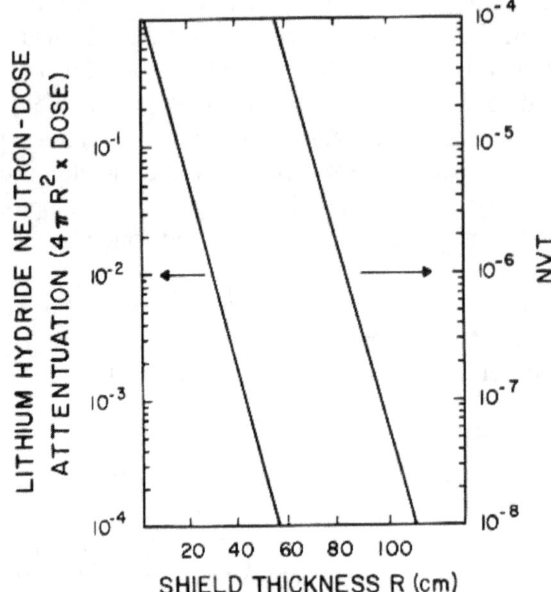

Fig. 38. Typical lithium hydride neutron attenuation. *Courtesy of Los Alamos National Laboratory.*

Tight containment of the lithium hydride represents a shield design complication in high neutron flux regions, if natural lithium is used as the shield material. In natural lithium, approximately 7.5 percent of the lithium has a very high thermal neutron cross section--that is, 945 barns at 2200 m / s for the ^{6}Li$(n, \alpha)^{3}$H reaction.

$$\,^{6}_{3}\text{Li} + \,^{1}_{0}\text{n} \rightarrow \,^{3}_{1}\text{H} + \,^{4}_{2}\text{He}. \tag{36}$$

The helium generated by this reaction would build up significant pressures within the LiH shield containers during the lifetime of the space power system.   Consequently, to avoid this problem, LiH isotopically enriched in $^{7}_{3}$Li, (generally 99.99 weight percent $^{7}_{3}$Li) can be used in shield regions where the thermal neutron fluence is expected to be high.

Lithium hydride must be loaded, processed, and when at room temperature, maintained in a very dry atmosphere to prevent reaction with water or water vapor. At elevated temperature, a hydrogen overpressure must be maintained to prevent hydrogen loss.

If shielding is required to protect against gamma radiation, to minimize overall shield volume and system mass, high density materials are used. Table 13 lists properties for some of the high density, high melting point materials considered as suitable for gamma ray shielding materials.

Table 13. Thermophysical properties of candidate gamma ray shielding materials.

Property	Uranium-8 wt% Molybdenum (U-8Mo)	Tantalum-10 wt% Tungsten (Ta-10W)
Density (g/cm³) (at room temperature)	17.40	16.83
Thermal expansion (10^{-6} cm/cm-K)	14.4 at 673 K	6.3 at 373 K 6.9 at 773 K
Thermal conductivity (J/s-cm-K)	1.34 at 298 K 2.89 at 773 K	0.762 at 298 K 0.695 at 773 K
Specific heat (kJ/kg-K)	0.138 at 298 K 0.176 at 773 K	0.150 at 298 K 0.147 at 773 K
Melting point (K)	1403	—

Chapter 2

Design Requirements

All nuclear power system to be useful must satisfy a range of requirements. These cover the mission needs for power, the lifetime of the mission, reliability, spacecraft radiation levels, survivability, operational environments, performance related to startup, power demand and shutdown, and pre-flight testability and repairability. In addition, the launch vehicle establishes certain constraints on the nuclear power system that also must be considered in establishing design requirements. These encompass size, length, volume and mass constraints and launch environments such as launch vibrational modes. Safety considerations further are part of all design requirements. These include such items as operational constraints to prevent reactor criticality until the operational orbit is achieved, multiple independent reactivity shutdown features, subcriticality in water immersion or land impact, reactor designs with a significant negative reactivity coefficient, methods to ensure heat removal after shutdown, and safety features in ground transport. Cost, schedule and development add further design requirements. The power system must meet a given schedule, within cost to support a given mission. Development risk enters into the design requirements as a measure of whether the power plant can support a given mission. Sometimes these requirements will seem in conflict with each other. For instance, the mass requirements may make it difficult to meet the reliability requirements. However, the design requirements are an all encompassing set of needs and thus all requirements must be satisfied. This chapter will expand on the various aspect of establishing nuclear power plant design requirements.

Mission Requirements

Power levels and lifetimes are usually the initial requirement to be defined for a given mission. The power system maybe designed for a wide range of missions, such as those given in Table 1 of Chapter 1. The reasons for performing a given mission defines what instrumentation, electric propulsion, ground processing power, etc. are needed to fulfill the objectives of the particular mission. These then define the power levels and lifetimes that the power system must provide.

Over time, the power levels have increased and the lifetime needs have become longer. SNAP 2, the first space reactor power system, was designed to provide only 3 kW_e and a lifetime of 1 year in the 1950's. During the 1960's, SNAP 8 was being designed to provide 35 kW_e for a year. In the 1980's, SP-100 was to deliver 100 kW_e for 7 years. In the 2000's, the Prometheus Program was to deliver 200 kW_e for up to 20 years.

Reliability is a function of how much risk that the mission organization is willing to accept. Since space missions are very expensive and often involve many years of operation before the desired data is obtained, the reliability levels are set very high (approaching 0.99). Maintenance is not usually an option. In addition to the high reliability design requirements, there is usually a requirement in the design specifications document to avoid single failure points. This is added as a realization that a fabrication, manufacturing or assembly error might occur despite all the precautions taken. If single point failure locations cannot be avoided in a practical manner, the design requirements usually specify that the power system must demonstrate sufficient robustness in the design to mitigate risk of failure. These requirements favor designs that have inherent redundancies, such as thermoelectric converters. Design using electric power conversion systems without inherent redundancy, such as Brayton cycles, must consider redundant loops with the added complexity of valves, controls, and instrumentation to detect a failed component and take appropriate action. Recovery actions may need to be automated because some mission require too long a communication time for ground control.

Components in the spacecraft are subject to radiation damage. Thus, design specifications include a radiation level that must be met at the spacecraft where radioactive sensitive components are located. An unshielded reactor will result in very high radiation levels in the spacecraft. Fig. 1 shows representative radiological levels surrounding a 1.6 MW_t reactor operating for 7 years.[1] The radiation can be attenuated by providing radiation shielding materials. Also, for system operating in the vacuum of space, radiation decreases as the square of the distance increases. A boom can be used to separate the nuclear power system from the rest of the spacecraft. In reality, shielding and separation are used to reduce the radiation levels in spacecraft to acceptable levels. The significance on the design of the radiological requirement is illustrated in Fig. 2.

Over time, scientific instruments and electronics have become more susceptible to radiation damage. SP-100 was being designed for radiation levels at the user interface plane not to exceed a neutron fluence of 1×10^{13} n / cm^2 1-MeV equivalent and gamma dose of 5×10^5 rads Si. The Prometheus program specified the need for shielding to reduce radiation levels at the science system hardware to a neutron flux of 5×10^{10} n / cm^2 1-MeV equivalent silicon damage and payload gamma flux to 25×10^3 rads Si.

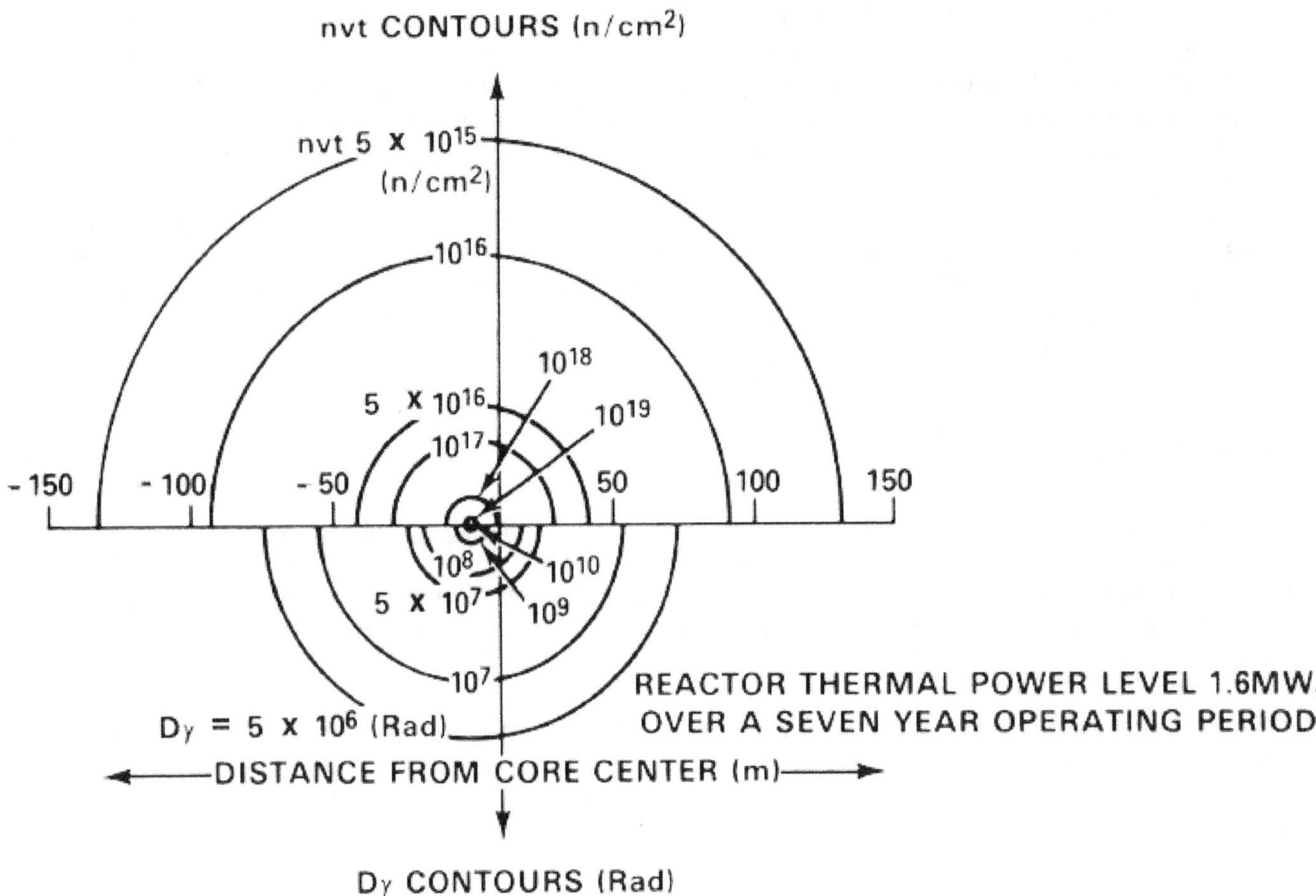

Fig. 1. **Radiation levels around an unshielded reactor based on a SPAR/SP-100 reactor operating at 1.6 MWt over a seven year operating period.** *From Joseph A. Angelo, Jr. and David Buden, 1982.*

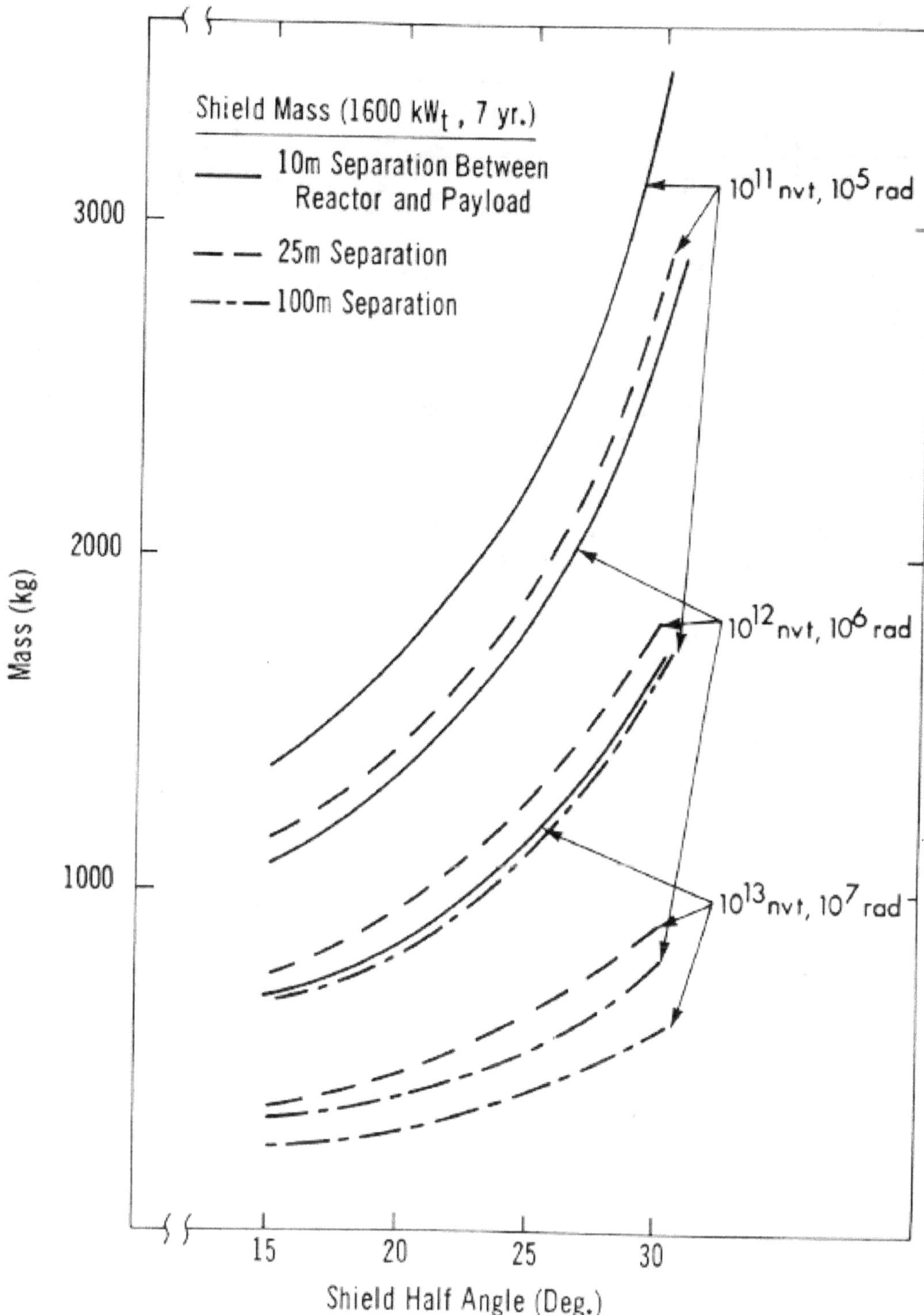

Fig. 2. Illustration of shield mass as a function of shield half angle and boom separation distance for various radiation levels at the spacecraft components. *From Joseph A. Angelo, Jr. and David Buden, 1982.*

Shielding considerations as a result of specifying radiation levels for manned missions is a greater challenge. The radiation exposure limits and exposure rate constraints for unit reference risk are given in Table 1.[2] A 4π shield, a radiation shield that completely surrounds the reactor, would have representative shield mass as a function of separation distance as shown in Fig. 3. For conservatism, 7.5 rem/quarter was used instead of 35 rem/quarter. Booms or tethered cables with separation distance of 100 m or greater leads to quite reasonable shielding masses.

Table 1. Radiation exposure limits and exposure rate constraints for unit reference risk.*

	REM**		
Constraints	Bone Marrow (5 cm)	Skin (0.01 mm)	Eye (3 cm)
1-year average daily rate	0.2	0.6	0.3
30-day maximum	25	75	37
Quarterly maximum	35	105	52
Yearly maximum	75	225	112
Career	400	1200	600

* For details, see "Radiation Protection Guides and Constraints for Space Missions and Vehicle Design Studies Involving Nuclear Systems, Report of the Radiobiological Advisory Panel of the Committee on Space Medicine, Space Science Board, National Academy of Science, 1970."

** REM (Roentgen equivalent man) is a unit of radiation dose equivalent. For details, see "International Commission on Radiation Protection," Publication No. 9, 1966.

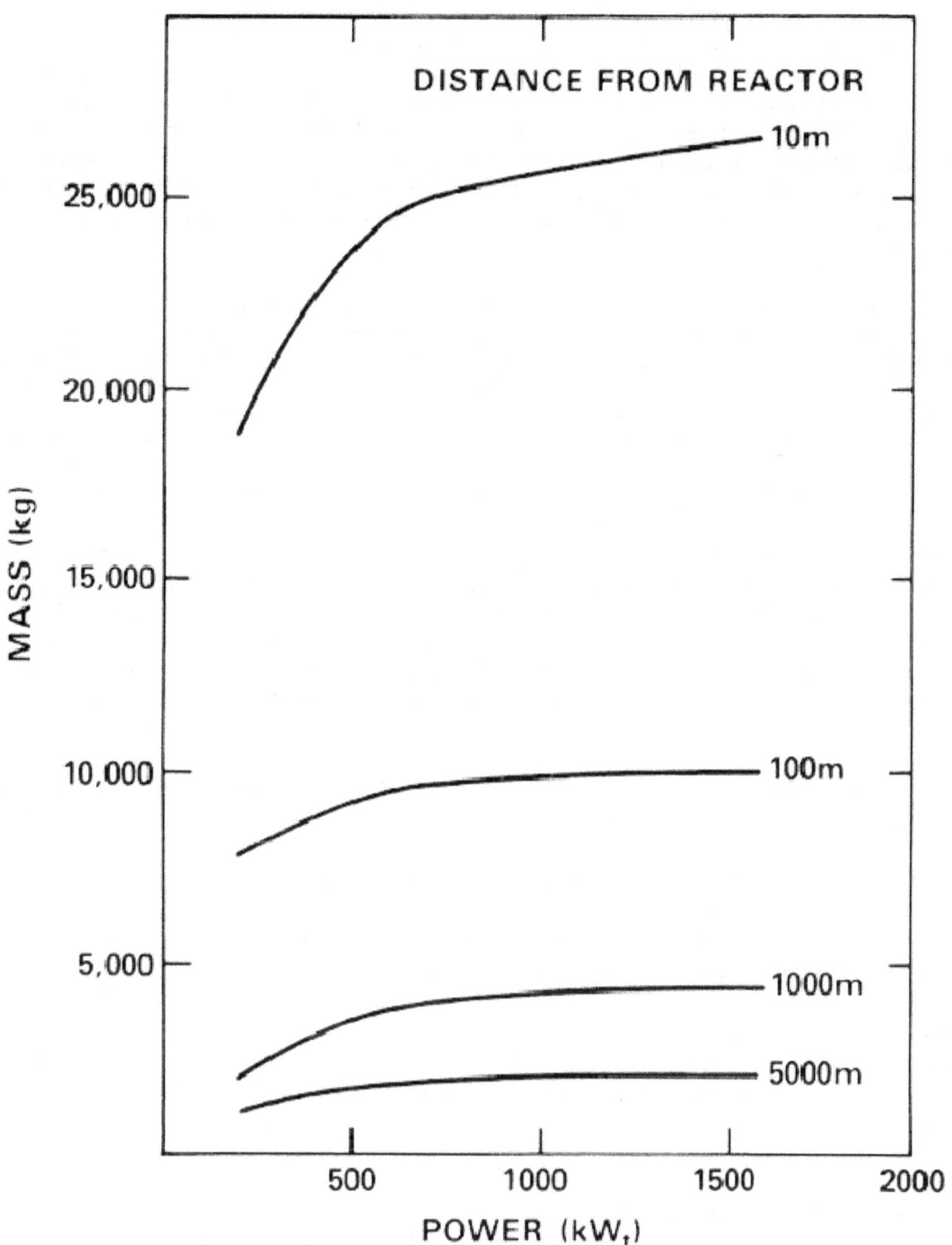

Fig. 3. Representative manned shield mass as a function of power level and separation distance.

Design requirements must accommodate space environments and, if operation is planned on planet or moon surfaces, these environments as well. Space operational environment means operating the power system in a virtual vacuum and near absolute zero conditions (about 3 K). This effect on the power system design concerns waste heat disposal, which must be rejected by radiation to space. The spacecraft must also accommodate the solid particle environments in space which includes protection against orbital debris and micrometeoroids. The crucial components here are pressurize components and moving assemblies. Also, the radiation environments in space must be accommodated. These include solar, galactic and planetary bodies and moons that might be associated with a given mission.

Book 3
Space Nuclear Fission Electric Power Systems

For lunar or Mars missions, surface environments will need to be considered in the development of nuclear power systems. The Moon has a surface gravity of 1.62 m / s^2 compared to Earth's 9.78 m / s^2. The Moon has a very thin atmosphere of helium, argon, sodium and potassium, but is too weak to provide protection against sunlight striking the lunar surface. During the month-long lunar day, equatorial temperatures can range from 100 to 400 K, with rapid (5 K / hr) temperature changes at sunset and sunrise. Radiation and micrometeorite impacts are of concern for equipment on the surface from galactic cosmic rays, solar flare particles, and solar wind particles (see Table 2).[3] The lunar soils, which might be used for shielding, have high concentrations of sulfur, iron, magnesium, manganese, calcium, and nickel. Many of these elements are in the form of oxides such as FeO, MnO, MgO, etc.

Table 2. Moon galactic cosmic rays, solar flare particles, and solar wind particles.

Radiation Source	Energy	Flux (cm^{-2}s^{-1})	Penetration Depth
cosmic rays	1-10 Gev/nucleon	1	few meters
solar flares	1-!00 Mev/nucleon	100	1 cm
solar wind	1000 ev/nucleon	!0^8	10^{-8} cm

Mars experiences temperatures between 186 K and 268 K. It's gravity is 3.69 m / s^2. The atmosphere, mainly of 95.72% carbon dioxide, 3.7% nitrogen, 1.6% argon, 0,2% oxygen, has a surface pressure of 0.7-0.9 kPa. The surface of Mars is thought to be primarily composed of basalt. Some evidence indicates the surface might be more silica-rich than typical basalt, perhaps similar to andesitic rocks on Earth. Much of the surface is deeply covered by dust as fine as talcum powder. The red/orange appearance of Mars' surface is caused by iron oxide (Fe_2O_3).[4]

Survivability is mainly a military concern and thus is a function of the threat scenarios that the spacecraft must be made to survive.

Operating requirements include startup, responding and maintaining the desired power profile, shutdown, and restart. Some systems will be operated too far from Earth to depend on Earth communications; thus, for these missions independent command and control will be needed. Reaction times to commands or changes in power plant loads will depend on the particular mission; these must be specified in the Design Requirements Document.

Pre-launch ground operations also must be considered. This includes any requirements to check that the system is ready for flight, such as a subcriticality tests of the reactor, electric heating end-to-end test prior to installing the reactor fuel, any desired ability to repair the system prior to launch, and transportation constraints on the power system.

Requirements Derived From Launch Vehicles
A most unique feature about nuclear space reactors compared to terrestrial reactors is that they must fit into a spacecraft and be launched by a rocket. This confines them in size and mass and results in using materials different than all other reactors, such as terrestrial power reactors or ship propulsion reactors. The reactor operating temperatures are much higher in order to make the size and mass of the nuclear power systems conform to the limitations of the rocket launchers.

The possible next generation of launch vehicles, Ares I (pictured in Fig. 4) and V (pictured in Fig. 5), was likely to be the main launch vehicles for the next generation of nuclear power systems--their development is in a state of uncertainty though some comparable launch vehicles will be necessary. Ares I is being designed to lift six astronauts or cargo to the International Space Station of approximately 21,820 kg; or up to four astronauts to low-Earth orbit with a mass of approximately 23,640 kg.[5] Ares V can carry nearly 188,200 kg to low-Earth orbit and with the Ares I launch vehicle to launch payloads into Earth orbit, Ares V can send nearly 71,400 kg to the moon.

Fig. 4. Artist picture of Ares I launch vehicle. *Courtesy of NASA Marshall Space Flight Center.*

Compare this to the Project Prometheus program that had a spacecraft total dry mass at launch not to exceed 25,000 kg or SP-100 designed to be launched on the Space Shuttle that could launch 24,400 kg to orbit. The use of Ares I or V (or their replacements) could reduce, but would not eliminate, some of the constraints on future nuclear reactor power system design. However, for many potential applications in the 10's of kilowatt range, the Ares launch systems may not be cost effective. These might require a smaller, less expensive launcher. Combining nuclear thermal propulsion with the nuclear reactor power system could replace the chemical rocket transfer orbit stage and be cost effective. Some of these concepts will be discussed in the chapter on Additional Space Reactor Power Plant Concepts.

Fig. 5. Artist picture of Ares V launch vehicle. *Courtesy of NASA Marshall Space Flight Center.*

Mass, length and size restrictions force the power plant designers to consider high temperature heat rejection systems since this subsystem dominates the power plant length and size. Heat rejection is a linear function of converter efficiency and an inverse forth power function of temperature. High temperature heat rejection leads to high reactor temperatures. The consequence is the need to consider fuels and materials that may be expensive or not well characterized. An example is the use of refractory metals in the reactor core.

The mass restriction also influences the design in that the reactor needs to be made as small as possible. Shield mass is a function of the reactor size and since shield mass is one of the heavier components, reactor size is very important. This means that the uranium fuel used in the reactor will be highly enriched. Also, using a moderator in the core increase the critical radius and drives shield weight up rapidly. Therefore, the reactors will tend to be designed with a fast spectrum. Reflector control drums tend to lead to smaller size reactors then control rods; thus these designs tend to also minimize shield mass.

In addition to the constraints imposed by the launch system on power system mass, length and size, the power system must be designed to withstand the vibrational forces during launch and rocket operations in space. These modes are unique for each launch system

Safety

Safety must be built into the power plant design from the start. It Is essential at the initiation of the design process, through final qualification and flight, that space nuclear reactors be designed, fabricated, and tested with public health and safety goals firmly established. The philosophy of designing for safety involving radioactive substances embraces the following approaches:

- Confine and Contain;
- Delay and Decay; and
- Dilute and Disperse.

Confine and Contain isolates the radioactive material from the population by barriers. For instance, in terrestrial reactors a large containment vessel is often used with the reactor fuel confined inside the barriers. In space, the barrier is the large separation distance from other satellites and the Earth's population. Delay and Decay provides for sufficient isolation time so that radioactive levels are reduced to meet radiation safety standards before exposure to the population. This technique is used in space reactors by insertion of reactors into very long-life orbits. Dilute and Disperse is used when there is a large medium for dispersal, such that elements in the population are not exposed to radiation levels greater than set forth in the radiological standards. This is also used in space reactors to disperse the fuel over a wide area upon reentry into the atmosphere. All of the methods can be used to meet radiological standards, and several combinations may be used in meeting various safety requirements in different regimes of a particular mission.

Safety standards are met by design and operational features. Design features include such hardware devices or elements that preclude, for example, unplanned reactor criticality until a satisfactory space orbit is reached. Operational features include procedures to avoid unplanned criticality such as not operating the reactor at substantial integral power levels prior to launch and not starting the reactor during launch or until the specified desired orbit is reached.

Various accident environments must be considered that take into account all phases of planned missions. Based on the launch vehicle and mission profile, actual environmental conditions for each postulated accident needs to be determined. Table 3 lists various postulated accident environments for space nuclear power sources to be evaluated.

Inadvertent reactor criticality, radiological hazards, ensuring end-of-life neutronic shutdown and final disposal of the reactor are the major nuclear safety design concerns associated with space reactors. Non-nuclear concerns may also exist such as beryllium toxicity and fragments from an explosion. Inadvertent criticality can occur by immersion of the reactor in a fluid, such as water or liquid hydrogen propellant; core compaction from being dropped or in a launch vehicle crash; control system malfunctions; or core meltdown. Associated radiological hazards include space reactors with radiation from a critical reactor or subcritical reactor that had been operating at significant power levels. Also, operated reactors contain actinides, fission products, and material activation products. Failure to shutdown at end-of-life must consider control system malfunctions or warpage of control elements. Final disposal relates to the elimination of the reactor as a potential long-term hazard to the Earth's biosphere. Proper consideration of safety from the initiation of the design process and adherence to prescribed operating procedures is needed to protect the public and biosphere.

Table 3. Accident environments be considered in nuclear power plant design.

<u>Launch and Ascent</u>

Explosion overpressure
Projectile impact
Liquid propellant fire
Solid propellant fire
Land or water impact
Sequential combinations

<u>Space / Earth</u>

Loss of control
Reentry
Land or water impact
Post impact environment (land or water)

Protection Against Inadvertent Criticality

Certain reactor designs if immersed in water or rocket propellant could become critical unless special design features are included in the reactor. Water is a moderator and, since many space reactors have a fast neutron spectrum, the moderation effect can cause the reactor to become critical or supercritical even with normal control reactivity elements in their shutdown positions. If a containment vessel or coolant heat pipes are included in the reactor, these may leak or break on impact with an ocean, lake, or river, or during inundation with liquid propellant. If they survive impact, erosion or corrosion could occur, thus allowing the fluid to enter the core. A violent reaction could occur between reactor coolants such as lithium or sodium and the water or rocket propellant. This could lead to a chemical explosion which would fragment the reactor and thus avoid criticality. However, some low probability exists that certain reactor designs can go critical on fluid immersion; therefore, it is necessary to make provisions in these designs for this possibility.

For thermal reactors, such as those fueled with uranium-zirconium hydride, methods have been investigated to make the core intrinsically subcritical upon water immersion, using highly spectrum-dependent thermal-neutron absorbers.[6] Isotopes were selected whose neutron absorption cross-section increases rapidly in energy just below the normal spectrum peak. Burnout of this poison in the operational mode would have the advantage of compensating for normal reactivity losses. Measurements have been made of Gadolinium (^{155}Gd), Samarium (^{149}Sm), and Europium (^{151}Eu) for reactivity loss, isothermal temperature coefficient, and prompt temperature coefficients. The penalty associated with use of gadolinium to provide intrinsic subcriticality results from reactivity losses due to residual (unburned) gadolinium. This increases the need for thicker reflectors which, subsequently, imposes a weight penalty. The following conclusions were derived: neither ^{155}Gd or ^{149}Sm can be used alone, due to the positive effect each has on the prompt temperature coefficients; ^{151}Eu should not be used by itself, due to its slow burnout rate and possible large increase in the temperature defect; and that a combination of ^{151}Eu and either ^{155}Gd or ^{149}Sm (or possibly both) could preserve the unopposed temperature coefficient.[7]

Fast reactors are more difficult to control by poisons in the core. Since neutron production is equal to neutron absorption plus neutron leakage, both terms of neutron production can be considered in design approaches to avoid inadvertent criticality. Neutron absorption can be increased by core control rods or a central control plug, such as boron carbide (B_4C). This idea was used in the heat pipe cooled reactor[8] and fuel pin reactor designs by incorporating central removable control plugs. The central plug is inserted into the core on the ground and would be removed once the desired orbit is reached. At that point, the general population is isolated from the reactor and is not, therefore, at risk. Leakage is increased using highly enriched $^{10}B_4C$ control drums.

Another scheme to prevent criticality is to insert the fuel into the core after orbit is achieved. Higher power reactors that might use fuel pellets, particles, or plasmas can possibly utilize this scheme. During launch, the fuel is stored in a geometry that could not become critical even if fluid immersion occurred.

Mechanical segmenting and separating the core is still another idea for preventing inadvertent criticality. Because of fluid containment and the need for pressure vessels in most reactor designs, this approach could be difficult to implement. However, the U.S.S.R. has successfully used this approach.

If water enters the reactor core, the core will either remain subcritical from one of the design features discussed; become a "chugging" reactor as it alternately boils the water forcing the fluid out and becoming subcritical, returning to the critical state when the water reenters the core, and then repeating the cycle; or violently disassembling. Several planned tests on reactors in the SNAP-10A and nuclear rocket programs showed that the reactor core would disassemble from rapid energy release if the core is made supercritical at a very rapid rate.[9, 10] In these tests, the cores disassembled from physical explosions, not nuclear explosions.

Inadvertent criticality by core compaction can occur on impact with the ground, such as accidentally dropping the reactor during installation into the launch vehicle or overpressure from an external explosion. Compressing a reactor into a critical mass is difficult since it will more likely break apart than be compressed. It is shaped as a cylinder, often surrounded by a pressure containment vessel and a large cylindrical layer of reflector materials about 10 cm thick. Mechanical design approaches to enhance reactor fragmentation or structurally strengthening the pressure vessel can be used to assist in preventing criticality in this type of accident.

Another form of core compaction is meltdown inside a confined space, such as the pressure vessel. This might occur by a control system malfunction, loss of heat removal capability, or fire. The first two can be avoided in the design by locking the reactivity control elements in place until a desired orbit is reached. In a manned launch vehicle, mechanical key locks could be used. The astronauts are present to unlock the control drums once the reactor is in space; for unmanned missions, electronic interlocks are feasible. If compaction does occur, the result would probably be a mild excursion or a violent non-nuclear disassembly. Core meltdown from an external launch pad fire does not appear to be a problem. However, this will need to be continuously evaluated, especially with larger launch vehicles.

Table 4 summarizes design approaches to avoid inadvertent criticality.

Protection Against Radiological Hazards
Radiological hazards must be considered from reactor construction to final reactor disposal. Since highly enriched uranium is used in the current reactor designs, handling of the fueled reactor will be under established Department of Energy (DOE) and Nuclear Regulatory Commission (NRC) procedures. These encompass physical security, materials control and accounting, and inspection to verify materials inventories. Physical protection measures are regulated by DOE Order 5632.2, and materials control and accountability measures by Order 5630.2. DOE regulations cover fabrications, assembly, testing, transportation and pre-launch storage. These regulations will probably not have a direct impact on the flight systems designs; external means to the flight power system are used to assure compliance. Under normal conditions on the launch pad, it is expected that reactor power plants can be handled the same as other cargo for the launch vehicles.

Table 4. Designing for safety to avoid inadvertent criticality.

Cause	Design Feature	Operational Feature	Effect of Safety Failure
Reactor fluid immersion	Increase neutron absorption	Lock control elements for launch	"Chugging"
	Increase neutron leakage		Core disassembly (non-nuclear)
	Mechanically separate core		
	Fuel reactors in space		
	Control element locks		
	Redundant reactivity controllers (drums and rods)		
Compaction	Mechanical design	Lock control elements for launch	Mild excursion by thermal expansion
	Fuel reactors in space		Core disassembly (non-nuclear)
	Neutron absorption in core		
Control system failure	Control element locks	Lock control elements for launch	Can not start reactor
	Redundant control elements		
	Safety override system		

Once on the launch pad, certain design features can be used to protect workers and astronauts. The radiological levels can be kept sufficiently low so that no special exclusion zones need to be established; non-operated fuel-enriched uranium-235 has radiation levels significantly below safety standard limits. Workers and astronauts are protected against radiation from a reactor prior to launch and achieving orbit by avoiding operation at a level to build up significant amounts of fission products. For flight test verification, the reactor can be made critical and operated at a sufficiently low level so that an insignificant amount of fission products are generated; and the neutronic behavior of the reactor can still be checked.

In order to avoid unplanned reactor operation and also to avoid a core meltdown, a number of safety features are usually incorporated into space reactors designs. These include having a significant number of independently controlled reactivity control elements. The control elements can be in the form of rotating cylinders with a segment of neutron-absorber materials. SNAP-10A used four control elements; SNAP-8 had six elements; and more recent designs for higher electrical power levels have twelve or more reflector drums.

Reactivity control elements can be made dual-functional. In these designs, a latching mechanism is included such that, if a loss of power to the mechanism occurred, the drum is spring-loaded to return to its shutdown position. Thus, positive action is needed to keep the reactor operating. The dual action plus having many separate control drums can be used to ensure that unplanned criticality does not occur. One control drum at a time can be safely tested to prove functionality. People working in the vicinity of the launch pad need not take special precautions because a reactor is present, as was demonstrated in the SNAPSHOT flight program.

The major concerns to a normally operating reactor in space are related to a manned spacecraft or a nuclear powered space station. Once the reactor has operated for any appreciable time, people must be protected. People are protected by radiation attenuation shields, and by separation and/or exclusion zones.

Biosphere protection is needed against a reactor reentering the atmosphere with high radioactivity levels. These high levels can be associated with the actinides, fission products, or activation of materials. Dispersing or burning up the reactor or intact reentry are the solutions recommended by the United Nations. A number of design approaches are possible to facilitate the dispersal mode. These include mechanical and chemical design features. Mechanical features might include means to separate the core from its containment vessel, such as using sections of materials which will readily oxidize at the start of atmospheric reentry. The atmospheric forces, plus chemical reactions, will cause separation without special explosives being part of the spacecraft--an alternate approach. Once the core has separated from the containment vessel, it needs adequate atmospheric exposure to enhance its dispersal However, many reactor designs use coatings around the fuels that will tend to retard this action. If insufficient dispersal is predicted to occur, it is possible to consider chemical reactions prior to reentry. This use of chemicals is a difficult design problem, because it involves a solvent that will react with the fuel materials and cladding, and yet can be pumped through the system without losing flow integrity. Also, the chemical reactants must be compatible with the normal core coolant; or this must be removed before insertion of the chemicals. A chargeable battery with solar arrays or an RTG might be used for pumping power; and the rate of circulation should be quite low.

Another approach, represented by a heat-pipe reactor or a reactor configuration with multiple independent coolant channels, allows the use of bands to hold the reactor together. The bands can have segments that quickly oxidize on reentry, such as molybdenum. The fuel need not be clad, because the heat transport channels perform this function.

A significant launcher and space station infrastructure exist currently. This infrastructure could be employed to attach booster rockets to any low-orbit reactor and dispose of them into higher orbits or place them on an Earth escape trajectory. If the initial booster rockets fail to work properly, replacement booster rockets could be attached to the reactor until safe disposal is achieved. The reactor would not reenter the atmosphere while there is a significant radiological source term present. This approach would not depend on atmospheric disposal upon reentry for biosphere radiological safety.[11]

Astronauts must also be protected against uncontrolled radioactive releases. This includes the need not only to ensure that the reactor will not operate in an uncontrolled state, but also that the reactor will not meltdown (thus resulting in unplanned radioisotope releases). Redundant reactivity control mechanisms have already been discussed. However, decay heat removal must also be considered, especially in the case of a loss of primary loop coolant. If there are many independent, redundant heat transport elements to cool the reactor, this is not a problem. Otherwise, parallel pumping elements would be one option. These require a very reliable containment system for the primary coolant loop. Another scheme would be to design a passive system that will conductively remove the decay heat from the reactor to radiating surfaces. Core heat pipes can also be used to remove decay heat to a radiating surface.

Table 5 summarizes design approaches to protect against radiological hazards

Table 5. Design goals and possible design approaches to protect against radiological hazards.

Goals	Reason	Design Approach
Radiation levels sufficiently low to launch to avoid special precautions	Protect workers and astronauts	Not operate reactor (except for zero power testing) until a stable orbit or flight path is achieved
		Two independent systems to reduce reactivity to a subcritical state
		Unirradiated fuel shall pose no significant environmental hazard
Prevent inadvertent criticality	Insure public not exposed to levels of radiation that exceed established standards	Subcritical if immersed in water or other fluid
	Protect Shuttle crew	Significant negative power coefficient
		Subcritical on earth impact accident
		Incorporate a reactor safety system
		Quality assurance standards
		Positive coded telemetry system for reactor startup
		Redundant control and safety system
		Independent source of electrical power for reactor control system, reactor protection system, reactor communication system
		Instrumentation to continuously monitor reactor status
Avoid release of radioactivity by-products in concentrations to exceed radiological standards	Insure public not exposed to radiation levels that exceed standards and protect biosphere against concentrations of radioactive elements about safety standards	Orbital boost system for short-lived orbits. Reactor dispersal if booster fails
		Spacecraft attitude controller for communications and boost system
Avoid unplanned core destruction	Protect space investments and avoid contamination of volumes of space environment	Independent shutdown heat removal systems for decay heat removal
		Two independent systems to reduce reactivity to a subcritical state
		Positive coded telemetry system for reactor startup
		Positive coded signal to operate reactor
		Fault detection systems for reactor protection

Safety Issues Relative to Orbit Lifetimes

To protect the Earth's population against undue risk, radiation levels at the time of a nuclear reactor reentering the Earth's atmosphere should be low. If the orbital lifetime of a satellite is sufficiently long, most of the fission products will decay away. A "sufficiently high orbit" has a lifetime of 300 years or more.[12]

The initial orbital altitude needs to be about 750 km. Figure 6 plots the orbital lifetimes as a function of altitude and ballistic characteristics of the system. The ballistic parameter is: $W / C_D A$ (where W is the weight of the reentry body; A is the drag area which is a function of the flight altitude with respect to the orbital path; and C_D is the drag coefficient which is influenced by the geometric characteristics of the body). Radioactivity dose calculations show that if the reactor reenters the biosphere after 300 years in orbit, the fission product activity will have been reduced from approximately 10^7 Ci to about 100 Ci.

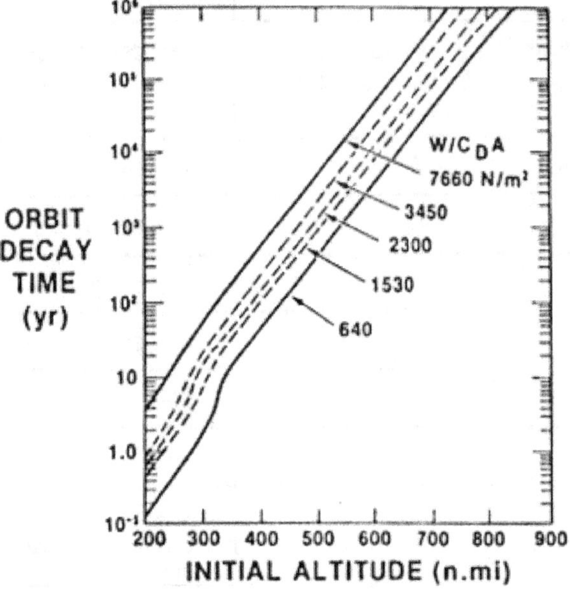

Fig. 6. **Minimum orbit decay time.**

Certain designs may use materials that are activated while in the reactor, such as Nb-1 Zr-0.1 C fuel cladding. The presence of these materials can result in the generation of some long-lived radioactive isotopes. For a reactor using niobium, activation of the fuel cladding results in an increase of 22 Ci at the end of 300 years, because ^{94}Nb is generated (half-life of 2×10^4 y). The total dose level after 300 y is 118 Ci.[13] It is derived mainly from long-lived isotopes. If the orbit time is increased to 600 y, the dose level decreases to 34 Ci; and in 2,000 y, to 28 Ci. If no Nb were used in the core design, the levels after 300 y would be 96 Ci; after 600 y, 12 Ci; and after 2,000 y, only 7 Ci.

Safety Issues Relative to Space Debris[14]

Collisions between spacecraft carrying nuclear power sources and space debris, and their possible consequences, is an area of concern that is recognized. Possible consequences of a collision is that the lifetime of the nuclear power system may be changed and the spacecraft could be destroyed with various pieces entering the atmosphere earlier than planned. Collision conditions will determine the potential effects. Small angles of impact will tend to bounce off while large angles of impact at the center of mass will go straight through or penetrate and stop. Large angles of impact on peripheral areas will tend to break off pieces of the spacecraft.

Most of the nuclear power sources in Earth orbit are located in a belt between 800 and 1100 km. Estimated debris in this same belt are: 6×10^{-8} / km^3 fragments larger than 1 cm, and 1×10^{-8} / km^3 fragments larger than 4 cm. Based on analyzing nuclear sources in the band between 900 and 1,000 km, there is a high probability of collision with space junk of approximately 1 cm diameter within 15 years, with impact damage. Penetrating impact due to debris from 5 to 10 cm in diameter will occur in the next 25 to 50 years, and catastrophic impact due to debris which is greater than 50 cm in diameter will occur 5 to 10 years later.[15]

Cost, Schedule and Development Risk

All projects have budget and schedule limitations and it is important to recognize this at the beginning of the design process! Uncertainty must be minimized if the design objectives are to reach a flight readiness by a certain date. Relying on known technologies where possible lessens risk, cost and makes adherence to schedule much more probable.

Unfortunately, because of the need to minimize power plant mass, space reactor designs operate in a temperature range much higher than in terrestrial power plants. Existing data on candidate core and structural materials are insufficient for longer lifetimes, so materials irradiation tests are necessary. If the reactor uses a fast neutron spectrum, tests must be in a relevant facility. The design, irradiate, and examination of material test specimens on an expedited basis takes approximately two to three years per cycle.[16] Several cycles are desirable.
Electric converter subsystems also must operate in environments significantly different than existing experience. With different materials and more difficult fabrication problems, further uncertainty is introduced into the design process. Also, the space environments are different than here on Earth, further complicating the development process. There has been a significant amount of work on converter subsystems in previous space nuclear power program to provide an excellent foundation of what works for thermoelectrics, thermionics, Rankine cycles, Brayton cycles and Stirling engines.

Another major consideration in establishing cost and schedules is the availability of materials, components, subassemblies and engineering system test facilities. Many of the facilities that were used in earlier space nuclear system programs have been decommissioned. New facilities will be needed to needed qualify the elements of the power plant and support the engineering design codes necessary for successful designs. The selection of a design concept must consider this aspect to develop realistic cost and schedules.

Chapter 3

Uranium-Zirconium-Hydride Reactor Power Plants

In 1957, active development of nuclear reactors commenced with the development of a uranium-zirconium-hydride (U-ZrH) reactor system.[1] This was in recognition that electric power needs would be significantly increased as more ambitious space missions developed. The initial design involved a reactor with a three kilowatt-electric mercury Rankine conversion system. This was designated as SNAP-2 and the program demonstrated the U-ZrH fueled reactor concept. This led to the start in 1958 of a 500 watt-electric power plant using thermoelectric converters, called SNAP-10A. The SNAP-10A became the only U.S. nuclear reactor to fly in space with a successful launch and flight testing in 1965. U-ZrH reactor development continued with the initiation of the SNAP-8 program in 1960, a 30-60 kW_e power plant, incorporating a higher power level mercury Rankine cycle. Major features of these early U-ZrH reactor systems are shown in Table 1. With a change in emphasis in the space program following the Apollo lunar landings, development activities on all these systems were terminated in the early 1970s.

Table 1. U-ZrH reactor power plant summary

	SNAP-2	SNAP-10A	SNAP-8
Power (kW_e)	3	0.5	35
Design lifetime (yr)	1	1	1
Reactor power (kW_t)	55	30	600
Efficiency (%)	9	1.6	8
Reactor outlet temperature (K)	920	810	975
Reactor type	U-ZrH$_x$ thermal	U-ZrH$_x$ thermal	U-ZrH$_x$ thermal
Primary coolant	NaK-78	NaK-78	NaK-78
Power conversion	Rankine (Hg)	Thermoelectrics (SiGe)	Rankine (Hg)
Boiling temperature (K)	770	—	850
Turbine inlet temperature (K)	895	—	950
Condenser temperature (K)	590	—	645
Hot junction temperature (K)	—	770	—
Cold junction temperature (K)	—	595	—
Radiator temperature (K)	590	595	575
Radiator area (m^2)	11.1	5.8	167.2
System unshielded weight (kg)	545	295	4,545
Specific weight (kg/kW)	182	590	130

SNAP-2A

Starting in 1957, SNAP-2 system development extended over approximately a decade and became the precursor for both the SNAP-10A and SNAP-8 programs. Major technical accomplishments of SNAP-2 system development included:

- Operational uranium-zirconium-hydride reactor concept heat source developed.
- Demonstrated the performance characters of mercury Rankine dynamic energy conversion system.

- Verified the feasibility of using a high-speed, single-shaft, working fluid-lubricated and hermetically sealed turboalternator assembly.

- Also, verified the safe operation of a Rankine cycle space power plant, including startup, steady-state operations and shutdown.

Fig. 1 is a schematic diagram for a 3-5 kW$_e$ SNAP-2 unit.[2] The nuclear reactor heat source was cooled by the primary coolant loop of liquid metal eutectic NaK-78. The coolant was pumped by a thermoelectric DC conduction electromagnetic (EM) pump. The EM pump obtained its electric current from a chromel-constantan themocouple that operated between the reactor outlet temperature and the boiler preheat temperature. Thermal energy was transferred to the power conversion loop through a boiler using mercury fluid. On the secondary loop side of the boiler the mercury became a superheated vapor and then expanded through the turbine. When the mercury vapor left the turbine it was condensed in the radiator and was then returned to the boiler by the mercury pump.

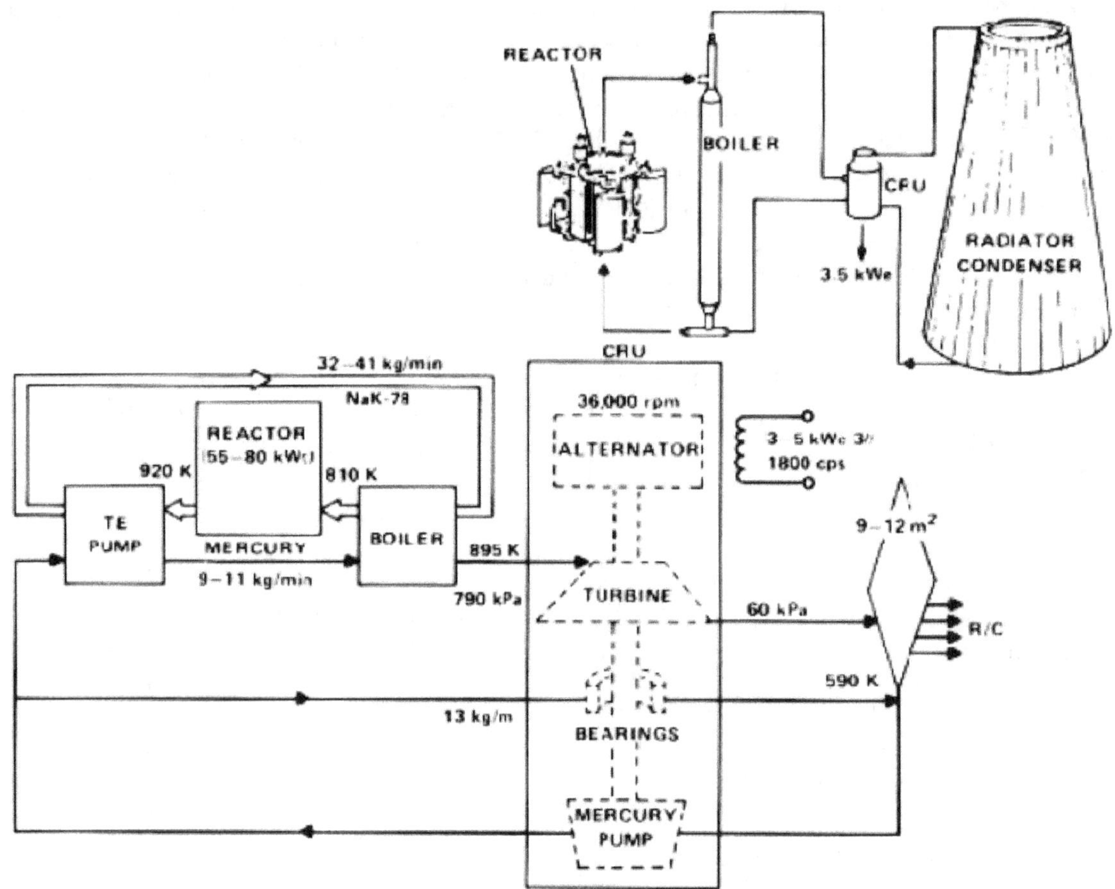

Fig. 1. SNAP-2 mercury Rankine system schematic *From A. A. Jarrett, 1973.*

A schematic of the reactor is shown in Fig. 2. These systems were thermal reactors fueled with fully enriched uranium-235, hydrogen-moderated, and containing a beryllium reflector. The power level was established by control drums that changed the relative position of a part of the beryllium reflector. The reflector also contained separate safety elements

The core assembly consisted of 37 fuel elements that were arranged in a triangular array on 3.2 cm centers. This arrangement resulted in a core configuration forming a right hexagonal cylinder about 20 cm across the flats, 23 cm across the corners, and 25 cm in length. The fuel material was zirconium hydride with approximately 10 weight percent (10 wt %) of enriched uranium alloy. This fuel mixture contained 90.08 wt % ZrH$_{1.87}$ and 9.92 wt % U. The hydrogen-to-metal ratio was 1.79. The core material volume was 7,000 cubic centimeters, with 3.12% by volume uranium metal and 96.88% by volume zirconium hydride.

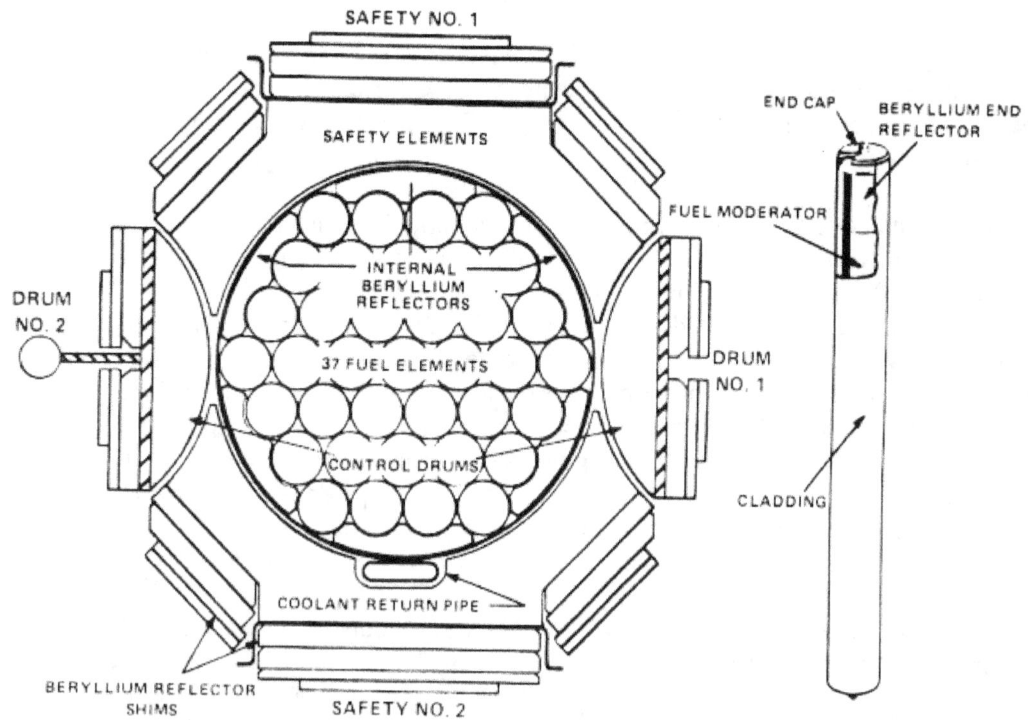

Fig. 2. Fuel rod assembly and core cross section for SNAP-2 reactor. *From A. A. Jarrett, 1973.*

The fuel material was fabricated in a solid cylindrical rod, 25 cm long and 3.1 cm in diameter. The fuel element was assembled by canning the fuel rod in a 3 cm outside diameter (0.3 mm wall thickness) Hastelloy N cladding tube. On each end of the fuel element there was a 3.8 cm long, 3.1 cm diameter beryllium oxide cylinder that served as an end reflector. The inside surfaces of the cladding tube and the end caps were coated with a ceramic barrier that inhibited the diffusion of hydrogen from the fuel element. This ceramic material contained samarium oxide to pre-poison the reactor with equilibrium samarium. The fuel elements were sealed by welding the end caps.

The fuel elements, some 33.6 cm in overall length, were supported and held laterally by top and bottom grid plates. Coolant flow holes (9.5 mm diameter) in these grid plates were located in line with the core coolant passages formed by each set of three adjacent fuel elements.

The reactor NaK coolant entered the core vessel through the lower grid plate at the center of the bottom head. NaK coolant then passed through the core into the plenum chamber between the top grid plate and the top vessel head. The coolant exit line originated at the center of the top head. After a U-bend, it went along the side of the core vessel through a slot machined in one of the beryllium safety elements. At a flow rate of 32 kg / min, the coolant experienced a total pressure drop of 1.2 kPa in passing through the core vessel. There was 86 kg of NaK in the primary coolant system.

The shape of the coolant flow passages were tricusp channels formed by three adjacent fuel elements--except for the outer row where the flow channels were formed by two adjacent fuel elements and the inner surfaces of the internal beryllium side pieces. The fuel element heat transfer area (over the 25 cm length) was 9,385 cm^2. At 50 and 65 kilowatts-thermal power level, the average heat flux was 53.3 and 69.3 kW$_t$/ m^2, respectively. The power density had a peak-to-average ratio of 1.29 in the radial direction and 1.26 in the axial direction. The overall peak-to-average power ratio was 1.63. The peak heat flux was 85.1 kW$_t$/ m^2 at 50 kW$_t$ power level and 112.9 kW$_t$/ m^2 at 65 kW$_t$.

An inside reflector, consisting of six radial reflector beryllium side pieces, was located within the core vessel. The outside reflector included two control drums with a total reactivity worth of 4.69$. Each control drum had individually driven, constant-speed, reversible motors capable of 0.56 degree per second or a maximum reactivity insertion rate of 2.1 ¢ per second. The safety elements were attached to the safety element drive yoke by electromagnets used for scramming the reactor. The SNAP-2 reactor could be scrammed in 200 milliseconds with 10.5$ negative reactivity insertion.

Two reactors were built and tested under the SNAP-2 program, the Experimental Reactor (SER) and the Developmental Reactor (S2DR). The SER was operated at 50 kilowatts-thermal for 1,877 hours with a core outlet temperature of 920 K; and at 50 kilowatts-thermal for 2,290 hours with a core exit temperature less than 920 K. In addition, SER was also operated at less than 50 kW_t and less than 920 K for 1,868 hours. The equivalent time at 50 kW_t operation was 4,493 hours or some 187 days. The reactor experienced both power-dependent and temperature dependent long-term performance degradation. The power-dependent reactor losses were attributed to fuel burnup and fission product poisoning, including samarium and xenon buildup. The samarium buildup was calculated to be the largest power-dependent factor. Temperature-dependent losses were attributed to the loss of hydrogen from the fuel rod, a phenomenon apparently caused by hydrogen dissociation from the matrix material and its subsequent diffusion through the cladding. The successful operation of the SER system permitted the S2DR reactor development to proceed with technical confidence.

The S2DR reactor was operated at 30.5 kW_t and 745 K for 1,150 hours to simulate SNAP-10 reactor conditions. It was also operated at 55 kW_t at a temperature of 920 K for 2,060 hours and at 30.5 kW_t at 810 K for 1,544 hours. During the 21 months of S2DR testing, there were no reactor failures. The success of this experiment supported the conclusion that SNAP-2 reactor operation for one year at 55 kW_t and a 920 K outlet temperature was feasible.

Turning to the power conversion Rankine subsystem, the SNAP-2 design had the rotating components of the power conversion machinery mounted on a single, common shaft. A major design feature resulted in the conversion system having one moving part, supported on liquid mercury bearings. The very heart of the power conversion unit was this combined rotating unit (CRU) (see Fig. 3). Mercury vapor entered at a temperature of 895 K and a pressure of 790 kPa, expanded through the two-stage axial flow impulse turbine that in turn drove the three phase, six-pole permanent magnet alternator. This alternator was a permanent magnet device with a sealed stator. It delivered 5 kilowatts-electric at 1,800 Hz. The shaft of the CRU rotated at 36,000 rpm. The mercury working fluid exhausted at a pressure of 60 kPa and cooled the alternator as it flowed through the unit and over the finned stator housing. The shaft itself was supported by two mercury-lubricated journal bearings, the lubricant for which was supplied by the on-shaft centrifugal mercury pump. The assembly was enclosed in a hermetic housing that prevented the loss of the mercury working fluid.

A boiler was used to transfer thermal energy liberated in the nuclear reactor to the mercury working fluid, which was superheated prior to entering the turbine. A single pass, counterflow design was used. Because the boiler needed to function in the microgravity environment of outer space, internal swirl wires were incorporated to accommodate better heat transfer. Table 2 summarizes the final SNAP-2 system boiler parameters. The final boiler configuration involved four 9.5 mm diameter tubes for the mercury. These were 0.5 mm wall thickness and 5.8 m long. The shell length was 1.4 m with a 10 cm diameter straight cylinder configuration. The boiler had a dry mass of 9.5 kg and a 13.6 kg wet mass.

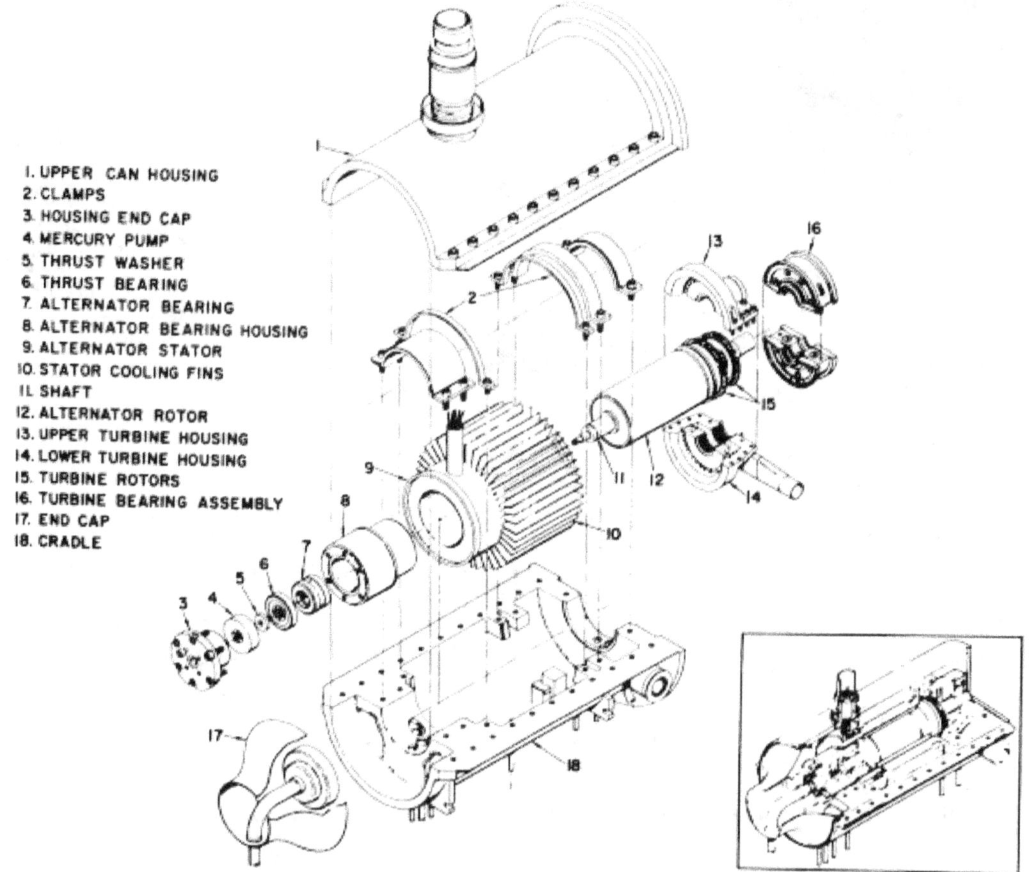

Fig. 3. Combined rotating unit components for the SNAP-2 system. *From A. A. Jarrett, 1973.*

Table 2. SNAP-2 system boiler parameters.

Secondary (Mercury) Loop Side	
Mercury flow rate (kg/min)	9.1
Outlet temperature (K)	895
Inlet temperature (K)	645
Outlet temperature (kPa)	827
Mercury pressure drop (kPa)	414
Primary (NaK) Loop Side	
NaK flow rate (kg/min)	32
Inlet temperature (K)	920
NaK pressure drop (kPa)	1.7

The NaK liquid metal pump was originally designed as a rotating member of the combined rotating unit However, this so complicated the design of an already complex unit that a separate electromagnetic NaK pump unit was eventually designed (see Fig. 4). This thermoelectric EM pump had the advantages of containing no moving machinery, supporting complete liquid metal sealing, providing a built-in electromagnetic flowmeter, and serving as a preheater to the mercury boiler. The pump combined a DC conduction pump and a thermoelectric (TE) generator in a single, all-metallic unit. The chromel and constantan TE elements extracted heat from the hot, circulating NaK and converted a portion of this thermal energy directly into electricity, which was supplied to the conduction pump. Since thermal energy was being rejected from the NaK loop to the mercury loop, the inherently low efficiency of the NaK EM pump proved to be acceptable.

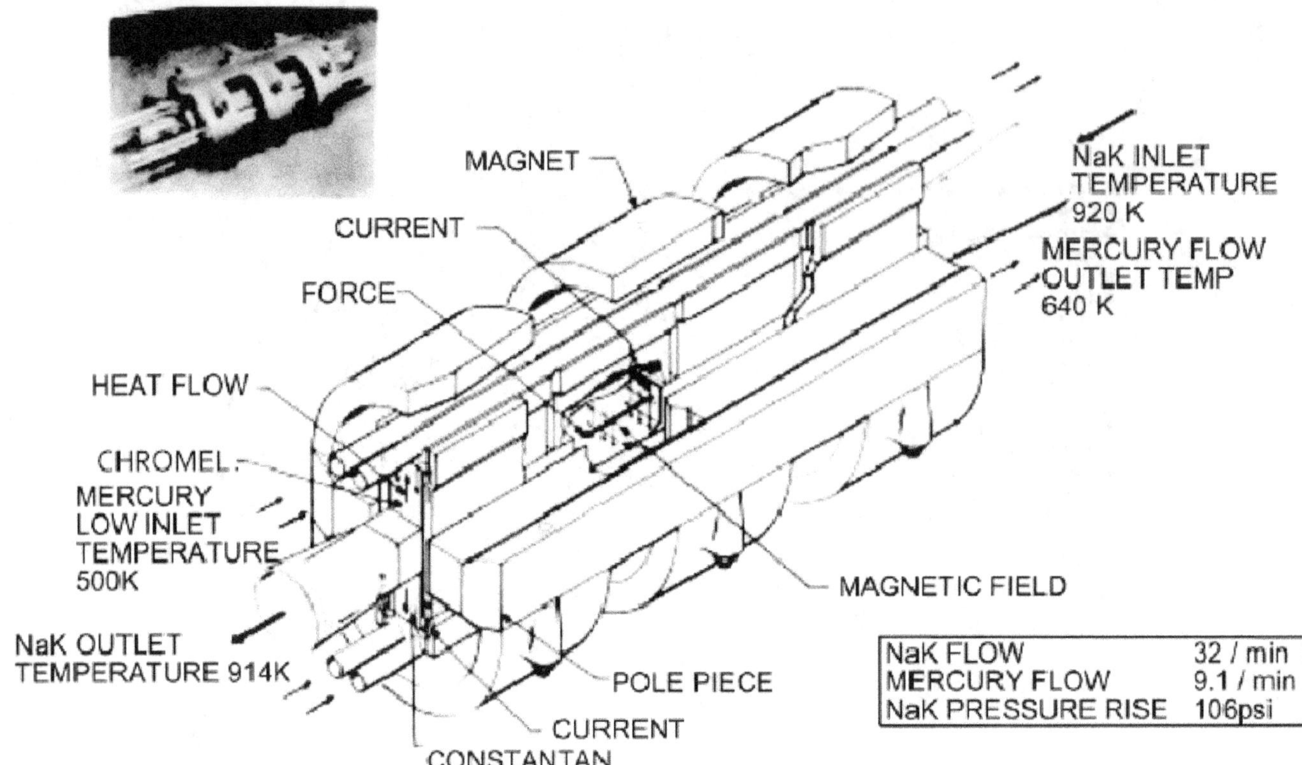

Fig. 4. SNAP-2 thermoelectric-electromagnetic NaK pump *From A. A. Jarrett, 1973.*

The SNAP-2 program faced many challenging technical issues. Table 3 summarizes some of these issues and the year in which they were resolved or satisfied. Engineering questions pertaining to the nuclear reactor included fuel fabrication, hydrogen retention, and coolant pumping.[3] In the power conversion subsystem, the mercury Rankine cycle mercury corrosion rate was regarded as one of the main problems. With mercury as the working fluid, special materials are needed throughout the power conversion system and the "crud" generated by such corrosion processes must be filtered or separated out. This was to prevent possible deposition in critical flow passages. Other technical issues included: (1) protecting the alternator stator windings from liquid mercury; (2) compensating for the poor lubricating properties of mercury so that the load-carrying capacity of the bearings would be adequate; (3) cavitation-erosion damage within the pump and bearings; and (4) the critical design tolerances and clearances associated with a dynamic conversion system for space applications.

Table 3. Technical issues associated with SNAP-2 system development.

Technical Problem Area	Primary Demonstration or Verification	Year Resolved
Criticality of U-ZrH Reactor	Zero Power Critical Experiment	1957
Fabrication of Enriched Uranium ZrH Fuel Elements	SER Fuel Elements	1958
Load Capability-Stability of Mercury-Lubricated Bearings	Prototype Bearing Component Tests	1958
High Speed Operability of Turbo-alternator Assembly	500-hr CRU[a]-1 Test	1959
920 K Power Operation of Reactor	SER Power Demonstration	1959
Zero-Gravity Boiling and Condensing	OFF-BASE Flight Test	1962
Protection of Alternator from Mercury Attack	CRU-V with Evacuated Stator Operation	1964
Long-Life Capability of Turbine, Pump, and Alternator	6,600-hr CRU-IVM-3 Test	1965
Mercury Corrosion Endurance Effects	2,500-hr CRU-V-1A Test (4,700-hr CRU-V-5C Test)	1963 (1966)
Stable Operation of Mercury Rankine	PSM[b]-1 Test	1964
Automatic Orbital Startup of Mercury Rankine System	PSM-3 Test	1965

[a] CRU = Combined Rotating Unit

[b] PSM = Power System Module

SNAP-10A

Design

On December 30, 1960 a decision was made to proceed with the design of a convection-cooled reactor system based on SNAP-2 technology for flight development and demonstration. The design goals were to provide 500 watts of electric power (at a nominal 28 volts) with a one year life duration. Major SNAP-10A design features included liquid metal heat transfer, a DC electromagnetic pump, and thermoelectric power conversion with a cold junction serving as the heat rejection area to outer space.

The major elements of the SNAP-10A system are shown in Fig. 5.[4] The reactor had a thermal fission spectrum using fully enriched uranium-235 fuel and a zirconium hydride moderator. Coolant was by circulating NaK. The electric power generation was by silicon germanium (SiGe) thermoelectric modules. The radiator surfaces used to reject waste heat to space were integrated into the thermoelectric modules. Approximately 5.8-m^2 of radiating surface was necessary. A thermoelectric-powered DC conduction pump was used to transport the liquid metal working fluid (NaK) to the TE elements. The arrangement of the components on the spacecraft is shown in Fig. 6.

A eutectic heat transport mixture, NaK-78, was chosen for thermal energy transfer because of its low melting point and low vapor pressure at reactor operating temperatures. Heated NaK exited the reactor core at 833 K and was circulated at a rate of 0.8 liter per second. The DC conduction pump consisted of a permanent magnet and an integral thermoelectric (PbTe material) power supply. This integral TE converter operated between the NaK outlet temperature and a cold junction temperature established by the heat rejection radiator. The temperature differential (ΔT) was 167 K. As shown in Fig. 5, the hot NaK was then circulated through the SiGe TE power conversion system and back into the reactor core. Bellows-type expansion compensators were attached to the return leg. These devices pressurized the orbiting system to 34.5 kPa and accommodated thermal expansion of the NaK working fluid during reactor startup.

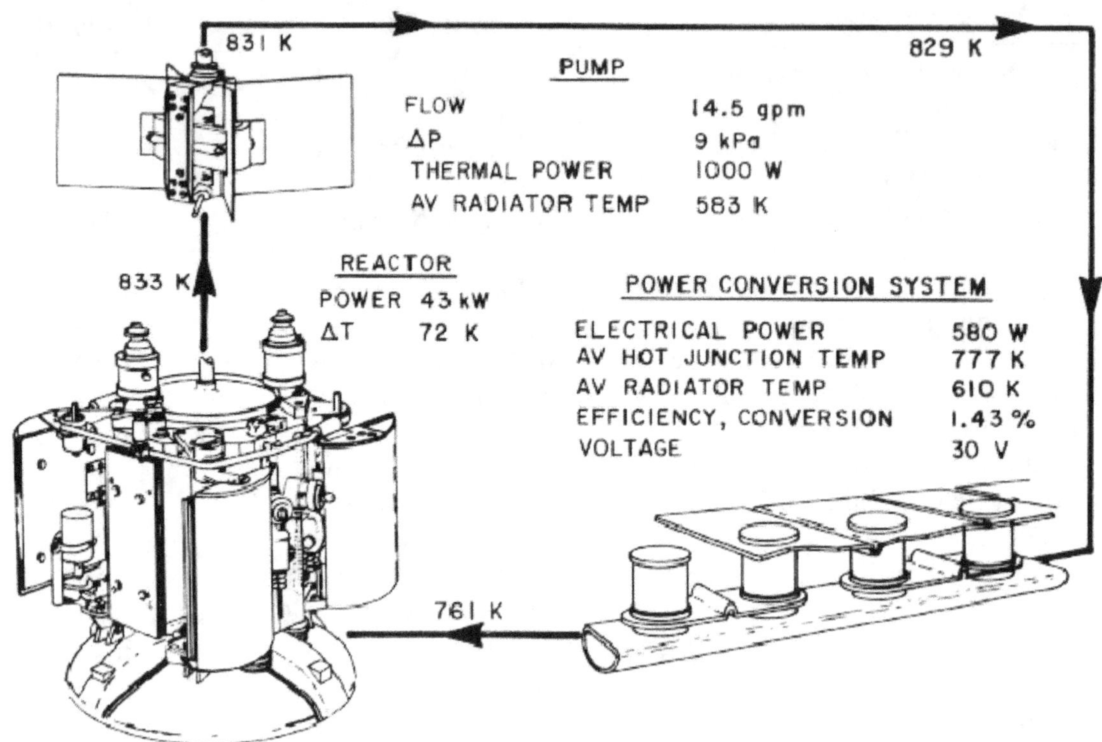

Fig. 5. SNAP-10A power system. *From H. M. Dieckamp, 196*

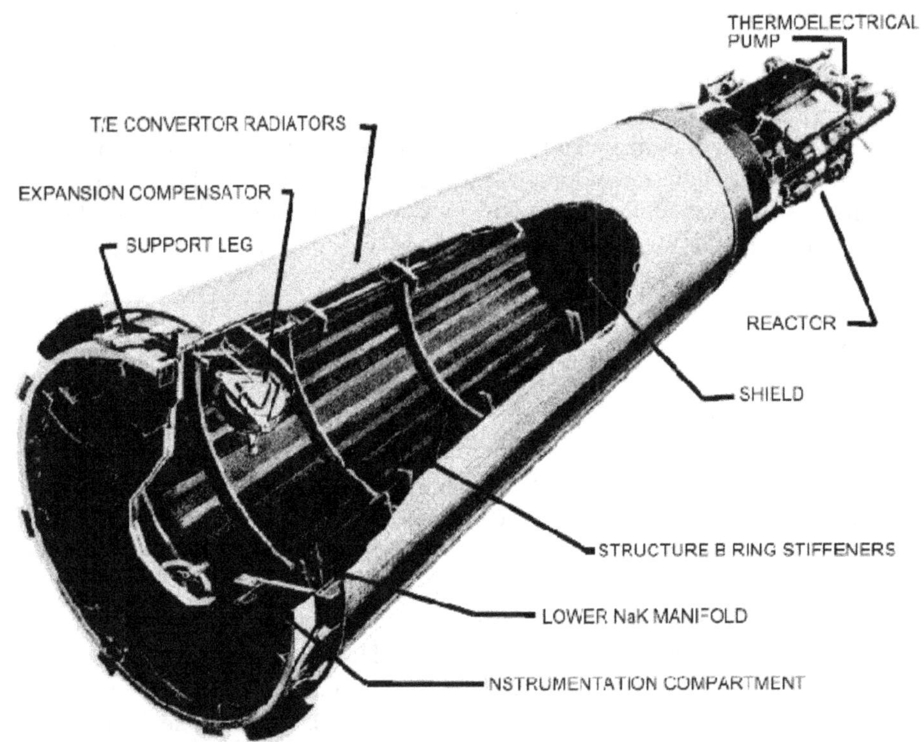

Fig. 6. SNAP-10A flight system configuration. *From H.M. Dieckamp*

Examining the reactor in more detail, the SNAP-10A reactor (see Fig. 7 and Table 4) was based on the SNAP-2A design. There were 37 uranium-zirconium fuel elements enriched with uranium-235. Each fuel element was clad in Hastelloy N tubing and had a ceramic hydrogen barrier applied to its inner surface. The fuel elements were positioned by upper and lower grid plates. The grid plates were positioned by a fixed ring at the bottom and hold-down springs at the top. The core size was 22.6 cm diameter, 41 cm high

Beryllium reflectors were used in the void space between the close-packed hexagonal fuel-element array and the circular reactor vessel. The top reactor vessel head served as both a vessel closure and restraining structure for the core hold-down springs. Four drums were located in the beryllium reflector surrounding the reactor to provide coarse and fine control. Reactor control was achieved by moving the cylindrical pieces outward to control the neutron leakage from the core.

Table 4. SNAP-10A system design parameters.

System Characteristics	Flight System (1965)
Electrical power, minimum (W)	533
Reactor thermal power, nominal (kW)	43.8
Reactor type	
Fuel	$ZrH\text{-}U^{235}$
Reflector	Be
Design life (years)	1
Radiator area (m^2)	5.8
Overall package length (m)	3.5
Mounting base diameter (m)	1.3
Coolant	NaK-78
Coolant flow rate (gpm)	13.3
Reactor outlet temperature (K)	827
Reactor inlet temperature (K)	790
Power conversion system	
Material	SiGe
Voltage	30.3
Mass (kg)	
Reactor	125
Shield	98
Converter	70
Pump	9
Expansion Compensator	13
Piping and NaK	20
Structure	38
Instrumentation and Compartment	48
Heat Shield	15
TOTAL	436

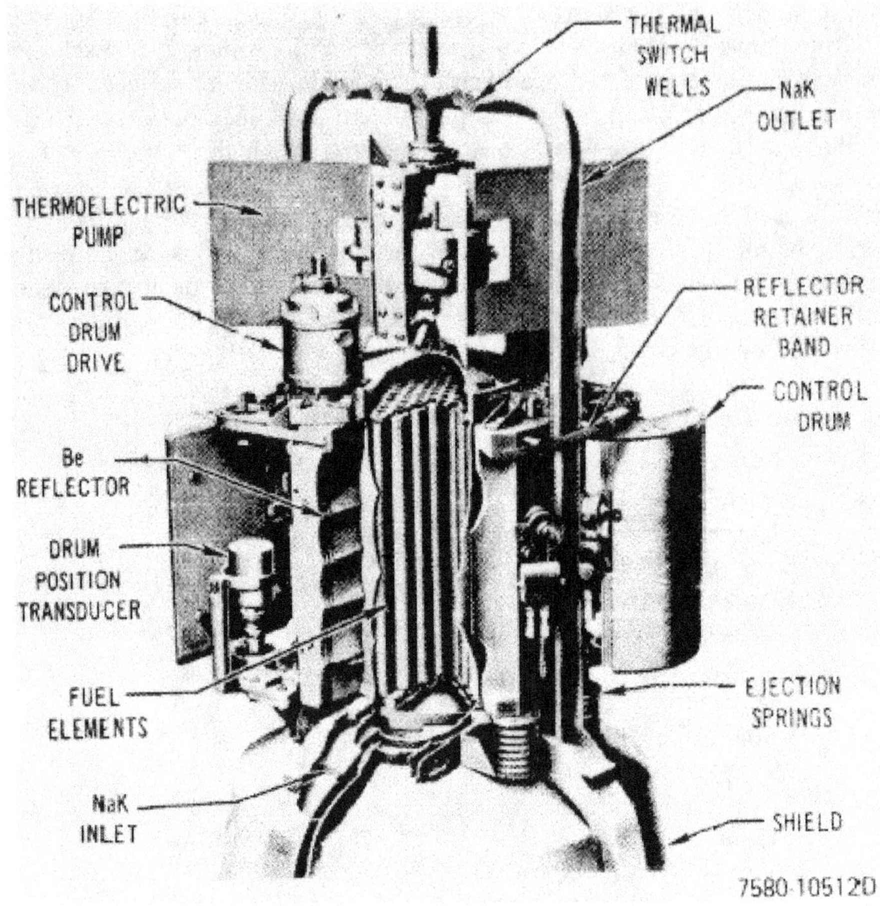

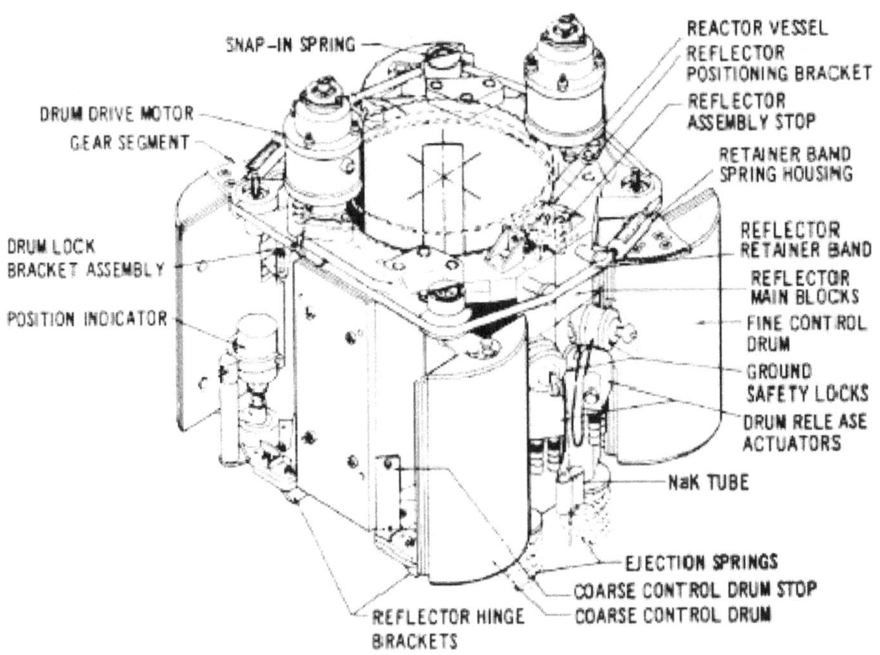

Fig. 7. SNAP-10A reactor (top) and control and reflector assembly (bottom). *From D. W. Staub (Ed.), July 1973.*

For the electric conversion system, silicon germanium (SiGe) thermoelectric material was selected over PbTe for the SNAP-10A system for the following reasons. SiGe is stable at temperatures above 1,255 K while PbTe TE material sublimes above 700 K; this would require that the PbTe material be encapsulant when operated in elevated temperature regimes. Second, stable low-resistance electrical contacts to SiGe can be made by metallurgical bonding. Fabrication experience with PbTe material resulted in unstable higher resistance mechanical contacting. Finally, the mechanical properties of SiGe alloy were more uniform and less design-restrictive

The thermoelectric couples were arranged along the NaK flow tubes. The NaK flow was divided equally among the 40 tubes. There were 72 SiGe thermocouple elements along each tube, for a total of 2,880 individual TE elements. Fig. 8 shows the arrangement and details of the converter. The temperature differential (ΔT) across the TE material was approximately 170 K from the hot to the cold SiGe junctions. The radiator had a total effective area of 5.8 m^2 and an average power dissipation capability of approximately 16.4 W$_t$/ cm^2 at a mean temperature of 590 K. All materials were brazed or metallurgically bonded to ensure a sound structure and a thermally conductive stack.

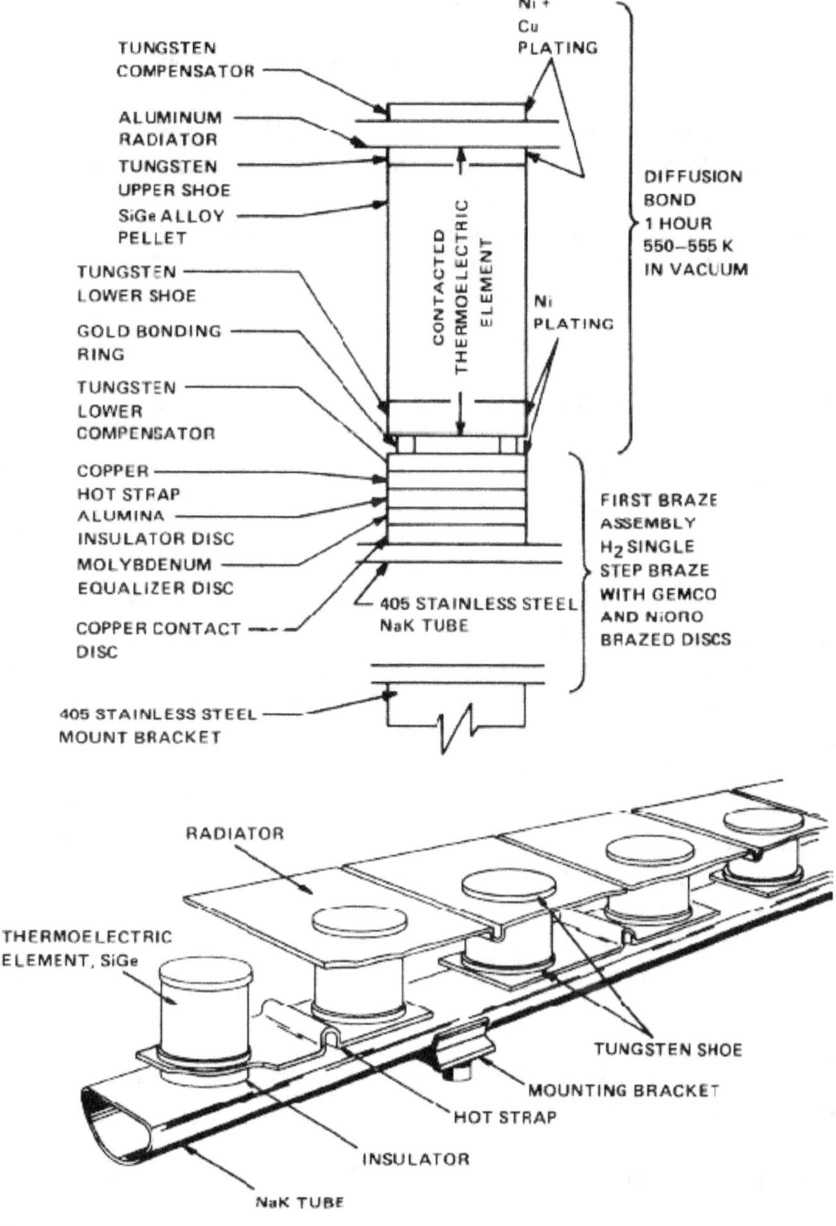

Fig. 8. SNAP-10A power conversion system. *From H. M. Dieckamp, 1967.*

The system was designed for flight operations to be initiated by a single command from the ground. This command was designed to release the locking pins holding the four control drums, apply power to the startup controller, and arm the heat shield eject temperature switch circuit. Two of the control drums were spring loaded and immediately drove to the full-in position; the two remaining drums were automatically inserted by the control system at a rate of $1.0^0 / 300$ sec. At this rate, approximately seven hours would be required to reach criticality, during which time the system remained with the heat shields in place to protect the liquid metal (NaK) working fluid from freezing in the space environment. After criticality was reached, a nominal power excursion occurred that was terminated by the reactor's inherent negative temperature coefficient of reactivity. With a continued insertion of reactivity, the reactor gradually approached full power. When the temperature of the liquid metal working fluid reached 408 K, the thermal heat shield switches would release the heat shield halves and preloaded springs caused their ejection from the power plant. With an increase in reactor temperature, the thermoelectric pump, located on the reactor pressure vessel head, caused immediate bootstrap flow. When the system reached full operating temperature, the ramp insertion of reactivity was terminated by the action of a temperature switch that sensed reactor outlet temperature. This temperature switch functioned like a thermostat by allowing an occasional reactivity insertion to maintain the reactor outlet temperature about the switch's setpoint. The SNAP-10A power plant was designed to have no moving parts except for the reactor's control system.

Driven by a system design requirement to minimize shield weight, a shadow shield configuration was chosen. The location is shown in Fig. 6. The shielding requirements specified limiting radiation doses at the power unit-spacecraft interface to approximately 5×10^{12} fast neutrons and 4×10^7 rads of gamma radiation in one year. All spacecraft components were located in the conical shadow of this shield to eliminate undesirable radiation scattering. The shield itself was located directly below the reactor and used a cold-pressed lithium hydride material that was reinforced with stainless steel honeycomb contained in a Type 316 stainless steel casing. The shield weighed 98 kg. It was designed to withstand a maximum temperature of 712 K at the attachment to the TE converter radiator and a minimum temperature of 579 K near the surface exposed to space.

SNAP-10A Flight Demonstration
SNAPSHOT, flown in 1965, demonstrated that a nuclear reactor could be successfully launched into and operated in the space environment; and that nuclear reactors could be used safely in space without exposing the terrestrial population or biosphere to undue risk. As such, the on-orbit testing of the SNAP-10A reactor in the SNAPSHOT mission achieved the SNAP-10A program objectives.

In preparation for launch, the entire system was thermally checked using electric heaters in a simulated space environment. The reactor was neutronically checked in a series of criticality tests that did not involve the creation of a significant quantity of fission product nuclides. Significant higher power reactor operations were not performed in order to avoid fission product buildup that would have inhibited prelaunch access to the unit and severely impaired launch pad activities. As a result, launch pad activities at Vandenberg Air Force Base in California were very much the same for the SNAP-10A payload as for other payloads. The SNAP-10A system was mated to the Atlas-Agena vehicle in the gantry complex, and all subsystems were given a prelaunch check. The reactor itself was not made critical on the pad. It should be noted that the time required to mate components, perform functional checks, and complete booster preparation closely followed patterns experienced with other, non-nuclear payloads. No unique operational constraints were introduced by the presence of the SNAP-10A system and no unusually elaborate aerospace ground handling equipment or procedures were required to support the SNAPSHOT launch. Finally, since the "cold, clean" reactor did not pose a radiation hazard on the pad, normal working access at the launch site to either the payload or its Atlas-Agena launch vehicle was not restricted.

SNAPSHOT was launched by an Atlas-Agena on 3 April 1965 at 1:24 P.M. Pacific Standard Time. The vehicle achieved an orbit with an apogee of 1,328 km and a perigee of 1,295 km-- the target orbit was a circular orbit at an altitude of 1,296 km. At 5:05 P.M. PST the start command was given to the SNAP-10A system. Criticality was reached at 11:15 P.M. PST and full power was successfully achieved at 1:45 A.M. PST on April 4.[5]

After approximately six days at full power, the on board automatic control system was deactivated by ground command.   From that time on, the SNAP-10A system was controlled by its inherent negative temperature coefficient of reactivity. This provided a significant safety feature in the SNAP-10A design.

However, on May 16, 43 days after startup, the SNAP-10A reactor system shut down. Analyzes indicated that the reactor performed normally up to that point and that the shutdown was caused by an apparent failure in the voltage regulator in the Agena vehicle. Because of built-in aerospace safety design features, no provision was made to restart the reactor system using ground based commands. Consequently, this reactor remains quiescent in its 4,000 year orbit.

The calculated radioactivity inventory for the SNAP-10A core as a function of time after shutdown is presented in Fig. 9.. Fifteen years after shutdown the fission product inventory is less than 100 curies, while after 100 years the core radioactivity level will be less than 0.1 curie.   When the SNAP-10A finally reenters the Earth's atmosphere several millennia from now, the radioactivity level of its core will be negligible. The reactor, as designed, should disperse harmlessly during a final fiery plunge through the Earth's upper atmosphere.[6,7,8,9]

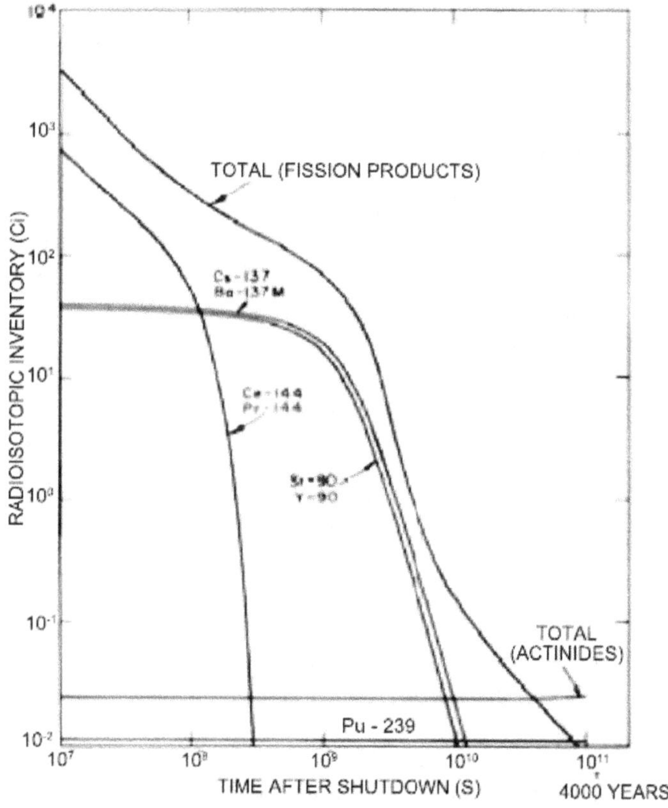

Fig. 9. Calculated radioisotope inventory of SNAP-10A reactor as a function of time after shutdown.

Despite this premature shutdown, SNAPSHOT represents an extremely successful experiment! The entire reactor subsystem performed properly and very predictably. Criticality was achieved at the reactivity levels anticipated, indicating that the launch environment produced no adverse effects on the reactor. Also, there were no signs of self-welding problems or failures in the high temperature bearings of the reactor control elements. The thermoelectric power conversion system proved reliable, and heat rejection to space performed a bit more efficiently than anticipated. This provided an extra power output of 3 percent. The launch environment did not adversely affected the liquid metal coolant loop. This included no unpredicted effects from the launch vibrations.   The launch shock and vibration loads were actually a factor of approximately 1.5 under the qualification levels. In addition, the system apparently did not experience any micrometeoroid punctures of its 0.5 mm thick tubing and there was no detectable deterioration of applied emissivity surfaces. Finally, the fast neutron radiation levels were about a factor of four higher on the vehicle centerline and a factor of four lower on the outer vehicle than predicted. The aft end of the Agena vehicle gamma radiation level was about two orders of magnitude lower than anticipated, because of scattering and attenuation by the vehicle structure and components within it.

As a complement to the SNAPSHOT mission, a duplicate of the flight system, called the FS-3, was operated in a shielded vacuum chamber on the ground for 10,000 hours. Post-test examination of this system revealed that there were no apparent wear out modes or incipient failures present.

SNAP-8

With enhanced launch vehicle capabilities that supported larger payloads in orbit, a second generation of SNAP nuclear power plants development was initiated. The SNAP-8 system was planned to provide 30 kW_e of power with capability to expand to 60 kW_e, a 10,000 hour operational lifetime, and a reactor outlet temperature of the working fluid of 975 K. The reactor was based on the SNAP-2 liquid metal (NaK) cooled, uranium-zirconium-hydride reactor design and was then coupled to a mercury working fluid Rankine cycle power conversion system. The SNAP-8 Experimental Reactor (S8ER) tested from May 1963 through April 1965 and the Developmental Reactor (S8DR) tested from January 1969 through December 1969 were developed and tested as part of the SNAP-8 program. Table 5 lists the reactor characteristics of these two systems.[10]

Table 5. SNAP-8 reactor characteristics.

	S8ER	S8DR
Design power level (kW_t)	600	600
Design life (hr)	10,000	12,000
Primary coolant	NaK-78	NaK-78
Number of fuel elements	211	211
Fuel loading (kg of 93.15% enriched U^{235})	6.56	8.2
Core vessel outside diameter (cm)	23.7	23.5
Reflector (thickness of Be, cm)	7.6	13.0
Control drums	6	6
Reactor outlet temperature (K)	977	977
Reactor inlet temperature (K)	866	866
Average power density (kW/liter)		37
Core mass flow (kg/sec)	6.1	6.1
Mean prompt neutron lifetime (μ sec)	6.7	8.4
Effective delayed neutron fraction	0.0077	0.0080
Thermal flux (n/cm^2/sec)	2×10^{12}	3×10^{12}
Medium fission energy (eV)	0.21	0.15
Clean, wet excess reactivity ($)		14.4
Control drum worth ($)		21.1
Max. cladding temp at design conditions (K)	1,055	1,015
Max. core temp at design power (K)	1,112	1,080

Fig. 10 is a schematic illustration of the four loop SNAP-8 power plant.[11] The reactor was coupled to the power conversion loop at the mercury boiler. The primary loop used the eutectic mixture of sodium and potassium (NaK-78) to transport thermal energy from the reactor to the boiler. The liquid metal working fluid was circulated by a specially designed NaK pump. The second loop, the power generation Rankine cycle loop, used mercury as the working fluid. Components of the loop included a boiler, turbine-alternator assembly, condenser, and boiler feed pump. The third loop was used to remove waste heat from the Rankine cycle second loop. It was composed of a pump-driven NaK loop which coupled the condenser and flight radiator assembly. The fourth loop used an organic fluid and low temperature radiator. It lubricated the bearings of the rotating components in the mercury loop and provided coolant for the alternator, pumps, and electrical controls. There was also an auxiliary loop (not shown in Fig. 10) for startup. It coupled the two NaK loops and provided an intermediate thermal load for the nuclear reactor prior to startup of the mercury loop.

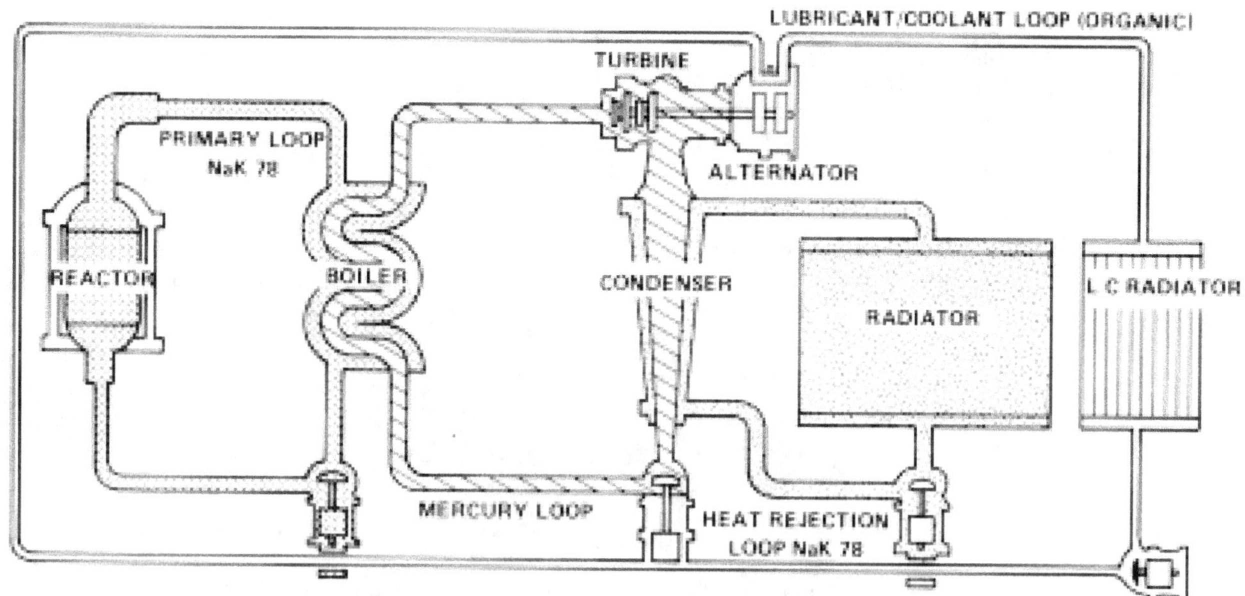

Fig. 10. SNAP-8 four loop power plant schematic. *From J. N. Hodgson, R. G. Germer, A. H. Kreeger, 1967.*

The S8ER system was designed to produce 600 kilowatts of thermal power, with a coolant outlet temperature of 975 K. During testing, this reactor generated over 5.1 million kilowatt-hours of energy--a performance that included 100 days of operation at 600 kW-thermal and 975 K outlet temperature, and 365 days operation at more than 400 kW-thermal and 975 K outlet temperature. These tests demonstrated static and dynamic stability, as well as tolerance to rapid transients. The core was subjected to some 115 rapid changes in power (i.e., 100 kW-thermal in one minute). However, post-test core disassembly and fuel element analyses indicated that 80 percent of the 211 fuel elements experienced cladding cracks. Detailed examination of selected fuel elements revealed that these cracks resulted primarily from high temperature swelling of the hydrated fuel, combined with the low ductility of irradiated Hastelloy N cladding material. This phenomenon was attributed to excessive fuel swelling experienced in the core at temperatures greater than 1,035 K.

The S8DR reactor was a modified S8ER design. The modifications of the core consisted of: (1) changing the fuel element hydrogen permeation barrier to a thermal cycle resistance coating; (2) increasing radial and axial fuel rod-cladding clearances; (3) redesigning fuel element end caps to alleviate stress cracking; and (4) modifying the coolant flow profile.

The S8DR reactor assembly (see Fig. 11) contained 211 fuel elements, each 1.34 cm in diameter and 42.7 cm long (see Fig. 12).[12] The center fuel region contained a hydrated zirconium-uranium alloy with 10.5 wt % U. The hydrogen concentration was 6.05×10^{22} atoms / cc. The axial hydrogen gap was 6.1 mm, while the radial gap was 0.09 mm. A 0.05 mm ceramic coating was included on the inside of the 0.29 mm thick Hastelloy N cladding and it served as a hydrogen diffusion barrier. Included in this ceramic was 58 mg of a burnable poison (Sm_2O_3), which compensated for reactivity loss due to fission product buildup. As shown in Fig. 13, the S8DR fuel elements were arranged in a triangular lattice on 14.6 mm centers. The core was cooled by the axial flow of NaK between the fuel elements.

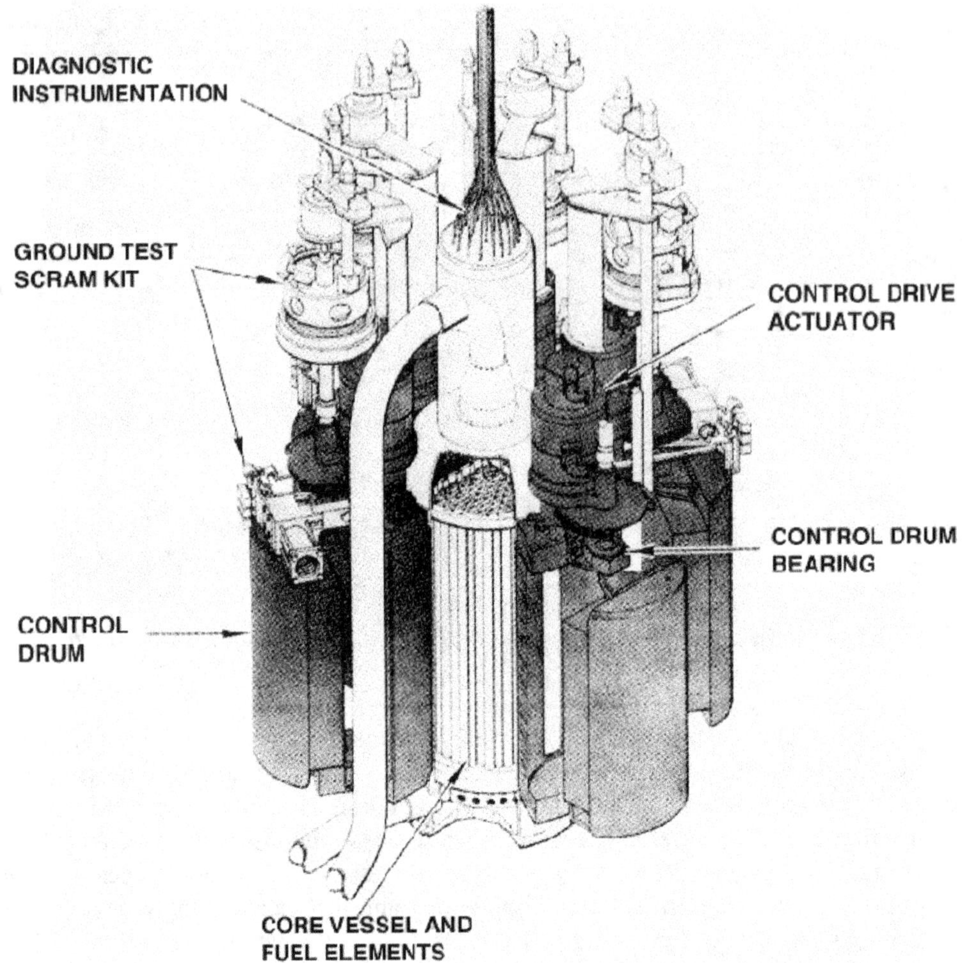

Fig. 11. S8DR ground test assembly. *From D. W. Staub (Ed.), July 1973.*

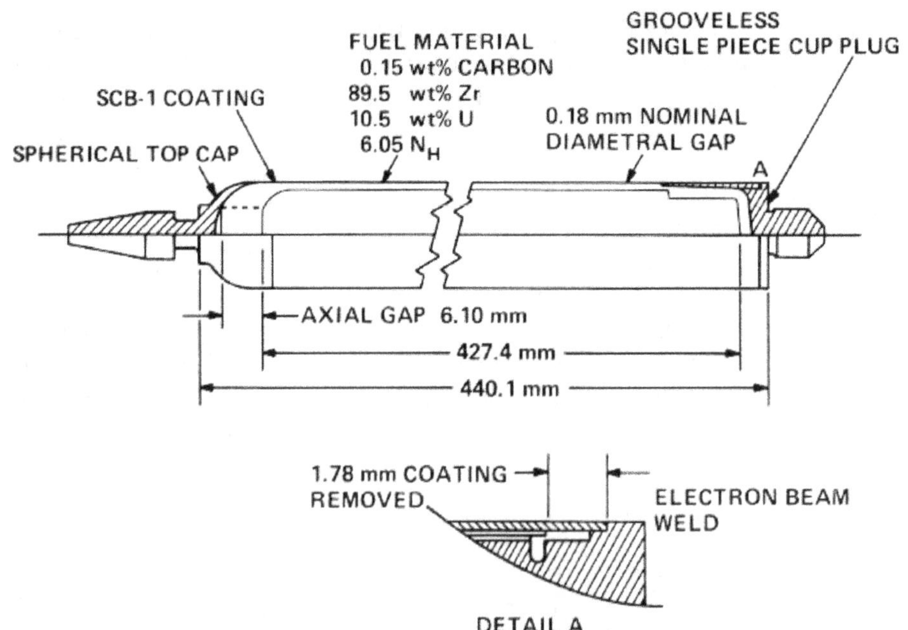

Fig. 12. Details of S8DR fuel element. *From D. W. Staub (Ed.), 1973.*

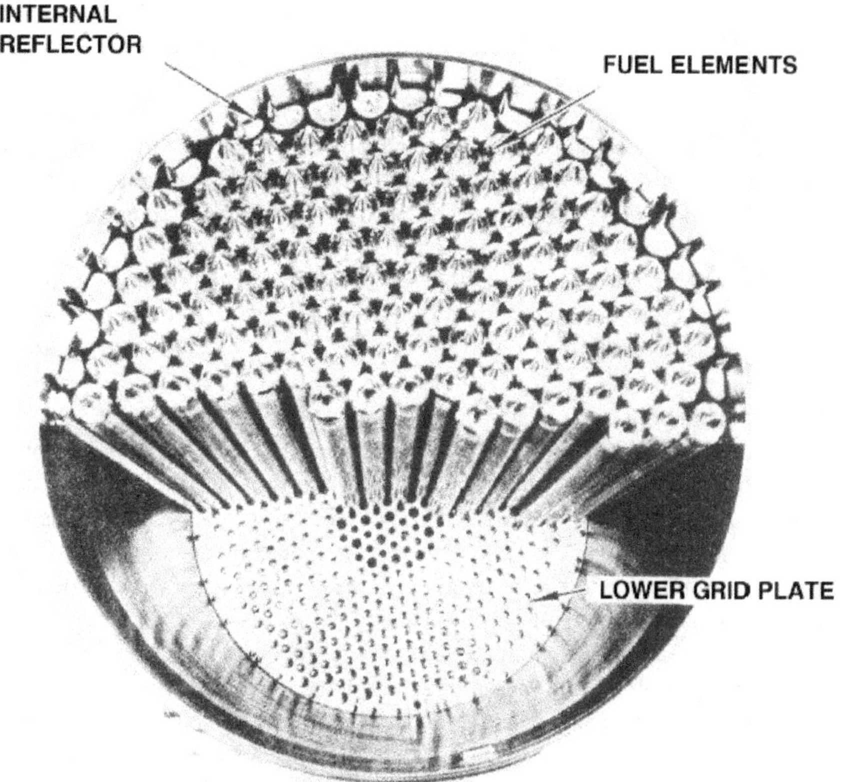

Fig. 13. Partial fuel array in SNAP-8 reactor core vessel. *Courtesy of Atomics International (AI)-AEC-13071.*

The reactor core was located in a right circular cylinder pressure vessel that consisted of a lower plenum head, an intermediate cylinder, and an upper plenum head. These components were fabricated with Type 316 stainless steel. The lower plenum head had a flat external bottom and a convex 45.7 cm radius internal bottom surface. This plenum head had internal brackets for mounting the flow distributor, the baffle plate, and the lower grid plate. An external mounting ring was used to couple the core vessel to the support structure. The intermediate cylinder, with a nominal wall thickness of 2.3 mm, was welded to the upper and lower plenum heads. This component had internal lugs to support and position the upper grid plate and external brackets on which were mounted the external reflector and control system. The upper plenum head contained a coolant outlet nozzle, 3 hold down lugs for the upper grid plate, 3 support lugs for the thermocouple bridge, and 27 thermocouples to measure coolant temperature at the reactor outlet. The core vessel was designed for operation at a pressure of 241 kPa with an inlet temperature of 894 K and an outlet temperature of 1,005 K.

Upper and lower grid plates were used within the vessel to position the fuel elements. A flow distributor and baffle plate were located beneath the lower grid plate to break up the flow coming into the lower plenum, to prevent vortexing, and to provide a constant pressure profile across the lower face of the flow baffle plate. There were 42 internal reflector pieces in the core to fill the void space between the hexagonal core array and the cylindrical vessel. Thirty were made of beryllium oxide, clad in Type 316 stainless steel, while the balance were solid Type 316 stainless steel.

The reactor vessel was surrounded by a beryllium reflector, as shown in Fig. 14. This reflector was divided into two stationary and six movable sections. Reactor control was provided through the movable sections. The reflector assemblies were attached to and supported by brackets that were an integral part of the reactor core vessel. The drums were shaped as segments of right-circular cylinders with 11.4 cm radii. The basic drum had a length of 38.9 cm and a maximum thickness of 5.4 cm. A 7.6 cm beryllium shim was attached to the back of each drum to make up the total thickness of 13.0 cm. All beryllium surfaces were anodized to provide a high emissivity. The reflector assembly was designed to limit the beryllium temperature to less than 1,005 K. Each L control drum rotated on a pair of upper and lower self-aligning bearings. These bearings were of a journal type, designed for operation at 894 K. The control drums were designed to compensate for the reactivity changes listed in Table 6.

The control drum actuator was an eight-pole reluctance motor that produced torque by the interaction of magnetic forces between the stationary teeth on the fixed poles and the notched teeth cut into the cylindrical rotor. The field coils for the poles were sequentially energized to produce a revolving magnetic field that turned the rotor. The stator-pole tooth alignment was such that, as each pole was energized, the rotor teeth were in a position to provide maximum torque. When not energized, the rotor was held in position by a brake mechanism. The control drum drives operated in conjunction with a scram mechanism that would rapidly rotate the control drums outward, when an electromagnetic clutch was released.

Operation of the S8DR system generated 4.3 million kilowatt-hours. This included 1,670 hours at 600 kW-thermal and an outlet temperature of 975 K, and 5,575 hours at 600 kW-thermal and an outlet temperature greater than 920 K. However, after disassembling the core following testing the fuel elements exhibited failure modes similar to those experienced by the S8ER fuel elements, with 72 out of 211 elements developing cladding cracks typical of stress-rupture phenomenon. These failures were attributed to excessive cladding strain caused by fuel swelling resulting from core temperatures in excess of design values. The probable source of this high temperature environment was entrainment of surge tank cover gas. This entrainment process was coupled with two standing, stable vortices in the inner plenum. The vortices tended to agglomerate small entrained gas bubbles and periodically release large bubbles that passed through the core in the region of maximum fuel element damage. Stable clustering also occurred, causing additional flow decreases and, consequently, fuel element temperature increases. This entire process resulted in fuel element temperatures that were estimated to be over 55 K hotter than the (maximum) design value. Once again, such over-temperatures were attributed to thermohydraulic phenomena within the core. Other reactor components operated satisfactorily.

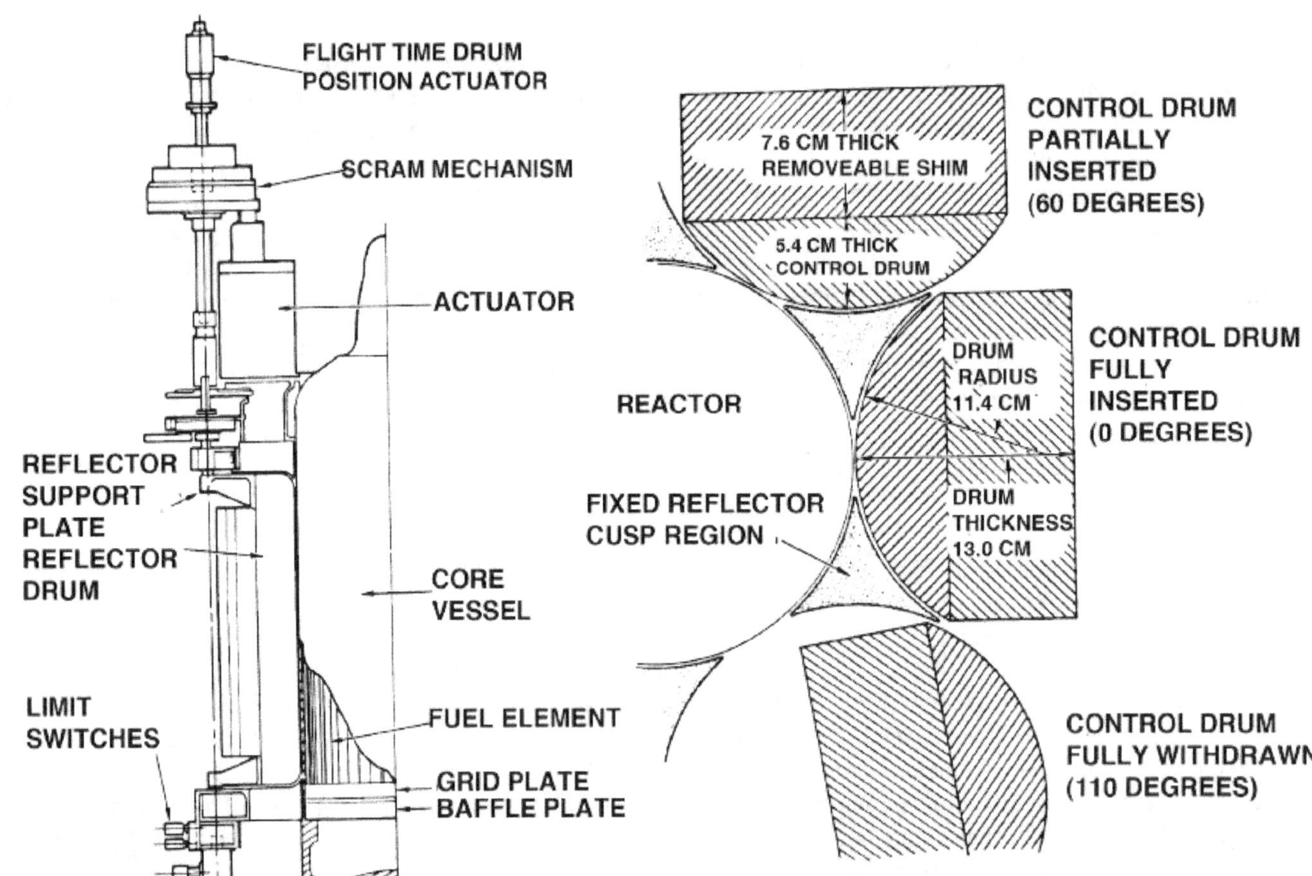

Fig. 14. Reflector and control drums for SNAP-8 development reactor. *From D. W. Staub (Ed.), 1973.*

Table 6. S8DR nuclear characteristics.

Clean, wet, excess reactivity ($)	14.4
NaK reactivity worth ($)	0.25
Total isothermal temperature coefficient of reactivity	$- 0.32¢/K$ at 295 K $- 0.36¢/K$ at 920 K
Power coefficient of reactivity	$- 0.035¢/kW$
Rise to 600 kW power and 920 K core average temperature ($)	$- 2.15$
Reactivity inventory (12,000 hours) ($)	
Hot, end of life excess (600 Kw)	5.45
Samarium burnout	2.90
Fission product poisoning	
Xenon	$- 1.05$
Samarium	$- 1.02$
Other	$- 1.17$
Axial hydrogen redistribution	$- 0.65$
Hydrogen loss	$- 5.15$
^{235}U burnup	$- 0.66$
Control drum worth ($)	
Total—6 drums	21.06
Single drum	3.90
Maximum differential worth	$5.75¢/deg$

The SNAP-8 boiler was a single-pass counterflow heat exchanger that consisted of seven mercury containment tubes in a single NaK tube outershell. The NaK part of the boiler was fabricated using Type 316 stainless steel. Tantalum, with its superior mercury corrosion resistance and good mercury "wetting" properties, was used as the mercury containment material. A non-flowing layer of NaK, contained in Type 321 stainless steel tubing, surrounded each tantalum tube. This non-flowing layer of NaK protected the tantalum from corrosion caused by interaction with the impurity materials in the primary loop NaK. The primary coolant loop working fluid (NaK) flowed over the outside of these Type 321 stainless steel tubes.

Two NaK pumps were used in the powerplant; one in the primary loop, the other in the heat rejection (third) loop. A special, single pump was designed and developed to suit both needs. The pump motor assembly (see Fig. 15) was a self-contained, hermetically sealed unit.[13] The unit combined on a single shaft: a centrifugal pump, a hermetically sealed motor, an internally lubricated coolant recirculating pump, and NaK-lubricated bearings. In addition, a recirculating loop was integrated with the pump motor assembly. This loop contained a "coldtrap" system, a heat exchanger, and a NaK filter. No static or dynamic shaft seals were used in the NaK motor. A close clearance annulus around the shaft between the pump and motor was used to isolate the NaK in the closed recirculating loop from the NaK in the primary loop. A feature of this design concept, a design in which a high thermal gradient existed between the pump and the motor, is that the recirculating fluid remained relatively uncontaminated by the primary loop working fluid. It should be noted that the primary loop NaK could contain a much higher oxide level than could be tolerated in the bearings and motor. Any oxides migrating through the annulus into the recirculating loop were trapped out.

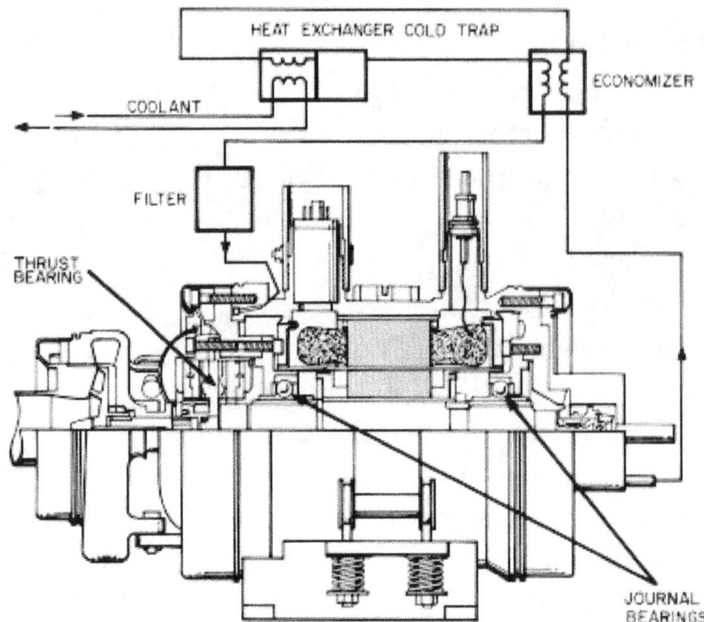

Fig. 15. SNAP-8 NaK pump-motor assembly. *From J. N. Hodgson and R. P. Macosko, 1968.*

A three-phase, 208 volt, 400 cycle squirrel-cage induction motor was used to drive the pump. The rotor and stators were canned with thin cylinders of Inconel. The rotor shaft was supported on two tilting pad journal bearings in assemblies of four pads per bearing. This design was selected because of its ability to accept misalignment, its inherent stability, and its low frictional characteristics.

The mercury turbine-alternator assembly in the secondary loop consisted of two subassemblies: the turbine and the alternator. These components were bolted together at a common flange. Turbine power was transmitted to the alternator by a splined drive on a flexible quill shaft (see Fig. 16). The turbine was a cantilevered, four stage axial-flow type. The first and second stages were partial admissions, and the third and fourth stages were full admissions. Preloaded, angular-contact ball bearings were used with an over-hung turbine, and a space seal system was interposed between the turbine exhaust duct and the outboard turbine bearings. The alternator was a hermetically sealed, three-phase, 220/208 V, 400 cycle, radial air gap, homopolar inductor. It featured a brushless solid rotor straddle-mounted on angular contact ball bearings. The turbine-alternator was lubricated and cooled by organic polyphenyl-ether.

The mercury pump motor assembly was a self-contained system with a "seal-to-space" (a mechanical shaft seal discharging to vacuum) (see Fig. 17). The pumping system consisted of a centrifugal pump, low-leakage dynamic shaft seals, and an induction motor. All of these components were located on one shaft, supported by angular-contact ball bearings lubricated with polyphenyl-ether. The motor and shaft seals were cooled with the lubricant fluid. Integral with this assembly were a jet-booster pump and lubricant coolant valves along with their associated plumbing. The jet-booster pump was used to improve the suction performance of the centrifugal pump. The centrifugal pump was a conventional design. The volute was a partial emission design and the back vanes of the pump impeller were used to balance the axial thrust.

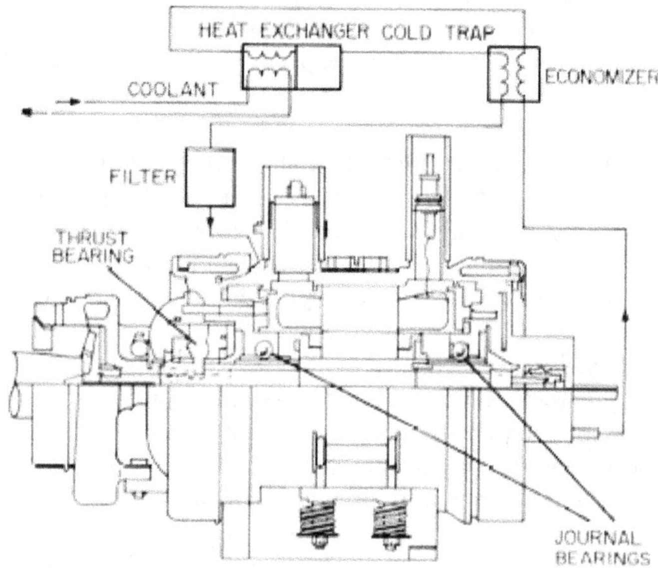

Fig. 16. SNAP-8 turbine-alternator assembly. *From J. N. Hodgson and R. P. Macosko, 1968.*

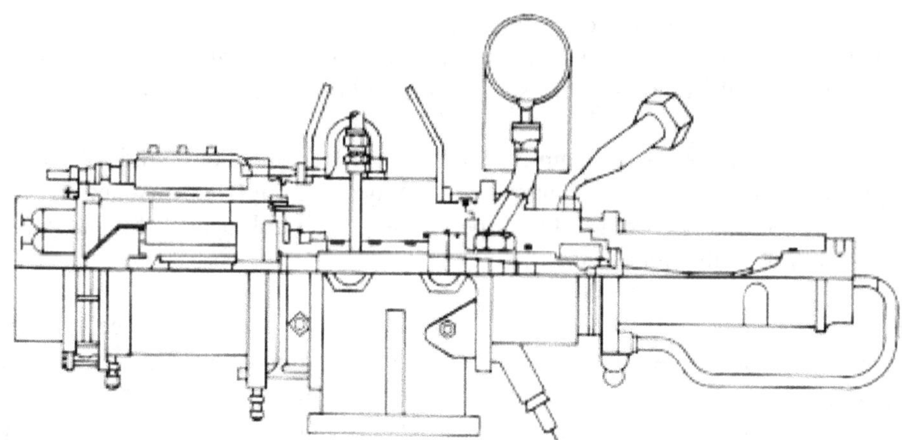

Fig. 17. SNAP-8 mercury pump-motor assembly. *From J. N. Hodgson and R. P. Macosko, 1968.*

A counterflow, tube-in-shell heat exchanger was used for the condenser. The mercury working fluid was condensed in 73 tapered tubes. The tapered tubes were designed to maintain a high vapor velocity through the length of the condenser, sustaining a continual movement of condensate droplets toward the liquid-vapor interface in a microgravity environment. NaK was the shell-side coolant and it flowed through the condenser as part of the heat rejection system.

In summary, the SNAP-8 ground based testing of the S8ER and S8DR reactors provided considerable operating experience and testing data. The S8ER was operated for over one year at a power level between 400 and 600 kW-thermal and a reactor outlet temperature of 975 K. The S8DR reactor was operated for 1,670 hours at 600 kW-thermal and a reactor outlet temperature of 965 K and for 6,688 hours at 600 kW-thermal and reactor outlet temperatures greater than 810 K. Even though the S8DR reactor developed a significant number of cracked fuel elements, appreciable operating life still remained when it was shut down. In fact, it has been predicted that this reactor could have operated for more than 16,000 hours.[14]

The cause of the fuel elements cracking was attributed to fuel element mechanical instability, coupled with larger than anticipated core coolant flow bypass. These caused excessive fuel element temperatures in selected core locations. This condition, in turn, resulted in greater than expected fuel swelling, which subsequently led to fuel element cladding cracking.

Since the U-ZrH reactors used integral moderated fuel, the fuel element cladding must be lined with a ceramic coating that acts as a hydrogen permeation barrier. In this case, by an integral moderated fuel, the fuel is a mixture of uranium with hydrided metallic zirconium that is equilibrated by a low hydrogen overpressure (approximately 330 mm of H_G at temperature). Excessive deformation or cracking of the cladding degrades or fractures the hydrogen barrier. This permits a higher than predicted loss of hydrogen that ultimately results in a far-shortened operating lifetime.

At the termination of the SNAP-8 program, power conversion system components were well on their way toward being proven items.[15, 16] All the major components were operated at least 10,000 hours without maintenance. A "breadboard" configuration had also been run successfully for 8,700 hours.[17] The turbine-alternator was operated for a total of 10,823 hours at system-rated load conditions and 35 start-stop cycles. During the last approximately 5,000 hours of this test, a 0.5 percent loss in turbine-alternator aerodynamic efficiency was observed. This loss was attributed primarily to a reduction in the third-stage nozzle area from deposits of mass transfer products and also from rotor blade leading edge erosion and corrosion damage. These conclusions were based on post-test inspection of the turbine-alternator, flow-testing of the nozzle assemblies, and measurements of blade erosion.

The mercury boiler feed-pump was tested at rated conditions for 12,227 hours and 109 start-stop cycles. Hydraulic performance tests were conducted at intervals to determine whether changes had occurred. The hydraulic-electric efficiency of the pump was the same at the end of the test as at the beginning.

The SNAP-8 NaK pump was tested as a complete assembly for 10,363 hours and 786 start-stop cycles. The hydraulic and electric performance of the pump was identical throughout this endurance test. Significant cavitation damage, however, had occurred in the impeller back vane hub area. This damage was so severe that some local penetrations of the hub wall had occurred.

In conclusion, the life expectancy of rotating components based on these performance tests indicated a five year lifetime could be achieved with the exception of the mercury pump and impeller. The condition of seals must also be carefully evaluated, and some improvement in the turbine rotor might also be necessary. As shown in Fig. 18, the lifetime of ZrH reactor systems can be increased considerably by reducing the temperature.[18]

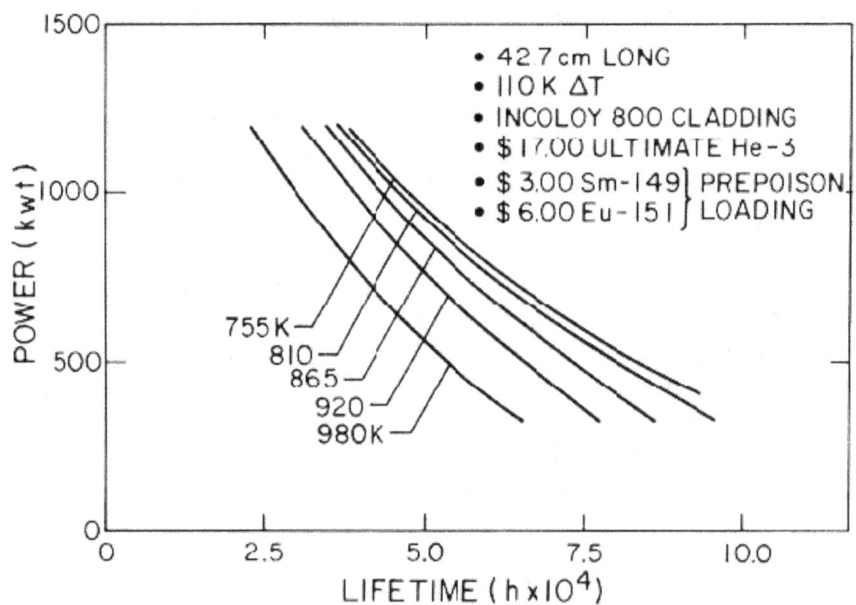

Fig. 18. Reference ZrH reactor performance. *From D. W. Staub (Ed.), 1973.*

Summary of U-ZrH Reactor Power Plant Development

Integral testing of the reactor technology was demonstrated in a series of six U-ZrH reactors. Table 7 [19,20] summarizes this reactor test experience. The uranium-zirconium-hydride reactor power plant programs demonstrated the technology for long-life U-ZrH reactor systems, such as five year lifetime at 920 K, are obtainable. Improved fuel elements and the coolant flow distribution in the core is needed. The use of fins on the fuel elements in order to reduce temperature differences across single fuel elements and maximize channel-to-channel coolant mixing is one idea.

Ideas to improve the fuel elements themselves were formulated; such as the substitution of Inconel 800 for the Hastelloy N. Inconel 800 would provide improved strain capability without a great loss of relative strength. In addition, a method was developed to produce a highly adherent hydrogen barrier coating on the Inconel 800, using a much simpler process than possible on the Hastelloy N material. The hydrogen content of the uranium-zirconium-hydride fuel material can also be increased to provide greater lifetime capability.

Fig. 19 and Table 8 describe major power system component experience also achieved.[21, 22] These now-terminated programs still provide a rich technical heritage for future nuclear power systems planning and the only U.S. reactor flight experience. The SNAPSHOT flight program demonstrate that nuclear reactors can be successfully launched and operated in the space environment without exposing the terrestrial population or biosphere to undue risk. It also demonstrated that such a launch could be achieved in the same manner as other payloads.

Table 7. SNAP reactor test experience.

	SNAP-2		SNAP-10A		SNAP-8	
	EXPERIMENTAL REACTOR (SER)	DEVELOPMENTAL REACTOR (S2DR)	FLIGHT SYSTEM		EXPERIMENTAL REACTOR (S8ER)	DEVELOPMENTAL REACTOR (S8DR)
			FS-3	FS-4		
Critical	September 1959	April 1961	January 1965	April 1965	May 1963	June 1968
Shutdown	November 1960	December 1962	March 1966	May 1965	April 1965	December 1969
Thermal Power (kW$_t$)	50	57	38	42	600	600–1000
Thermal Energy (MWh)	225	273	383	41	5,100	4,350
Electric Power (watts)	—	—	402	650 BOL 535 EOL	—	
Electric Energy (MWh)	—	—	4.0	0.6	—	—
Time at Power and Temperature	1,800 hr at 920 K 3,500 hr above 750 K	2,800 hr at 920 K 7,700 hr above 750 K	417 days at approx 800 K	43 days above 785 K	100 days at 600 kW$_t$ at 975 K; 1 yr above 400 kW$_t$ at 975 K	429 hr at 1,000 kW$_t$; 1,670 hr at 965 K; 6,680 hr above 810 K

Table 8. Major component experience for U-ZrH reactors.

	Number Units Tested	Total Test (Hrs)	Maximum Temperature (K)	Maximum Duration Single Test (Hr)
ZrH Reactor	6	43,000	975	10,000
NaK Pumps	59	212,000	920	20,000
NaK Volume Accumulators	50	100,000	670	10,000
Control System Actuators	68	241,000	880	24,400
Bearings	93	241,000	1,090	15,500
Radiation Shields	13	91,000	755	10,000

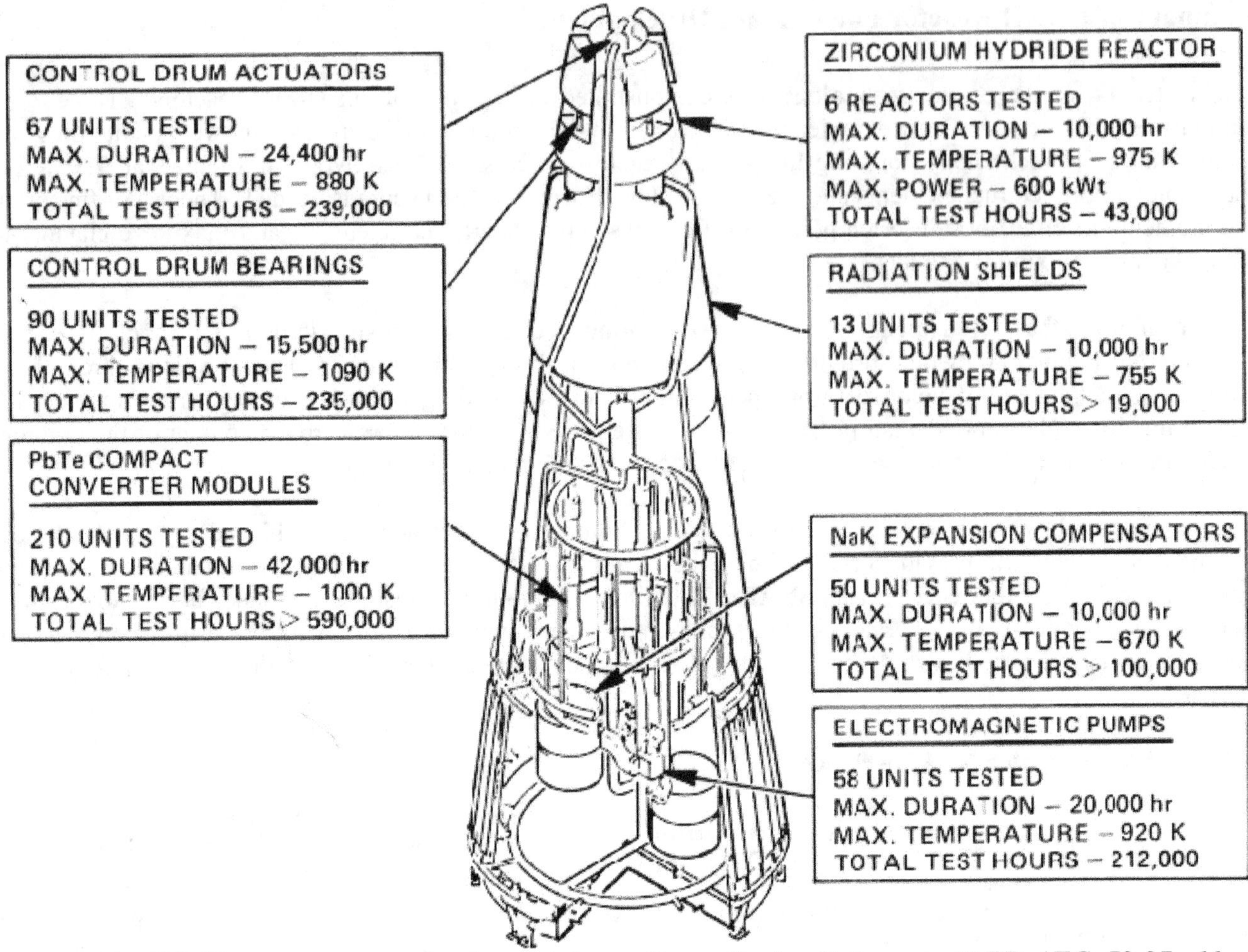

Fig. 19. U-ZrH component technology status. *From Rockwell Report AI-BD-AEC 73-27, 11 September 1973*

Bibliography

G. Asquith, D. G. Mason, and Stamp, "Zirconium Hydride Space Power Reactor Design," Intersociety Energy Conversion Engineering Conference Proceedings (1972), p. 572-580.

G. E. Berg, "Final SNAPSHOT Performance Report," Rockwell International Report NAA-SR-11934 (15 August 1966).

G. F. Burdi, "SNAP Technology Handbook, Vol. I, Liquid Metals," Rockwell International Report No. NAA-SR-8617, Vol. I (II August 1964).

L. E. Donelan, "Space Nuclear System Sheathed Electrical Cables Summary Report," Rockwell International Report No. AI-AEC- 13089 (June 1973).

L. E. Donelan, "Zirconium Hydride Reactor Control Drive Actuator Development Summary Report," Rockwell International Report AI-AEC-13080 (June 1973).

J. P. Dooley and J. P. Beall, "The Disassembly and Post Operation Component Examination of the SNAP-10A FS-3," Rockwell International Report NAA-SR-12504 (15 September 1967).

R. E. Dyer, "Disassembly and Post Operation Examination of the SNAP-8 Experimental Reactor," Rockwell International Report NAA-SR-11520 (1965).

R. R. Eggleston (Ed.), "SNAP Programs Bibliography," Rockwell International Report AI-AEC-13066 (30 June 1973).

L. M. Fead, L. D. Felton, and V. L. Rooney, "SNAP-8 Experimental Reactor Operations and Test Results," Rockwell International Report NAA-SR-10903 (28 June 1965).

J. P. Hawley, "Water Immersion Safety for SNAP Reactors," Rockwell International Report NAA-SR-I1361 (30 September 1967).

J. P. Hawley and R. A. Johnson, "SNAP-l0+A FS-3 Reactor Performance," Rockwell International Report NAA-SR-11397 (15 August 1966).

P. H. Horton and W. J. Kruzeka, "Zirconium Hydride Reactor Control Reflector Systems Summary Report," Rockwell International Report AI-AEC-13078 (June 1973).

H. G. Hurrell, F. Boecker, Jr., and K. S. Jefferies, "Startup Testing of the SNAP-8 Power Conversion System," Intersociety Energy Conversion Engineering Conference Proceedings (1970), p. 11-28 to 11-34.

S. Mak and J. R. Pope, "Operation Experience with NaK Lubricated Journal and Thrust Bearings for a NaK Pump-Motor Assembly," Intersociety Energy Conversion Engineering Conference Proceedings (1970), p. 11-71 to 11-73.

J. M. Otter, K. E. Buttrey, and R. P. Johnson, "Aerospace Safety Program Summary Report," Rockwell International Report AIAEC-13100 (June 1973).

Staff, "SNAP-8 Summary-S8ER and S8DS," Rockwell International Report NAA-SR-Memo-12099 (30 August 1966).

O. P. Steele, III, and E. J. Donovan, "Design Summary Reflector Drive System" Rockwell International Report TI-652-220-006 (10 October 1972).

A. W. Thiele and S. Berger, "SNAP-8 Summary Report," Rockwell International Report AI-AEC-13070 (June 1973).

G. M. Thur, "Design Changes in SNAP-8 for Reduced Reactor Operating Temperatures," Intersociety Energy Conversion Engineering Conference Proceedings (1970), paper. no. 709137.

G. M. Thur, "SNAP-8 Power Conversion System Assessment," Intersociety Energy Conversion Engineering Conference Proceedings (1968), p. 329-337.

J. H. Van Osdol, "Design and Integration of Zirconium Hydride Reactor Power Systems," Rockwell International Report AI-AEC- 13072 (June 1973).

F. H. Welch, E. C. Phillips, J. G. Asquith, and N. F. Davies, "Lithium Hydride Technology, Vol. I, Properties of Lithium Hydride and Corrosion Studies," Rockwell International Report NAA-SR-94 00, Vol. I (8 May 1964).

Chapter 4

SP-100

Design Requirements

The SP-100 program was active from 1983 - 1993. The program was being designed to provide tens-to-hundreds of kW_e power for 7 y at full power and 10 y overall operation. Power plant components were to provide this wide range of power without significant requalification of the basic building blocks. These requirements were originally derived from projected DoD needs for robust surveillance systems, survivable communications including anti-jam capabilities, electric propulsion systems for reusable orbital transfer, and weapons applications,[1] and from projected NASA needs for nuclear electric propulsion for planetary missions, earth orbit tug, civilian anti-collision aircraft radar, advanced direct broadcast satellites, growth space station power, and planetary bases.[2] A specific reference system design point of 100 kW_e was selected for demonstration of the power system technology. The total system mass of < 4,000 kg and safety requirements were key design drivers. In addition, the design was to decrease radiation at 25 m from the reactor to 1×10^{13} n / cm^2 and 5×10^5 rad. These criteria were to be met without servicing or maintenance in space. The size of the power plant was to be compatible with the Space Shuttle. Launch vehicle requirements were modified for launch on a Titan IV with the operating orbit above 2,000 km. Survivability requirements for defense missions had been put on hold indefinitely. Table 1 summarizes the key requirements.[3]

Table 1. SP-100 Key Generic Flight System design system requirements.

Operational Life	**10 years total with a maximum of 7 years at full power**
Launch Vehicle	**Titan IV / Centaur**
Mission Operation Orbit	**2000 km altitude circular earth orbit at 28° inclination**
User Interface Plane	**4.5-m diameter separated by 22.5 m from the base of the reactor vessel**
Self-generated Radiation at User Interface Plane	**Neutron Fluence: 1×10^{13} n/cm^2 (1 MeV equivalent)** **Gamma Does: 5×10^5 rads (Si)** **Thermal Power Density: 0.14 W/cm^2**
Main Bus Electrical Power To User	**99.7 kWe at 200 Vdc $\pm$ 7 Vdc** **300 W at 28 Vdc $\pm$ 7 Vdc**
Maximum Systems Mass	**4600 kg**

Power Plant Description

The Generic Flight System (GFS) design was completed in May 1988 and found in an independent System Design Review to be acceptable as the basis for technology and component development. The selected design is a fast spectrum reactor coupled to thermoelectric energy conversion units by liquid lithium, which is magnetically pumped through niobium alloy piping. Since then there have been major refinements to enhance reliability and to reduce mass. The major elements in the system shown in Fig. 1 and 2 are;[4] (1) a fast reactor using the niobium base alloy PWC-11 (niobium-1% zirconium with 1,000 ppm of carbon) for structures with a uranium nitride (UN) ceramic fuel, (2) control system using sliding neutron reflectors and safety rods with appropriate actuators, sensors, and controllers, (3) radiation attenuation shadow shield of lithium hydride and depleted uranium, (4) heat transport loops that transfer heat from the reactor to the power conversion units by means of pumped liquid lithium using self-actuated thermoelectric electromagnetic pumps, (5) solid-state thermoelectric power conversion units using silicon germanium/gallium phosphide (SiGe/GaP) to convert thermal power to electricity, and (6) the heat rejection system consisting of a carbon/carbon matrix structure and armor with titanium/potassium heat pipes.[5,6]

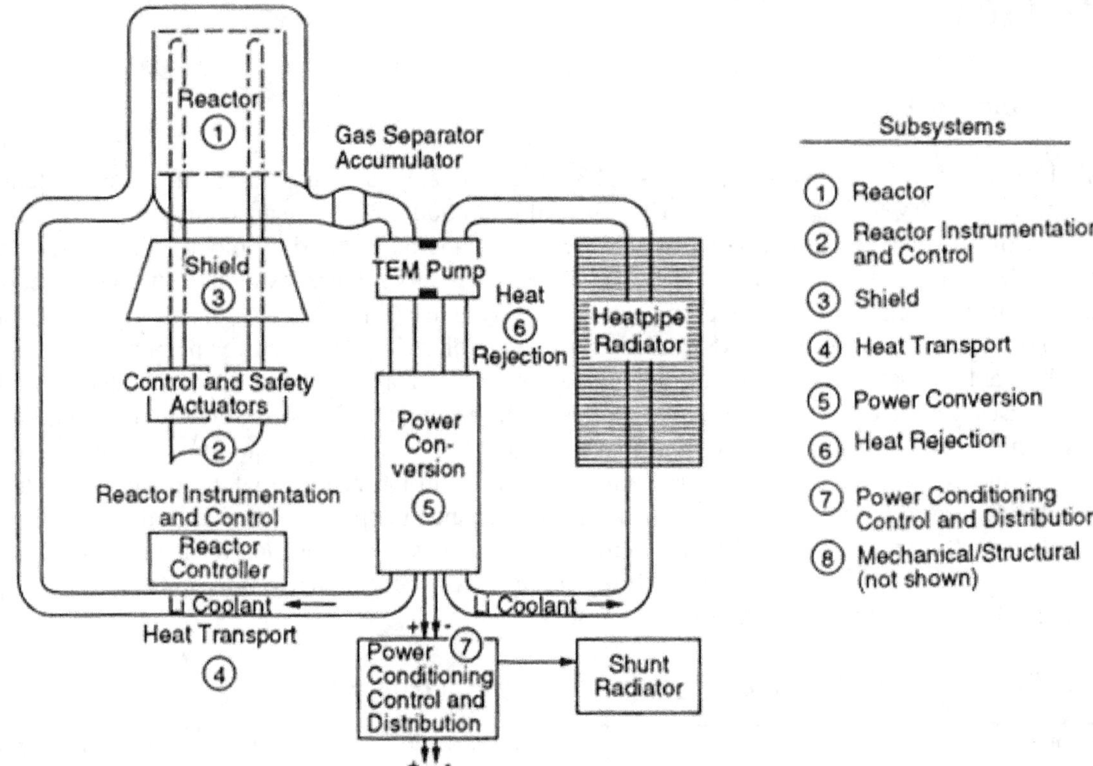

Fig. 1. SP-100 Generic Flight System schematic. *From* **Truscello, V. C. and L. L. Rutger, 1992.**

The building blocks that form SP-I00 can be configured in any number of arrangements depending on mission needs and desired power levels. The baseline configuration arrangement for a 100 kW_e power plant has the reactor at the forward end of the power plant away from the payload. The largest segment of the power plant is the heat rejection area. The total length of the deployed power plant without the boom is 12 m. The boom, used to minimize the amount of shielding needed, makes the overall length 25 m. This can be stowed for launch.

The heart of the reactor design is the fuel pin, shown in Fig. 3. The UN fuel is fabricated in the form of pellets, having a density of 94.5 percent theoretical and uranium enrichment of 97 percent. Approximately 50,000 pellets are needed for a 100 kW_e thermoelectric system core. Cladding provides structural strength for the fuel pin, while a layer of rhenium acts as a barrier between the fuel and lithium coolant, prevents loss of nitrogen from the UN fuel, allows thinner cladding, and acts as thermal poison if the reactor is immersed in water. The fuel pin cladding is the niobium alloy PWC-11 refractory material tubing with rhenium (Re) tubing bonded to the internal surface. The UN fuel operates at a peak surface temperature at the beginning of life of 1,400 K and end of life of 1,450 K and a peak burnup of 6 atom per cent. Surrounding the fuel pin is a wire wrap spacer that provides space for the lithium coolant to flow. The fuel pins have an upper plenum region to contain fission products.

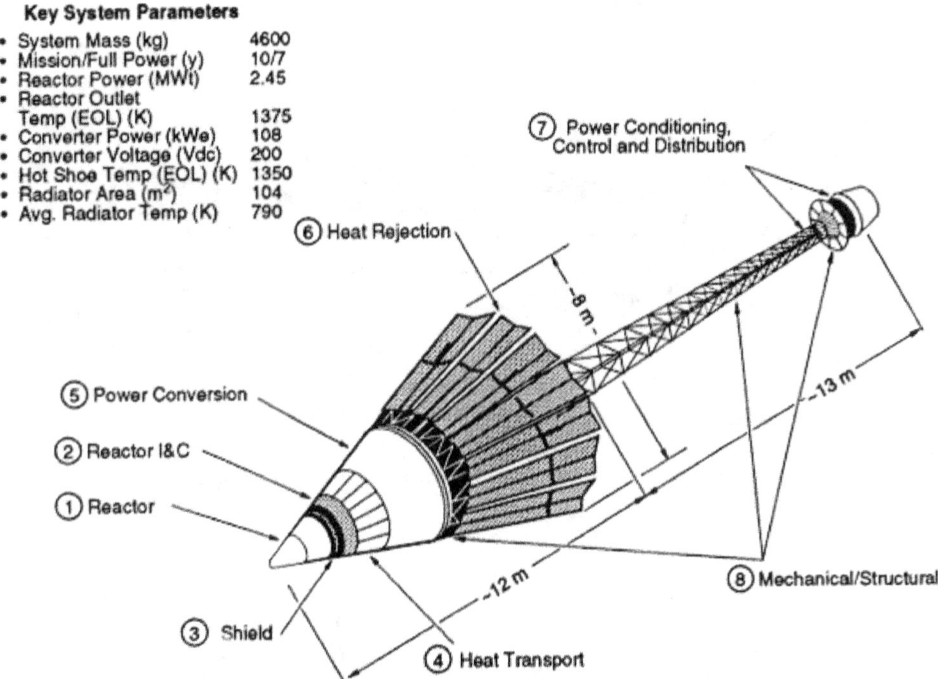

Fig. 2. SP-100 Generic flight System configuration . *From* **Truscello, V. C. and L. L. Rutger,1992.**

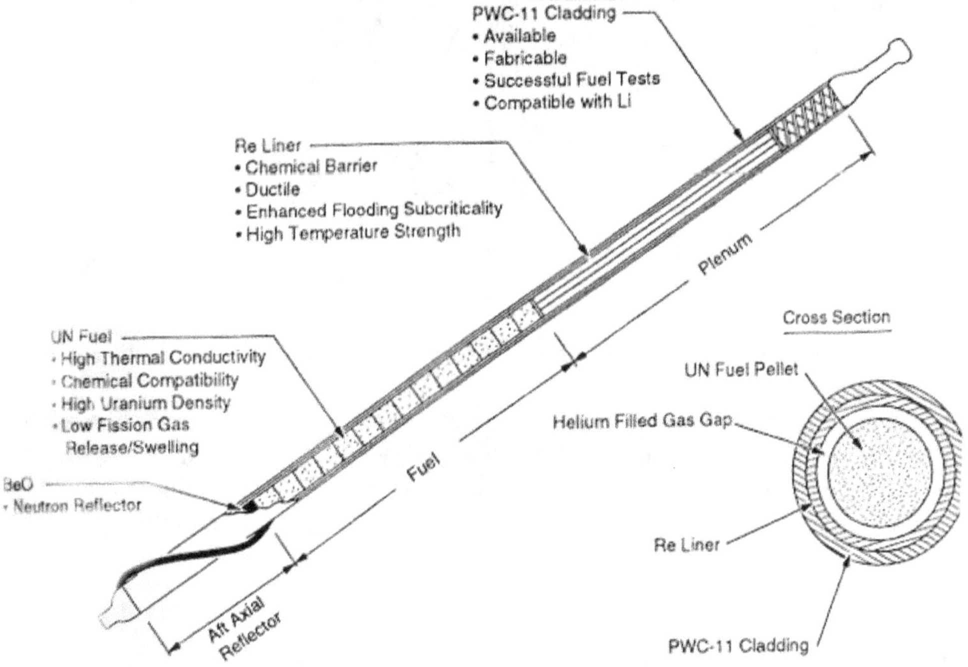

Fig. 3. SP-100 fuel pin. *Truscello, V. C. and L. L. Rutger, 1992.*

The reactor core is separated into 10 hexagonal-shaped assemblies with approximately 61 fuel pins per assembly and six partial assemblies, each with approximately 50 pins, around the outer edge (similar to the configuration shown in Fig. 4). The partial assemblies result in the hexagonal shape being closer to a circular form. The core also contains three safety rods for accidental criticality mitigation and, in case of loss-of-coolant flow, an auxiliary cooling system to prevent core melt down. The in-core safety rods provide a redundant shutdown system and help maintain the core subcritical in case of fire, explosion, water submersion, or compacting accidents. The spacing between the fuel pins is used as coolant paths for the lithium which transfers heat to the converters.

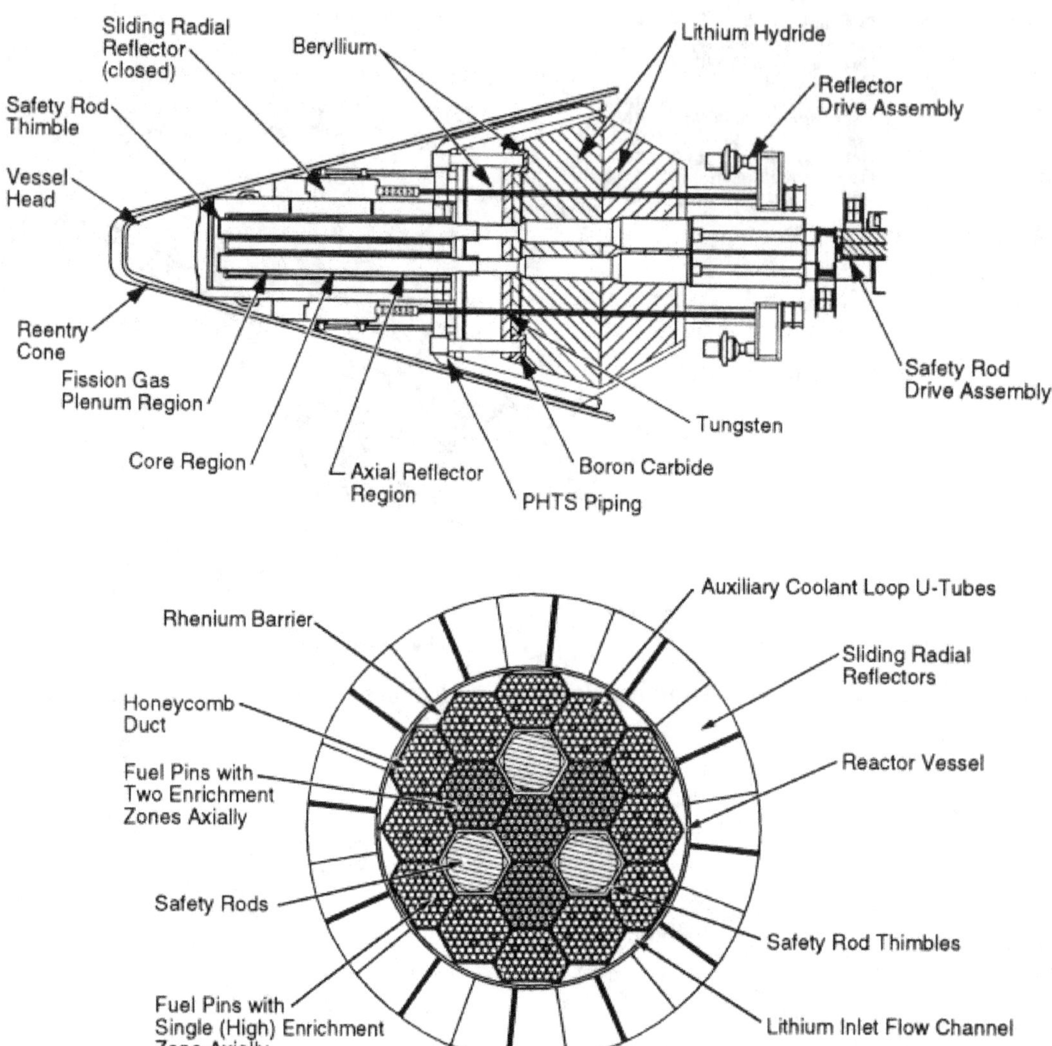

Fig. 4. SP-100 reactor subsystem components. *From* **Truscello, V. C. and L. L. Rutger, 1992.**

The auxiliary coolant function option is designed to limit reactor fuel temperatures to less than 2,000 K in case of a loss-of-coolant accident. It provides coolant loops totally independent of the normal heat removal system. Heat removal is by 42 u-tubes located throughout the core pin bundles.[7] The auxiliary lithium coolant passes through the core into a collection manifold, from which it is pumped to an independent radiator for rejection of heat to space.

The in-core safety rods are of boron carbide (B_4C) neutron absorbing material inside structural thimbles that can be located in or out of the core. The rods have a follower segment of beryllium oxide (BeO) that acts as moderator during operation.

Normal control is by means of twelve sliding tapered reflector segments. The segments control neutron leakage from the core. The positioning of the reflector segments is used to bring the reactor critical once the in-core safety rods are removed and to compensate for fuel burnup and swelling. The reflector segments consist of BeO contained within a Nb-1 % Zr shell.

The radiation shielding approach for minimizing system mass resulted in a conical shadow-shield with a cone half-angle of seventeen degrees.[8] Within the shield, the design must thermally isolate high performance/low mass shielding material with limited temperature capabilities from the adjacent high temperature components.

The shield is fabricated from lithium hydride (LiH) pressed into a stainless-steel honeycomb to attenuate neutrons. Depleted uranium plates are added primarily to attenuate gamma radiation. Beryllium is used to transfer heat to the outer surface of the shield, where it is radiated to space. Stainless steel is used as a structural element and to contain the shield.

Surrounding the reactor and butting against the radiation shield is a carbon-carbon reentry heat shield. The purpose of the reentry shield is to ensure confinement of the core fission products if the reactor reenters the atmosphere. It is conical in shape, and the geometry protects the reactor vessel from overheating during reentry, keeping the temperature to 300 K even through the reentry shield might reach 3,200 K.[9]

Heat is transferred to the converter by six interrelated pumped lithium loops.[10] Thermoelectric-electromagnetic (TEM) pumps are used to circulate liquid lithium in each loop (Fig. 5). Each pump also circulates lithium coolant on the cold side of two heat rejection loops. These are self-energized, DC conduction, electromagnetic TEM pumps. Hot and cold ducts run parallel to each other along the active length of the pump. Sandwiched between them is a series of thermoelectric elements connected on each side by compliant pads. The temperature differential imposed across the thermoelectric converters generates an electrical current that travels at right angles to the lithium flow in the ducts and in closed paths around a magnetic center iron; as the current circulates, an induced magnetic flux is generated in the center iron that is directed through the lithium ducts perpendicular to the current. The interaction of the magnetic flux and the current produces a force on the molten lithium that drives it along the ducts. Hence, flow responds to the temperature in a self-regulated way.

During reactor operation, helium gas is generated in the coolant loop by neutron reactions with lithium. Enriched lithium-7 (99.9% ^{7}Li, 0.1 % ^{6}Li in contrast to natural lithium, which has an isotope ratio of 92.6/7.4) is used to minimize production of helium. A gas separator/accumulator (Fig. 6), used to separate the helium from circulating lithium, utilizes the principles of surface tension and centrifugal force.

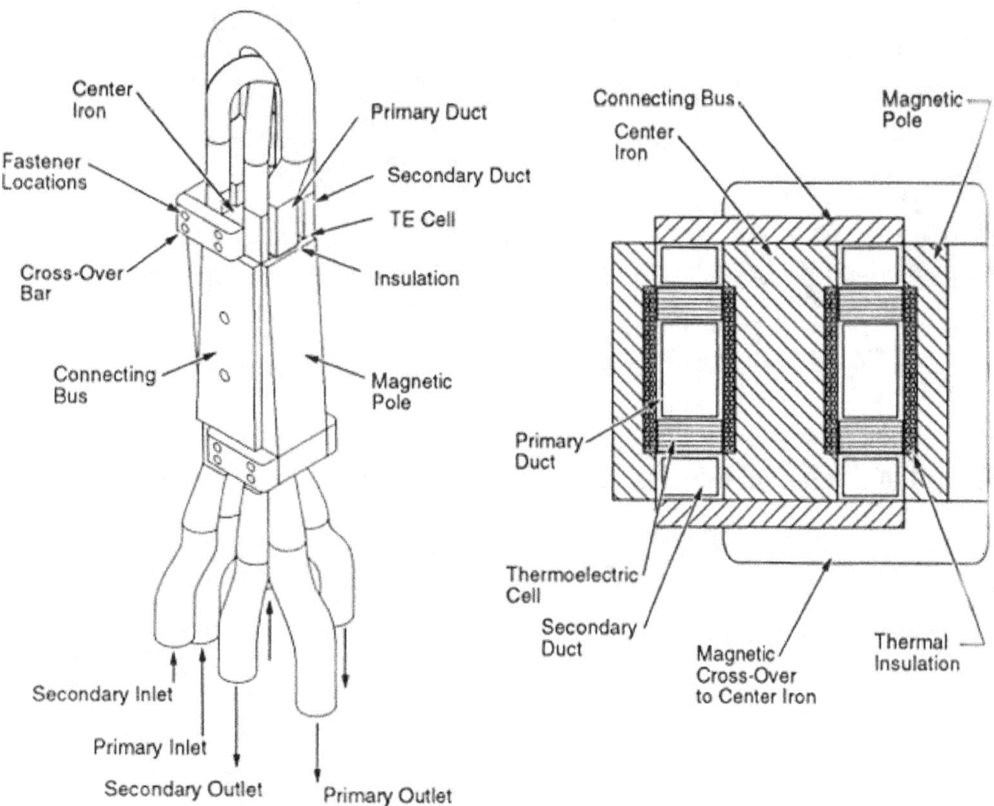

Fig. 5. SP-100 TEM Pump development *From* Truscello, V. C. and L. L. Rutger, 1992.

The reactor is controlled based on temperature and neutronic measurements. Flow is inherently maintained by the TEM pumps (for example, higher reactor temperatures increase TEM pumping). Sensors are multiplexed in an analog multiplexer/amplifier located behind the reactor shield. These are designed to operate for 10 y in a radiation environment of 4×10^{15} n / cm^2 and 2.4×10^8 rad (gamma). N type junction field effect transistors (JFET) semiconductor devices were being used.

Conversion of thermal energy from the reactor to electrical power is accomplished through the Power Conversion Assembly (PCA) (see Fig. 7).[11] Heat is conducted from hot lithium in a central heat exchanger, through TE cells on either side (which convert some of the heat to electricity), and then to a pair of heat exchangers where cooler lithium carries the waste heat to the heat rejection subsystem. The PCA building blocks can be packaged in any combination to generate the desired output. These building blocks included the thermoelectric cells and thermoelectric converter assembly (TCA). Each assembly had two arrays of 60 TE cells. The TCA consisted of two cell arrays, one hot side heat exchanger, and two cold side heat exchangers. A typical TCA configuration had 6 x 10 cell arrays. The thermoelectric (TE) converter assemblies were in a stack of six plate-and-frame configuration, each of which produces 1.5 kW$_e$ at 34.8 VDC. A total of 8,640 cells was required for a 100 kW$_e$ system.

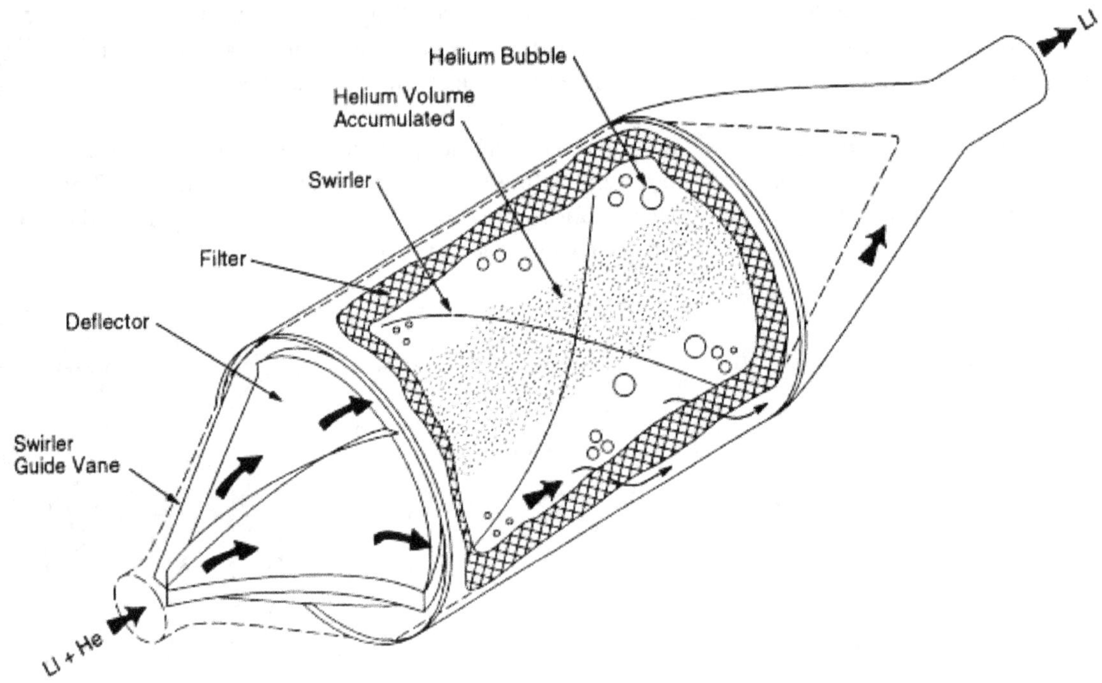

Fig. 6. SP-100 Lithium/Helium Gas Separator-Accumulator. *From* **Truscello, V. C. and L. L. Rutger, 1992.**

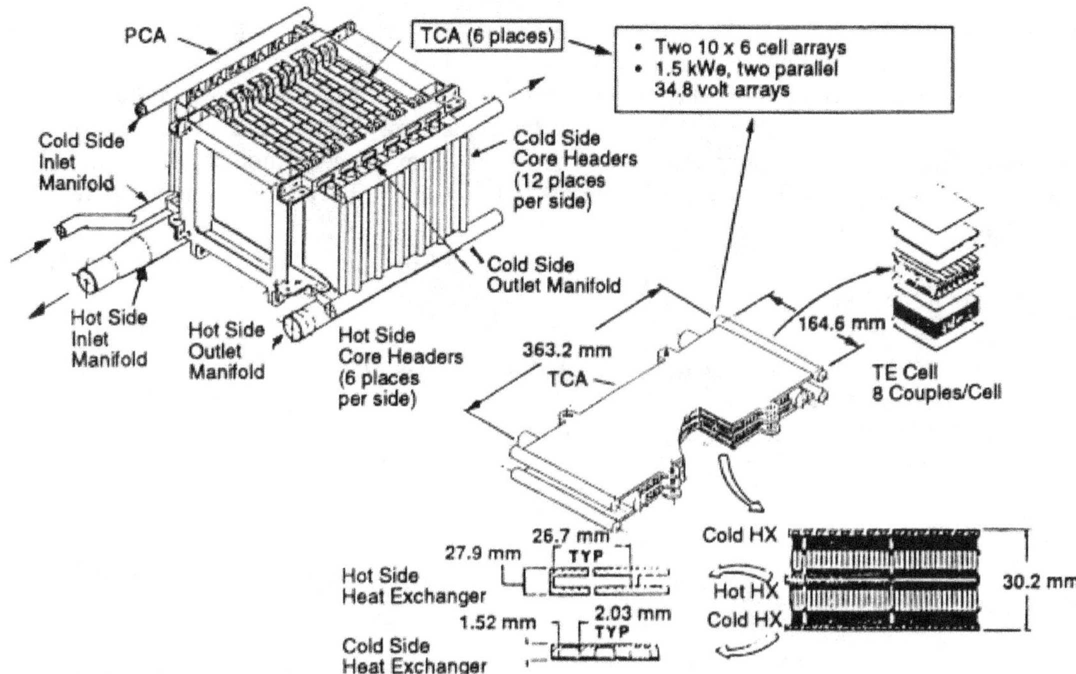

Fig. 7. SP-100 Power Converter Subsystem components. *From* Truscello, V. C. and L. L. Rutger, 1992.

The thermoelectric cells (see Fig. 8) used silicon germanium/gallium phosphide (SiGe/GaP) materials. Each cell consisted of a thermoelectric module, compliant pads to accommodate thermal stresses, electrical insulators to isolate the electrical power from the spacecraft, and conductive coupling to the heat exchangers. The cells were arranged in a parallel/series electrical network to provide the desired 200 V output. The thermoelectric material figure-of-merit was being increased to 0.85×10^{-3} K^{-1} by the addition of GaP to 80:20 SiGe from 0.67×10^{-3} K^{-1} for SiGe. A graphite-electrode SiGe bond was used to make the electrical contact resistivity of that joint less than 25 $\mu\Omega$ / cm. The compliant pad was used to prevent cracking of the TE elements from thermal expansion. It consisted of niobium fibers bonded to niobium face sheets on both sides. The niobium face sheet matched the thermal expansion of both the heat source material (PWC-11) and the heat sink material (Nb-1% Zr). The insulators were single-crystal alumina, with 4,000 V / cm voltage gradient at 1,375 K.

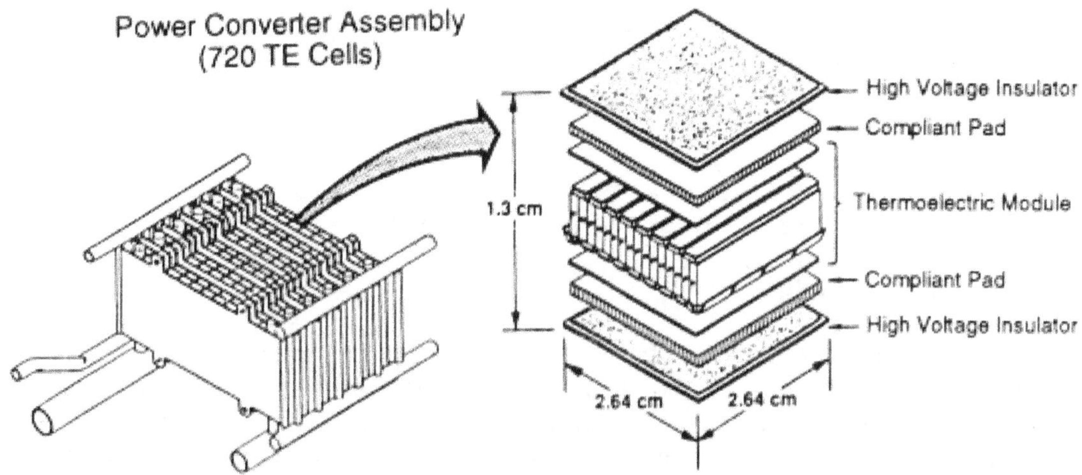

Fig. 8. SP-100 Thermoelectric Cell. *From* Truscello, V. C. and L. L. Rutger, 1992.

The radiator was designed to be tailored to the particular application. In the GFS, the heat rejection subsystem included twelve radiator assemblies, each constructed of an array of varying lengths of potassium heat pipes brazed to a lithium duct-strongback structure at the center (see Fig. 2). The lithium duct-strongback structure consisted of the lithium supply and return ducts. Each of the twelve radiator panels had flexible joints for the supply and return ducts so that the radiator can be folded for launch and deployed for reactor operation. Accumulators were included within the heat rejection subsystem to accommodate the variable volume necessary to compensate for expansion and contraction of the lithium coolant. The accumulators were fabricated out of titanium. Heat rejection area was 104 m^2.

For restarting the power plant in space, the reactor was used to heat an auxiliary liquid metal loop to melt the lithium throughout the system.[12]

Table 2 summarizes some key power plant performance parameters.

Table 2. SP-100 GFS design performance parameters.

Key System Performance Characteristics	
Rated Electrical Power Output (kWe)	100
BOM Reactor Outlet Temperature(K)	1350
EOM Reactor Outlet Temperature (K)	1375
Reactor Thermal Power Required (kWt)	2400
Average EOM Radiator Temperature (K)	791
Key System Design Characteristics	
Launch Vehicle	Titan IV/Centaur
Shield Half-cone Angle (deg)	17
Separation Distance (m)	22.5
Deployable Boom	Yes
Thermopile Area (m^2)	7.08
Radiator 1-Side Physical Area (m^2)	106
Power Distribution Voltage (Vdc)	200
Number of Thermoelectric Elements	
Power Conversion	8640 Cells
Auxiliary Cooling and Thaw (ACT)	180 Cells
Thaw Provisions	NaK Tracelines
Mass by Subsystem (kg)	
Reactor	700
Shield	960
Heat Transport (Includes thaw battery, if required)	520
Reactor Instrumentation and Control (I&C)	320
Power Conversion	450
Heat Rejection	1040
Power Conditioning, Control and Distribution (PCC&D)	390
Mechanical/Structural	220
Total	4600
System Power-to-Mass Ratio (W/kg)	21.7

Performance and Scaleability

SP-100 was designed as a set of building blocks. Thus, the reactor power level could be varied by changing the number of fuel pins, changing the number of thermoelectric converters, or using different power conversion units. The radiator area could be adjusted according to the power level. This flexibility was one of the major reasons for selecting the concept in 1985. The mass of the system was a function of the various combinations selected. Concepts of 8, 10, 20, 30, 40, 50, 100, 200, 300, 1,000, 5,000, 10,000 and 15,000 kW$_e$ had been configured. Fig. 9 shows the specific power difference as a function of thermoelectric figure-of-merit and using Stirling engines at various efficiencies and peak operating temperatures. The same 2.5-MW$_t$ reactor, used in conjunction with thermoelectric conversion to generate 100 kW$_e$, could be coupled to the Stirling conversion system to generate nearly 600 kW$_e$.

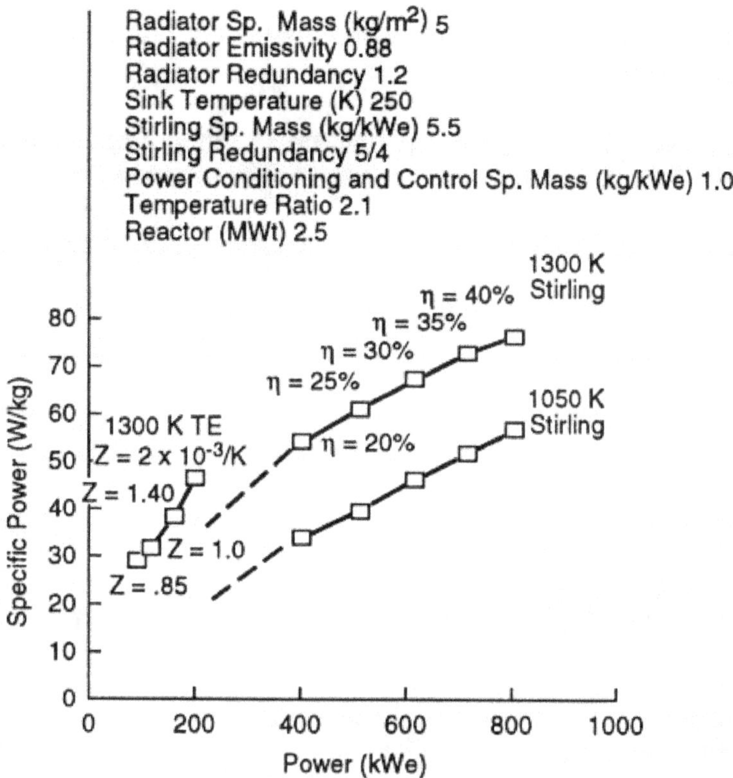

Fig. 9. Extending SP-100 reactor power systems capability.

Nearer term options are given in Fig. 10 and 11.[13] The mass is shown to be a trade-off at the design thermoelectric conditions with Brayton or Stirling conversion systems. The Brayton conversion work was performed in the 1960s where 46,000 h of test data was accumulated on the rotating machinery. Despite their higher efficiencies, the Brayton and Stirling conversion systems require much larger radiators than the thermoelectric conversion systems. This is a result of the lower radiator temperatures associated with them.

The SP-100 reactor can be reconfigured for higher thermal power levels and mated with a high-temperature Rankine cycle to provide even higher powers, such as several megawatts for nuclear electric propulsion. Also, SP-100 can be configured in a number of arrangements for lunar and Mars surface applications.

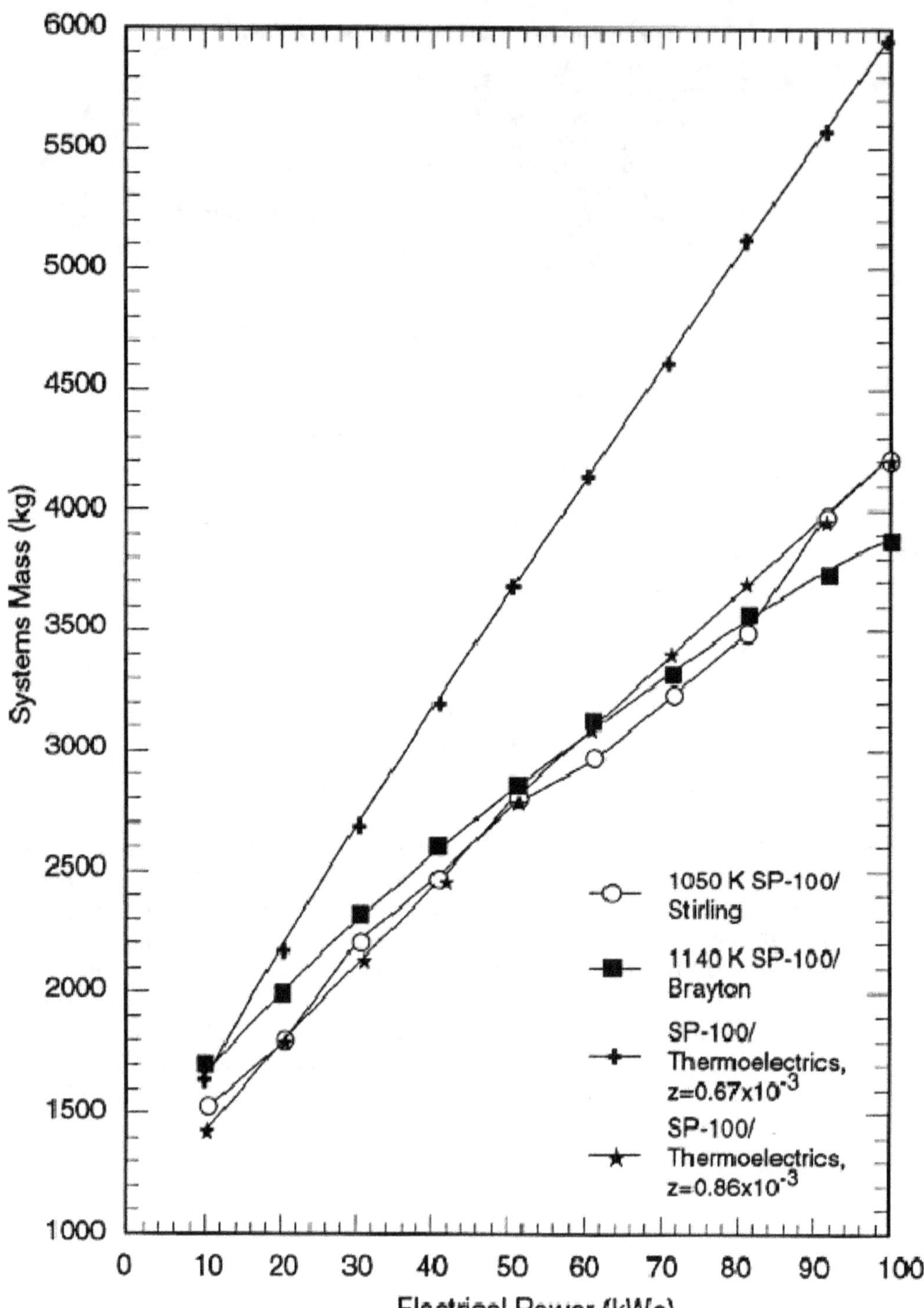

Fig. 10. SP-100 system mass comparisons for near term options. *From* **Schmitz, P., H. Bloomfield, and J. Winter, 1992.**

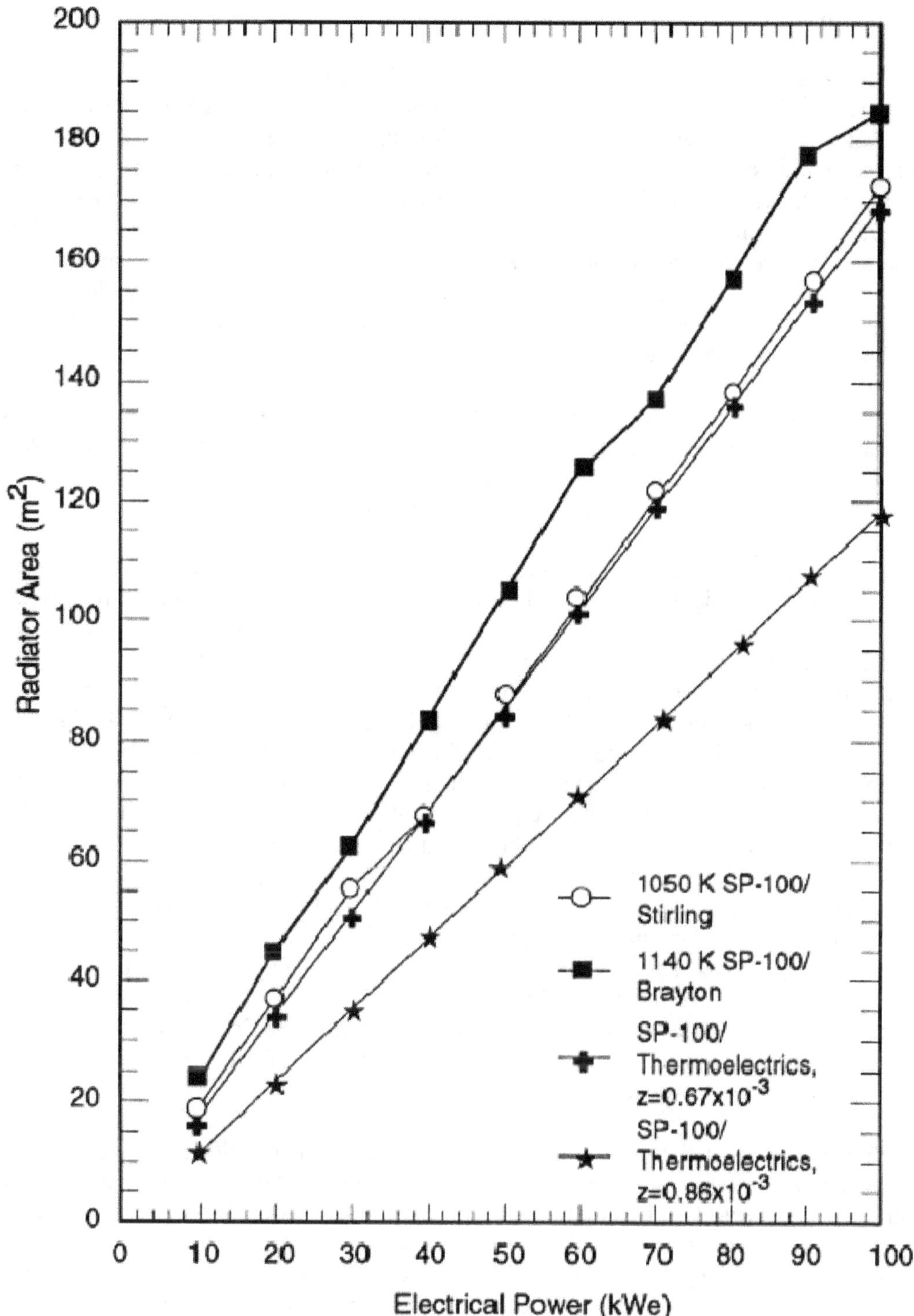

Fig. 11. SP-100 radiator area comparisons for near term options. *From* Schmitz, P., H. Bloomfield, and J. Winter, 1992.

Development Status

From 1985 until 1994, substantial progress had been made in all key technology areas, and early feasibility issues have been resolved.[14, 15] Fuel development included the fabrication of the fuel pins and testing in an environment sufficient to develop high confidence of meeting the 7-y full-power operational lifetime requirement. The fabrication processes necessary to produce high quality UN fuel pellets had been successfully developed and the fuel irradiated in the Experimental Breeder Reactor II (EBR-II) and Fast Flux Test Facility (FFTF) test reactors. Approximately 75 fuel pins had been tested with some tests performed at three times the nominal power and at fuel pin surface temperatures as high as 1,500 K. All tests have met or exceeded expectations, and the goal of 6 atom-percent fuel burnup had been achieved. The pins tested used the Nb-1% Zr alloy cladding because PWC-11 was not yet qualified. This alloy was predicted to provide similar design margins, but at less mass. The tests showed the fuel pins can meet the requirements of a 7-y system.

A dual material tube had been qualified with an outer tube of niobium alloy and an inner tube of rhenium for strength as well as a barrier between the fuel and niobium alloy. Tubes 0.6 m long had been tested that show the bond is sufficiently rugged for 10 y of operation. Approximately 50 tubes had been fabricated with Nb-1% Zr cladding and another 10 were produced with PWC-11 cladding material.

The nearly 50,000 fuel pellets needed for a reactor core had been fabricated and placed in storage awaiting the fabrication of the cladding. Production of the fuel pins for a reactor was being delayed until they were needed.

Materials are fundamental to the success of the development of a reactor that must operate for 7 y with a maximum coolant outlet temperature of 1,375 K. Sufficient materials data did not exist on the alloys of choice at the start of the program. Now, suppliers had been qualified for refractory alloys fabrication including Nb-1% Zr, PWC-11, and rhenium. Fabrication procedures included electron beam and gas tungsten arc welding, cold forging, drawing, hot isostatic pressing, diffusion bonding, vacuum sputtering, chemical vapor deposition, and high-temperature heat treatment. Compatibility testing with lithium had been performed in Nb-1% Zr and PWC-11 test loops at 1,350 K for thousands of hours without failures.[16]

Reactor physics behavior had been experimentally verified in critical assembly tests. Measurements performed on the assembly showed close agreement with predictions under both normal design and postulated accident conditions. These tests provided assurance that the reactor would meet safety criteria imposed to protect the environment. The experimental results verified that: (1) the internal shutdown rod reactivity met the shutdown requirement with margin, (2) reflector control worth versus position confirmed necessary control margin, (3) flooding reactivity worth was confirmed and subcriticality assured, and (4) buried reactor reactivity worth was confirmed and subcriticality was assured. [17]

Control and safety drives were the only mechanical moving parts in the SP-100 power plant design. Low-temperature development tests had been completed to confirm the key mechanical design features. Environmental tests of control drive motor, clutch, and brake assemblies had demonstrated predicted performance of these components at 700 K. The self-aligning bearings of the reflector control drives had been successfully tested for 30,000 cycles at temperatures up to 1,170 K, and tests of safety rod bearings were underway at 1,570 K. Tribological coatings necessary to protect against self-welding, friction, and wear had been tested using refractory borides, carbides, and nitrides. These must operate at temperatures > 1,600 K. Test data at 1,700 K indicated refractory carbides were the best material. Accelerated testing had been completed demonstrating that the equivalent life at SP-100 conditions is 50 y.

Long-life temperature and pressure sensors were needed to measure lithium coolant conditions. Temperature sensor concepts using a Johnson Noise Thermometer and W/Re thermocouples had been developed. Fabricated units were found to be mechanically robust. Sensor lifetime, accuracy, and stability were being established in a series of tests. In addition, the key features of a pressure transducer hydraulically coupled to the lithium coolant had been tested.

Multiplexers were located near the reactor where the nuclear radiation and temperature environments would be severe (1.2×10^8 rad gamma, 1.6×10^{15} n $/$ cm^2, and 800 K over 10 y). The temperature would be reduced to 375 K by the use of insulating blankets and radiators. Gamma testing indicated that radiation damage annealing will prevent unacceptable levels of drift in the circuit. Neutron testing was still underway.

Shadow shielding was to be used to attenuate the neutron and gamma radiation to acceptable levels for the spacecraft. Characterization of the LiH shield material for thermal conductivity and expansion, material compatibility experiments to confirm the long term behavior of the materials, and irradiation of LiH to establish swelling rates indicated that these were no long-term issues with the current design. This included irradiation testing at temperature. Material compatibility testing showed that LiH is not compatible with beryllium (Be). Therefore, the shield now has a stainless steel barrier between the Li and Be.[18]

Work had been done on the major issues associated with the heat transport subsystem. For the gas separator, a feasibility experiment using air and water indicated that the design was sound. Experiments of a prototypical gas separator with helium and lithium were planned. A magnetic bench test had been performed on the TEM pump to demonstrate the ability to accurately predict the three-dimensional magnetic field. An electromagnetic integrated pump test had been performed to verify the calculated pumping forces.

Power conversion major issues included: (1) thermoelectric material figure-of-merit, (2) bonding of the cell between the heat exchangers to accommodate critical problems of thermal stress and electrical isolation, and (3) electrical insulation at 1,350 K while sustaining a voltage gradient of 8,000 V $/$ cm for a period of 7 y in a deoxidizing environment created by the proximity of molten lithium. The SiGe material used in radioisotope thermoelectric generators (RTGs) had been improved so that the figure-of-merit was increased from 0.65×10^{-3} K^{-1} to 0.72×10^{-3} K^{-1}. The design goal of 0.85×10^{-3} K^{-1} must still be achieved.

A major technical achievement had been the development of a compliant pad to connect the thermoelectric cell to the heat source and sink. This was accomplished by use of a brush-like design that carries heat conductively while absorbing mechanical stresses due to the large temperature difference across the cell. The pad fibers were coated with a thin film of yttria to prevent self welding. Pads have performed under prototypic conditions for thousands of hours satisfactorily. Tests were not completed to demonstrate lifetime performance.

High voltage insulators were positioned at the top and bottom of the thermoelectric cell. Single crystal Al$_2$O$_3$ insulators, equipped with oxygen permeation barriers made of molybdenum, had been developed that will maintain the necessary electrical isolation for more than twice the lifetime required.

Very high electrical conductivity was needed to interconnect the TE couples. A multilayer electrode consisting of tungsten or niobium sandwiched between graphite layers had been developed. The tungsten provides good intercouple conductivity and strength, while the graphite isolates the tungsten from the TE material with which it reacts. Initial experiments had been performed, indicating that low resistivity bonds can be achieved, but long-term stability had yet to be demonstrated.

The process for fabricating the thermoelectric converter heat exchanger with the integral headers had been successfully demonstrated.

Cells can now be routinely fabricated using validated processes. The development had progressed through three phases. These all used SiGe thermoelectric materials. Progress in each phase included:[19]

- PD-1 first demonstrated the basic concept of conduction coupling of thermoelectric cells to their heat source and heat sink heat exchangers. The PD-1 cell contained certain features (such as low temperature braze) which prevented driving the cell to full prototypic temperature levels, and the test fixture limitations resulted in thermal inefficiencies which prevented the cell from reaching maximum potential power output. The PD-1 cell delivered 4.0 W$_e$ at 500 K temperature change for prototypic conditions and 4.8 W$_e$ for 545 K ΔT at peak power.

- The PD-2 cell improvements allowed operation of the cell at prototypic temperatures (1335 K hot side), but still was constrained by certain thermal inefficiencies. The PD-2 cell produced 4.0 W_e at 500 K ΔT prototypic conditions and 8.7 W_e peak power at 730 K ΔT .
- TA Cells: New analytical techniques[20] were developed and applied to design a "fracture safe" configuration. The TA cells are near prototypic. These were subjected to prototypic and higher thermal conditions. The measured cells were 8.8 W_e (versus 8.9 W_e predicted) at the 500 K ΔT prototypic temperature conditions and 13.7 W_e under 660 K ΔT peak power temperature conditions. The two TA cells tested had efficiencies of conduction coupling of 75% and 79%, respectively, almost identical to pretest predictions. Lifetime testing had now reached 3,670 h on one cell with the cells still functioning.

Only limited work had been performed on the heat rejection system. A half-dozen titanium heat pipes, with potassium as the working fluid, were fabricated and had been life tested. A 0.9-m section of radiator duct was fabricated and tested using a low melting point liquid-metal (Cerrobend) that substituted for the lithium to demonstrate the ability of the lithium in the radiator to thaw during startup. Test results showed that the actual thaw rate was twice as fast as predicted.

The early SP-l00 development issues are summarized in Table 3.[21] These challenges were mainly resolved. To meet the safety challenges, an auxiliary coolant loop had been designed to maintain the fuel pin clad below 2,000 K. This maintains the fuel structural integrity for disposal. The disposal location will be in space either at high Earth orbit or at its planetary destination. Thermoelectric cell design issues were resolved with development of the compliance pad and validation of the electrical insulator and electrical contact resistance. The fuel pin design and performance had been confirmed in reactor radiation testing. Heat transport hermiticity had been demonstrated in a lithium loop test. The major elements of the pump, including the magnetics, were demonstrated in element testing. The nitrogen loss that could limit lifetime had been resolved in testing that verified the success of the rhenium liner in the fuel pin to contain the nitrogen. System mass continues to be a challenge. The mass was at 4,600 kg for the GFS design and 3,900 kg for outer planet design; the specification is 4,000 kg. The gas separator to remove helium from the lithium loop had been demonstrated in air/water tests in Earth's gravity with further tests planned using lithium/helium. Enriched ^{7}Li was being used to minimize helium generation. To meet the low cost, low mass radiator heat pipe challenge, a design had been developed and successful operated for limited periods of time. For the shield temperature control, the forward LiH material was replaced with Be and B_4C to reduce the radiation dose by a factor of 200 and the temperature of the LiH to 700 K. This resolved the shield temperature control challenge. Remaining challenges for flight readiness will be discussed under the Flight Readiness section.

Table 3. SP-100 top ten challenges as of 1987

1. Safety	2. Thermoelectric Cell Technology
<ul><li>Core coolability with loss of coolant</li><li>Reactor control and safety drives</li><li>End of mission disposal</li></ul>	<ul><li>Electrical insulator development and performance</li><li>Electrical contact resistance</li></ul>
3. Fuel pin design and performance validation<ul><li>Fuel pellet development and scheduled</li><li>Fuel clad liner development</li><li>Fuel pin clad creep strength</li></ul>	**4. Thawing coolants**<ul><li>Startup from frozen lithium</li></ul>
5. Highly reliable heat transport loop<ul><li>Hermetic</li><li>TEM pump development and performance</li></ul>	**6. System lifetime**<ul><li>N_2 loss from fuel elements</li></ul>
7. System Mass<ul><li>Compliance with specification</li></ul>	**8. Gas accumulator/separator**<ul><li>Li^7 versus natural lithium</li></ul>
9. Heat pipe design and manufacture<ul><li>Transient performance/rethaw</li></ul>	**10. Radiation shield temperature control**

Flight Safety

The key safety requirements included:[22]

1. Reactor operation not started and operated (except for zero power testing) until operational orbit achieved.

2. Remain subcritical to environments associated with credible failures or accidents during assembly, transportation, handling, prelaunch, launch, ascent, deployment, orbit acquisition, shutdown, and transfer to high permanent storage orbit. Minimum situations include:

   A. Core internal structure and vessel generally intact, all exterior components removed for: (1) all possible combinations of soil and water surrounding core; and (2) reactor vessel exposed to solid propellant fire for 1,000 s.

   B. Core internal structure and vessel generally intact, compaction along the pitch line of the pins to produce pin-to-pin contact, and for: (1) all exterior components removed and all possible combinations of soil and water filling and surrounding the core; (2) normal exterior components and reflectors compressed around the core; exterior absorber material, if any, in its normal shutdown position; and core containing its original coolant or any possible combination of soil and water; and (3) all exterior components removed and aluminum surrounding the core containing its original coolant.

   C. Reactor vessel as designed, core fuel pins spread radially apart to the maximum distance allowed within the fuel channel lattice design. All exterior components removed:
      - Water filing and surrounding the core.
      - Saturated soil filling and surrounding the core.
      - Dry soil filling and surrounding the core.

   Calculated effective reactivities for these conditions with margin for modeling and calculation uncertainties shall be < 0.98.

3. In response to fires, reactor, without reflector elements, neutron shield and reentry heat shield, remain subcritical in the liquid and solid propellant fire environments. Limited melting and creep deformation allowed; however, the as-built geometry essentially maintained.

4. Structural response to explosions, the reactor shall remain subcritical in launch vehicle explosion environments.

5. For inadvertent reentry following reactor operation, reactor designed to reenter through the earth's atmosphere sufficiently intact to prevent the dispersion of fuel and fission products. Essentially intact defined as: (1) reentry structural and thermal loads not cause loss of effective fuel/safety rod alignment; (2) if after reactor operation, the reactor structure remains sufficiently intact to allow effective burial on impact; and (3) reentry not breach the reactor vessel nor impair the predictability of its structural response on impact.

6. For burial intact reentry through the earth's atmosphere, the reactor capable of producing effective burial as it impacts on water, soil, or pavement-grade concrete. Effective burial defined that the fuel, reactor vessel, and internal components are within the formed impact crater and below normal grade level.

7. Designed that reactor can be transferred to high permanent storage orbit at end-of mission. Reactor designed to prevent such core disruption and structural degradation during and following operation that compromise structural integrity and predictability of desired reactor behavior during final shutdown and transfer to high permanent storage orbit.

8. At final shutdown, reactor designed to ensure high-confidence permanent subcriticality at the final shutdown to preclude further production of fission products and activated material and ensure subsequent reduction of radioactive inventory. Final shutdown activated automatically, irreversible, and not initiated or rendered inoperable by any credible single failure or initiating event.

9. Final shutdown clock irreversibly interrupt the supply of power to all in-core safety rods and control reflector clutches when preset final shutdown time is reached.

10. Loss of primary coolant reactor designed to accommodate an instantaneous complete loss of main loop primary coolant followed by scram during operational phases without: (1) exceeding fuel design cladding temperature limits; (2) impairment of capability to achieve final shutdown; and loss of structural integrity sufficient to impair the capability to boost to permanent storage orbit.

11. For core heat removal capability, coolability assured with high confidence for all credible accident conditions to maintain the structural integrity and thereby the predictability of desired reactor behavior during final shutdown and transfer to high permanent storage orbit.

12. Nuclear feedback calculations, reactor core and associated coolant systems designed over entire power operating range, net effect of the prompt inherent nuclear feedback characteristics tends to compensate for the rapid increases in reactivity.

13. For power oscillations, nuclear subsystems designed to assure that thermal power oscillations which can result in conditions exceeding acceptable fuel design limits are not possible or can be reliably and readily detected and suppressed.

14. Reactivity control redundancy included two means with suitable independence and diversity to assure adequate protection against common cause failures. Each of these means capable of performing its nuclear safety function with a single active failure.

15. Inhibits included means that until operational orbit achieved, startup shall be precluded by three independent inhibits, one of which precludes startup by radio frequency energy.

Significant design features of SP-I00 had resulted from satisfying the nuclear safety requirements Fig. 12 and 13 summarize key safety features built into the SP-I00 design.[23] The reactor was designed to prevent inadvertent criticality during handling or in accident situations. This was accomplished by including two independent control elements that were physically locked in their shut down positions during ground transport, handling, launch, ascent, and final orbit acquisition (assuming Shuttle launch). These could not be released until two independent signals were given. A large reactivity shutdown margin was provided by the safety rods to ensure that the reactor remains shutdown should any accident occur. This included accidents involving severe fires, core compacting, projectile impacts, over-pressure, and immersion/flooding environments. At the beginning of the mission, opening of any 7 of 12 radial reflectors or the insertion of any 1 of 3 safety rods would have shutdown the reactor. As the mission progresses, the number of reflectors required for shutdown reduced, becoming 3 after 7 y of full power operation. The reactor was designed with a prompt negative reactivity coefficient to ensure stable reactor control and enhance shutdown if a loss-of-coolant should occur.

The reactor fuel pins included a rhenium poison that acts as thermal neutron absorption in case of water flooding.

While in space, the reactor was protected against impacts of micrometeorites and orbital debris by bumpers. If the debris would cause a loss-of-coolant, the reactor would automatically shut itself down. A loss-of-coolant auxiliary cooling loop was included in some configurations to ensure adequate core heat removal under accident conditions.

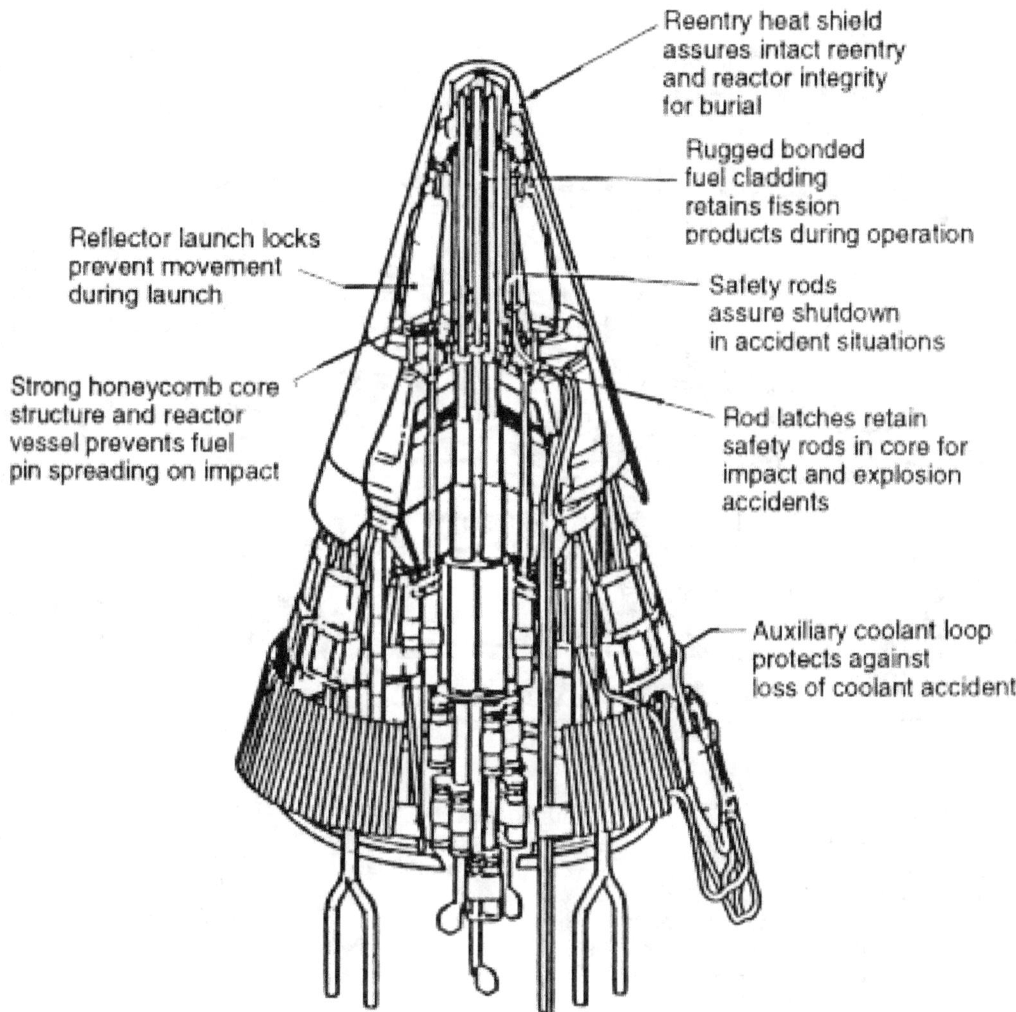

Fig. 12. SP-100 key safety features. *From* General Electric Co.,l989b.

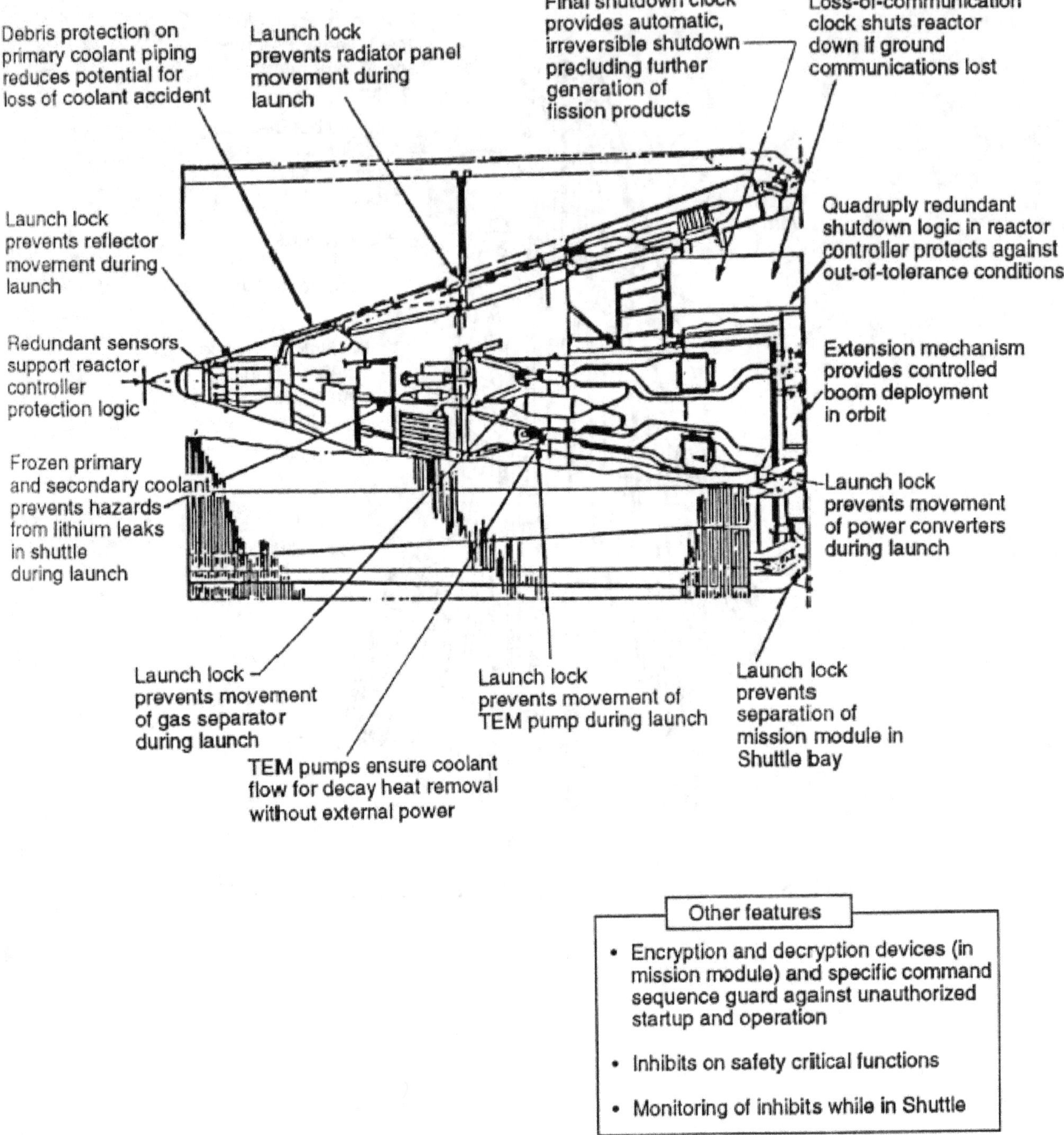

Fig. 13. Additional SP-100 safety features. *From* **General Electric Co., 1989b.**

The reactor was protected by a reentry heat shield to ensure it remains intact if the reactor should reenter the atmosphere. It was also designed to bury itself on impact, and to remain subcritical if it falls into the ocean.

The SP-100 used lithium as the working fluid. During launch, it was to be in the frozen state for added safety so that any launch-induced accident that might have cause piping rupture would not endanger crew or equipment. Reactor energy was used for thaw once the operational altitude was reached.

In the event of a loss of electrical power to the control system during operation in space, the reactor would have automatically shut down. Reflector elements and safety rods were spring actuated to their shutdown positions upon a loss of power.

As a result of the stringent design features, the reactor was predicted to offer much less radiation hazard to the public than transcontinental airline flights, diagnostic medical examinations, or therapeutic medical services. A mission risk analysis performed indicates no outstanding public safety issues. The analysis quantifies risk from accidental radiological consequences for a reference mission. The total mission risk based on expected population dose is estimated to be 0.05 person-rem based on a 1 mrem / y as the threshold for radiological consequences; this is a negligible amount in absolute terms and relative to the 1.5 billion person-rem / y that the world population experiences from natural radiation sources. The design had been subjected to several rigorous independent safety reviews with no negative findings.

Flight Readiness

A summary of the technical status and challenges as seen in 1991 is given in Table 4.[24] The progress on each technical challenges provides an assessment of flight readiness.[25, 26]

System Level
The power-versus-lifetime prediction codes were available though the verification of these codes was dependent on the completion of all component lifetime performance tests being conducted under the subsystems. With regards to verification of a 10-y system, all of the critical components, their failure modes, and the failure mechanisms had been identified. The failure mechanisms had been well defined and analyzed. The failures have been placed in three categories: i.e. 1) design margins determined to be adequate based on existing data, 2) additional test data needed, which is included in the SP-100 planned effort, and 3) additional test data needed which was planned to be obtained during the flight development phase. System start-up and restart from frozen lithium in zero G had been conceptually designed based on component thaw tests and still needed to be verified with system level ground tests. Flight system acceptance tests were still being defined.

Reactor Subsystem
The fuel pin swelling and fission gas release predictions had been verified up to the 10-y mission lifetime (6% burn up) based on in-pile accelerated fuel tests. The uranium nitride fuel pins were irradiated to burnups greater than that required for a ten-year mission. There were no fuel pin failures while operating at clad temperatures of 1,500 K, 100 K hotter than the hottest fuel pin design temperature in the operational system. Cold critical testing was performed in the Idaho ZPPR zero power physics reactor to verify the neutronic design codes for operation, shutdown, and safety under simulated operational and accident conditions. Hydraulic tests were performed on reactor mockups to establish the core's fluid flow resistance and flow paths. Uranium nitride fuel pellets for a 100-kW$_e$ space reactor have been fabricated and placed in storage at Los Alamos. Approvals had been completed for fabricating the bonded cladding and the fuel pins.

Irradiation effects on niobium alloys had been completed. The 10-y creep strength of Nb-1Zr had been verified by uniaxial and biaxial long term creep test. The very high creep strength of PWC-11 is dependent on the material processing and requires the results of additional long term creep tests . The reactor transient behavior would have been verified by the first operating SP-100 reactor. The SP-100 reactor component development was completed and technology ready for a flight system.

Fabrication development for the reactor subsystem components had been completed and technology readiness demonstrated for building a nuclear subsystem. An end-to-end electrically heated system test would be desirable as part of a flight hardware development program. While the design and fabrication of all reactor subsystem components have been successfully demonstrated, the final development of PWC-11 to demonstrate the effect of controlling the carbon precipitates to obtain maximum creep strength was yet to be demonstrated.[27]

Table 4. 1991 technical challenges

System Level

- Verified power versus lifetime prediction codes
- Verified reliable 10-y system-design margin codes
- Startup from frozen lithium in zero-gravity
- Flight system acceptance tests

Subsystems

1. Reactor
   - Verified prediction of fuel pin behavior
   - Verified 10-y creep strength of PWC-11
   - Verified transient behavior

2. Reactor Instrumentation and Controls
   - Reflector control drive actuator insulators and electromagnetic coil lifetime
   - Temperature sensors lifetime
   - Radiation hardened multiplexer amplifiers lifetime

3. Shield
   - Verified LiH swelling properties

4. Heat Transport Subsystem
   - Gas separator performance and plugging lifetime
   - TEM pump (TE/Buspar) bond performance and lifetime
   - TEM pump (Cu/Graphite Bus Duct) bond performance and lifetime

5. Converter Subsystem
   - Electrodes and bonds to TE legs lifetime
   - TE cell assembly low cost fabrication
   - High figure-of-merit TE material performance and lifetime ($Z=0.85 \times 10^{-3}$ K^{-1})
   - Cell to heat exchanger bond performance and lifetime

6. Heat Rejection Subsystem
   - Carbon-carbon to titanium bond performance and lifetime
   - Low cost and low mass heat pipe lifetime

Reactor Instrumentation And Controls Subsystem

The actuator insulator and EM coil lifetime tests and analyses verified that these components would operate for the 10-y mission lifetime. The temperature sensors were verified for 5-y missions. The lifetime of the multiplexers in a very high radiation field still needed to be verified. The control/safety drive subassembly (motor/brake/clutch/gears/spring drive/position sensor) was fabricated and performance tested. The balance of the control/safety drive assembly (CDA) was developed and designed, but needed to be fabricated and tested. The Control Drive Assemblies, the only active components in the SP-100 system, was considered adequate for five year missions, but had not completely demonstrated the full ten-year capability.[28]

Shield Subsystem

The LiH swelling properties as a function of temperature, radiation dose, and radiation dose rate were verified for ten years operation. Neutron irradiation of LiH swelling behavior at temperature was required to confirm that it is the same as that for gamma irradiation. The SP-l00 shield development was completed and technology ready for a flight system. The shield subsystem needed to be fabricated and tested in an end-to-end system test prior to flight.

Heat Transport Subsystem

The heat transport subsystem development was complete for near-term systems except for final validation tests of the thermoelectric electromagnetic (TEM) pump and the gas separator/accumulator (GSA). The gas separator performance and plugging still needed to be verified in a flowing lithium test. The TEM pump bond performance had been verified and the pump lifetime was primarily dependent on the TE cell lifetime. The pump TE cell performance and the lifetime verification must be completed. A development pump was fabricated using near-term technology, but needed to be tested in a lithium loop. The long-term testing of a prototype TEM pump was the only significant development activity remaining.[29]

Converter Subsystem

The TE cell electrodes and bonds had been developed and incorporated into a prototype TE cell. Three of these prototype TE cells were being tested at design conditions. The first cell operated at prototypical system conditions for 3,400 hours as predicted. Then, the internal electrical resistance began to increase. At 4,100 hours the unit was shutdown to determine the cause of the high internal electrical resistance, now about three times operational value. The second cell operated over 5,000 hours with only a slightly greater increase in electrical resistance. The third cell started to experience increase electrical resistance after 500 hours and was shutdown after 2,300 hours when the resistance reached three times normal. The first and third cells were found to have debonded at one of the graphite-to-SiGe "N' leg bonds. This SiGe-to-graphite bond is the weakest bond in the cell and is located on the hot side of the thermoelectric cell where the load is the greatest. The program ended before an improved bond could be tested. Complete cells need to be fabricated and tested as well as complete thermoelectric converter assemblies (TCA).

The low cost fabrication of TE Cells had been factored into the design with the manufacture of a large number of cells. The current converter manufacturing process used hot isostatic pressure bonding of niobium-to-niobium for the cell-to-heat-exchangers bond. Small scale fabrication had shown this bonding to be very successful To meet the long-term performance goals, additional work was needed to improve the integrity of the metallurgical bonds between the current collectors and the thermoelectric material to meet the life requirements.[30] Also, the high figure-of-merit TE material development was progressing, but needs additional work.

Heat Rejection Subsystem

An integral carbon-carbon tube and fin had been developed and successfully bonded to a thin wall (0.085 mm) Nb-1Zr potassium heat pipe, 25 mm diameter by 376 mm long and operated as a radiator heat pipe. Since the Nb-1Zr tubes can be manufactured and are leak tight with such a thin wall it may not be necessary to go to a titanium liner. A high conductivity integral carbon-carbon tube and fin 25 mm diameter by 1 m long has also been fabricated. Nb-1Zr potassium screen-wick heat pipes had been tested for long times (>10,000 hr) with no failures or degradation. Heat pipes of Nb-1Zr had been fabricated with high conductivity carbon-carbon armor and fins, but needed to be tested. A flight radiator panel of lower mass heat pipes, such as a 1.3 cm diameter titanium potassium heat pipe with bonded carbon-carbon fins and armor, needed to be built and tested.

Design Growth-Using Stirling Engine Development

As shown in Fig. 9, Stirling engines offer an attractive power conversion system in the hundreds of kW_e when mated with an SP-l00 reactor. For instance, a 600 kW_e power plant using Stirling engines that operate at 1,050 K has a specific power over 45 W_e / kg. Therefore, in 1985, it was decided to develop Stirling engine technology along with thermoelectric power conversion for SP-100. The Stirling engine development program[31] is based on a free-piston design that features only two moving parts (displacer and power piston), close clearance, non-contacting seals (no wear of mating parts), hydrostatic gas bearings for dynamic members (no surface contact of dynamic components and no oil lubrication necessary), dynamic balancing, and the potential for a hermetically sealed power module. The free-piston Stirling concept utilizes gas springs, which have hysteresis losses.

The program started with a 650-K Space Power Demonstrator Engine (SPDE) technology development and proceeded with the development of common designs for 1,050 K and 1,300 K.[32] The 1,050 K engines provide the means to demonstrate that the engine technology issues have been solved using easier to work with superalloy materials before demonstration of the refractory or ceramic materials version. The specifications for the 1,050 K design are summarized in Table 5. A schematic of the 1,050 K engine is shown in Fig. 14.

Table 5. 1,050 K Stirling Space Engine goals and specifications.

Balanced opposed configuration total power output (kWe)	50
End of Lift power (kWe/piston)	25
Efficiency (percent)	>25
Life (h)	60,000
Hot side interface	Heat Pipe
Heater temperature (K)	1,050
Cooler temperature (K)	525
Vibration - casing peak -peak (mm)	<0.04
Bearings	Gas
Specific mass (kg/kWe)	<6.0
Frequency (Hz)	70
Pressure (MPa)	15

In October 1986, the 650 K SPDE demonstrated 25 kW_e.[33] The SPDE was a dual-opposed configuration consisting of two 12.5 kW_e converters. After this successful demonstration, the engine was cut in half to serve as test beds for evaluating key technology areas and components, now called Space Power Research Engines (SPRE). The electrical output had been measured as 11.2 kW_e at overall efficiency of about 19%. The goal is 12.5 kW_e and efficiency greater than 20%. Sensitivity of the engine performance to the displacer seal clearance and the effects of varying the piston centering port area were being studied.

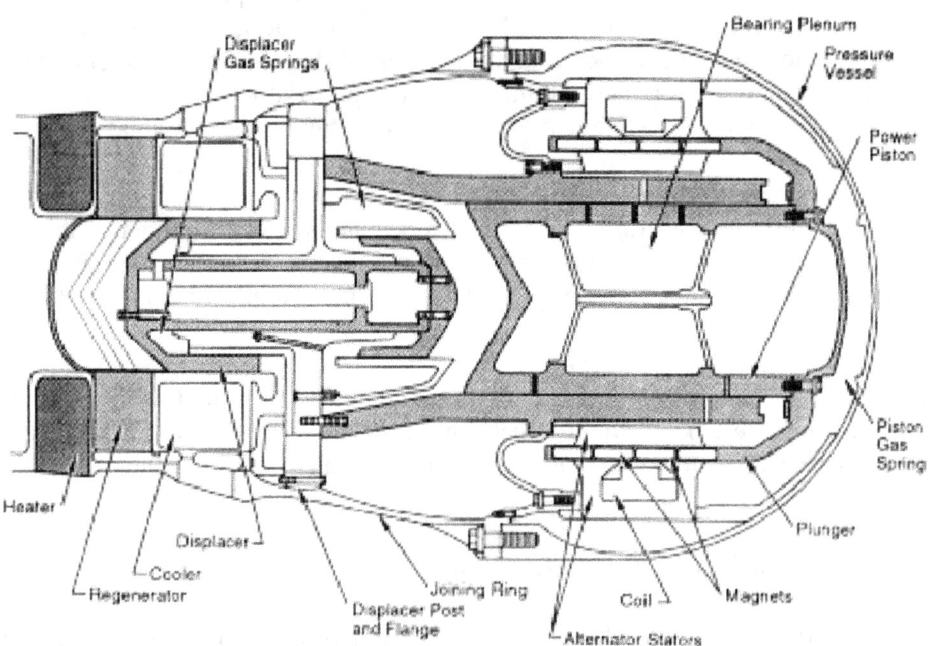

Fig. 14. Preliminary design of 1,050 K Stirling Space Power Converter. *From* **Dudenhoefer, J. E., 1990.**

The Component Test Power Converter (CTPC) was a 12.5 kW$_e$ cylinder technology engine for the Stirling Space Power Converter (SSPC). Inconel 718 was used as the heater head material to permit early testing for short terms (100 - 1,000 h at 1,050 K). This testing had demonstrated 12.5 kWe at a 1,050 K operation temperature and greater than the 20% goal. A heater for the SSPC of Udimet 40 L1 superalloy that will have a design life of 60,000 h still needed to be fabricated. The CTPC was being used to evaluate critical technologies identified as: bearings, materials, coatings, linear alternators, mechanical and structural issues, and heat pipes. The impact of temperature on close-clearance seal and bearing surfaces in the cold end of the power converter had been tested, with no problems observed. The CTPC linear alternator used an alternator that can reach a peak temperature of 575 K, close to the upper operating limits of samarium cobalt magnets. Test results indicated sufficient design margins.[34]

Endurance testing was underway on a 2 kW$_e$ free-piston Stirling engine called EM-2 that operates at a heater temperature of 1,033 K. At the end of 5,385 h, only minor scratches were discovered due to the 262 dry starts/stops, and no debris was generated. The heater head of Stirling power conversion systems is the major design challenge because heater head creep is predicted to be the life-limiting mechanism for Stirling engines. The difficulty in creep analysis stems from inadequate knowledge of elevated temperature material behavior and inadequate knowledge of inelastic analysis techniques. Typically, the high operating temperatures and long operating periods of Stirling engines are taxing the ultimate capabilities of even the strongest superalloys.

The free-piston Stirling engine has continued to be developed following the termination of the SP-100 program. Currently, the technology has progressed to being used as a power conversion system in the Stirling Radioisotope Generator.

Summary[35]

For near term applications of planetary missions employing nuclear electric propulsion, a three-to five year lifetime, 20 kW$_e$ power plant was designed based on existing SP-100 technology. These missions are not as aggressive as ones using a 100 kW$_e$ power plant, but offer very attractive near term options. The 20 kW$_e$ power plant configuration is shown in Fig. 15.

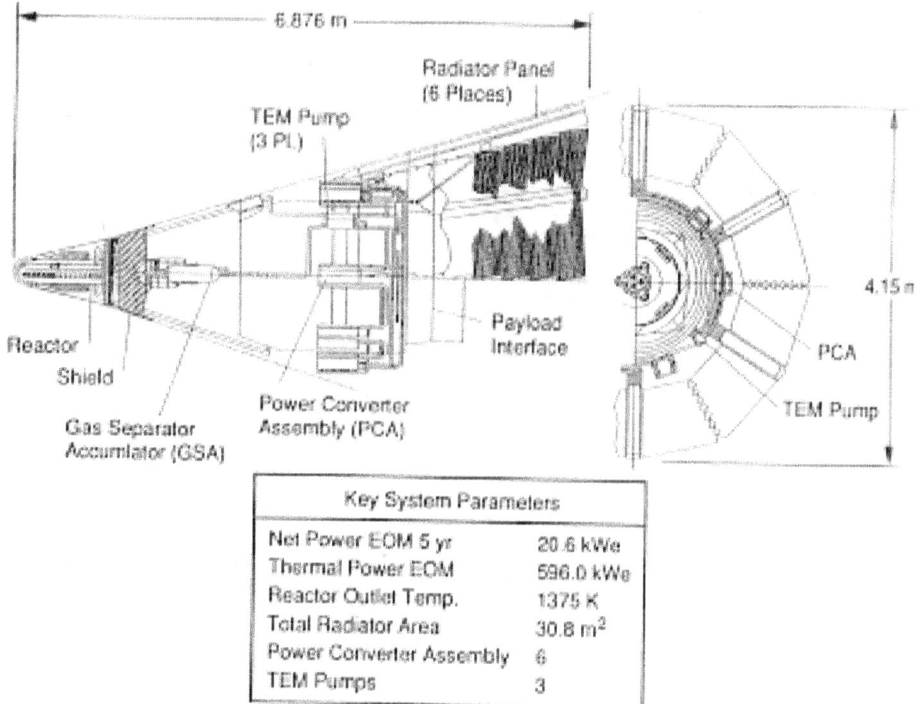

Fig. 15. 20 kW$_e$ space reactor thermoelectric power plant using SP-100 technology.

The SP-100 program had essentially completed its component performance development phase. This included validation and component fabrication of the critical technologies needed to build a reactor space power system. A generic flight system had been completed as well as plans for validating the overall power plant for flight. Table 6 summaries the technologies for near-term, intermediate and mature systems.

Table 6. SP-100 power plant technology

	Near-Term	Intermediate Term	Mature
System			
Power	10-40 kWe	10-60 kWe	10-300 kWe
Lifetime (Full Power/Mission)	3/5 Years	5/7 Years	7/10 Years
Reactor			
Fuel	UN	UN	UN
Coolant	Lithium	Lithium	Lithium
Clad	Nb-1Zr/Re	PWC-11/Re	PWC-11/Re
Structure	Nb-1Zr	Nb-1Zr	PWC-11
Outlet temperature	1350 K	1375 K	1400 K
Reactor I&C			
Mode	Dual	Dual	Separate
Safety	In-Core	In-Reflector	In-Core
Control	In-Core	In-Reflector	Reflector
Heat Transport			
Pump	TEM	TEM	TEM
Material	Nb-1Zr/PWC-11	Nb-1Zr/PWC-11	PWC-11
TE Material	SiGe--0.67X10-3/K	SiGe(GaP)--0.72X10-3/K	SiGe(GaP)--0.85X10-3/K
Converter			
Type	Multicell	Multicell	Multicell
Power	8.8 We/Cell	10.8 We/Cell	12.8 We/Cell
TE Material	SiGE-0.67X10-3/K	SiGe(GaP)-0.72X10-3/K	SiGe(GaP)-0.85X10-3/K
Radiator			
Heatpipe	K-Ti (2.5 cm Dia.)	K-Nb-1Zr (2.5 cm Dia.)	K-Ti (1.3 cm Dia.)
Fins	Ti	C-C	C-C
Armor	Ti	C-C	C-C
Duct	Ti	Nb-1Zr	Ti

Chapter 5

Thermionic Reactor Power Plants

Earlier Developments

Thermionic converters can be integrated into the reactor core or located outside of the core. The advantage of the in-core arrangement is that all high temperature elements are located in one unit--high temperature heat transport is minimized. Out-of-core thermionics however, separates the development problems of long life thermionic elements and reactor development.

The mid-1950s saw the initial development activities on thermionic reactors. By the 1963 to 1973 period, effort was being centered on in-core thermionic conversion concepts.[1,2,3,4,5,6,7,8,9,10,11,12] Reactor system designs were prepared for a variety of space applications including: a 5-10 kW_e system for unmanned satellites; a 40 kW_e system for manned space laboratories; and a 120 kW_e system for nuclear electric propulsion. These power system concepts employed the same basic thermionic fuel element entitled the in-core "flashlight" design. This is actually only one of several approaches to designing a thermionic reactor, some of which will be discussed later. The in-core "flashlight" thermionic reactor concept was the most advanced approach.

The thermionic diode shown in Fig. 1 is the basic building block on which the major thermionic development effort concentrated. The appearance is obviously similar to a flashlight battery. The arrangement is such that the fuel is in the interior of the diode in order to isolated it from the core structure. Surrounding the fuel is the thermionic emitter that emits electrons that travel across the inter-electrode gap to the collector. It is designed to operate at very high temperatures of up to 2,000 K. In the space between the emitter and collector is a cesium reservoir. The cesium atoms in the reservoir are ionized partially by electron collisions and partially by contacting the hot emitter surface. The positive cesium ions are used to help neutralize the space charge effect that would otherwise be built up by the flow of electrons. Also, the positively charged cesium ions prevent the electrons adjacent to the emitter surface from attaching themselves to both the emitter and collector surfaces. This has the beneficial effect of lowering the work functions. A sheath is used to encapsulated the thermionic configuration.

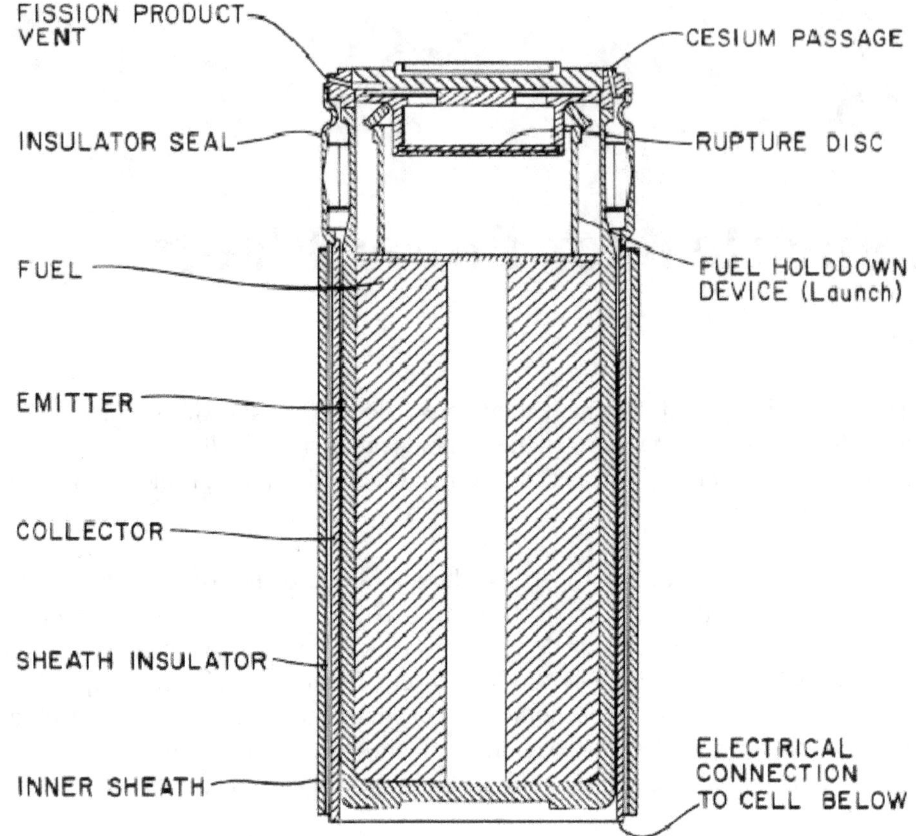

Fig. 1 Thermionic diode. *Courtesy of General Atomic.*

Similar to a flashlight with its batteries in series, the thermionic diodes were packaged six converters per fuel element (see Fig. 2).[13] These fuel elements, which were about 2.5 cm in diameter by 40.6 cm long, can then be configured into many different reactor core configurations and meet a varied of power requirements.

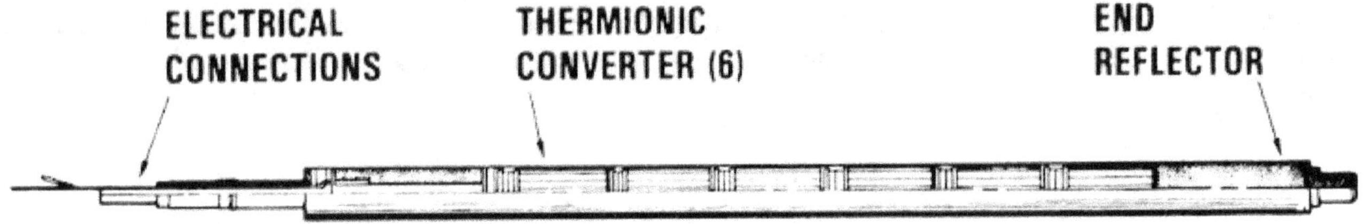

Fig. 2. Schematic of in-core thermionic fuel element. *From J. W. Holland et al.,1971.*

As examples, Fig. 3 shows some possible reactor core configurations using the thermionic fuel elements to provide 40-60 kW$_e$ and 100-250 kW$_e$. The solid circles in the figure represent the thermionic fuel elements. The 40-60 kW$_e$ moderated reactor contained 60 thermionic fuel elements that are coupled with U-ZrH elements (the open circles in Fig. 3) similar to those developed for the SNAP-8 program. In this configuration, the thermionic fuel elements generated the desired electric power, while the U-ZrH fuel elements were added to achieve criticality and to moderate the neutron spectrum. The core was surrounded by a radial reflector region, containing control elements in the form of drums. The reflector material was beryllium oxide; the control drums included segments of boron carbide to be position for neutron absorption. The control drums, contained in dry wells within the reactor vessel, were rotated to achieve reactor control. Positioning the boron carbide segments towards the core caused the reactor to reduce power; if the boron carbide segments were turned away from the core, reactor power level increased.

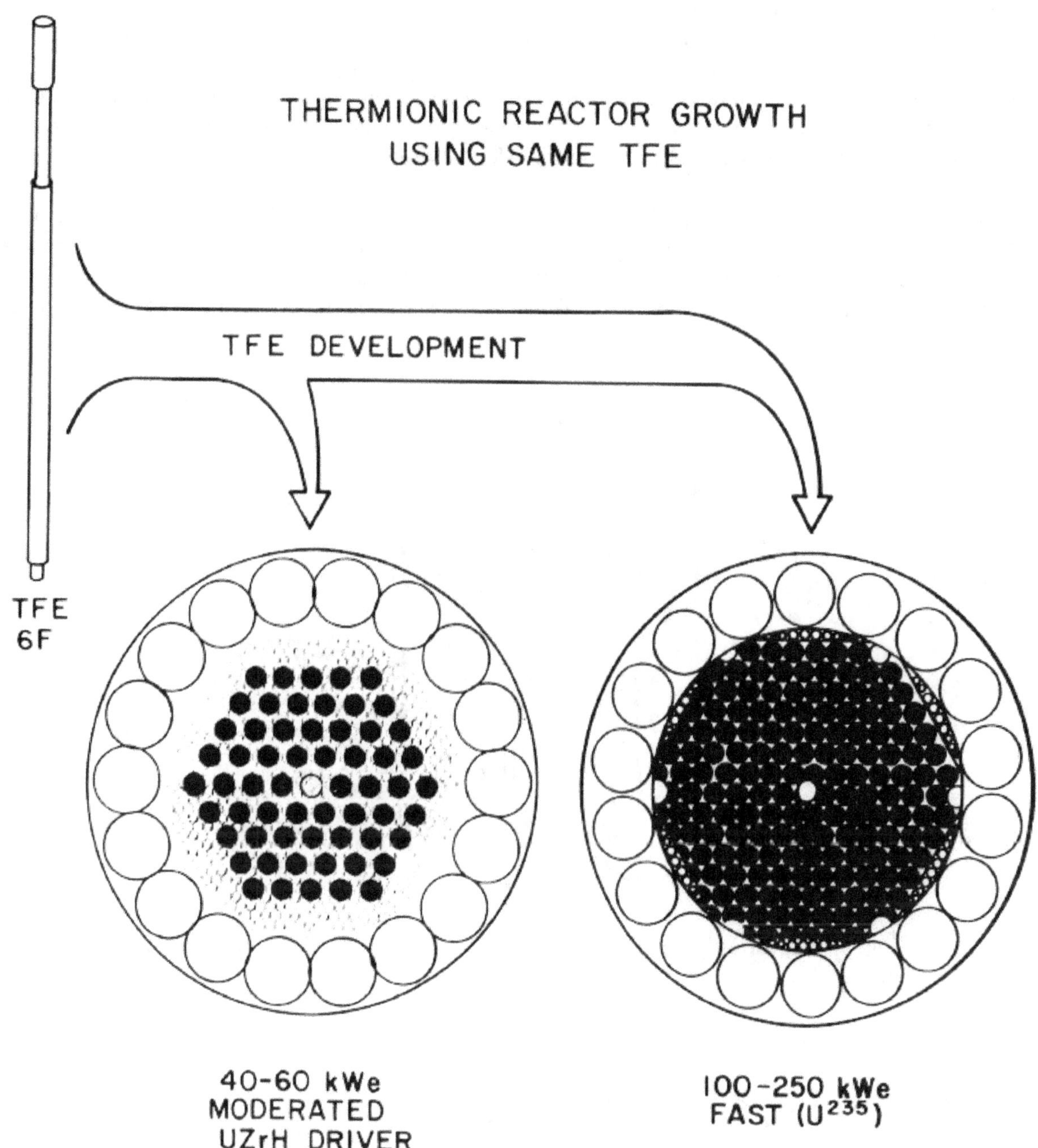

Fig. 3. Thermionic reactor core arrangements for 40-60 kWe and 100-250 kWe configurations. *From* **J. W. Holland et aI.,1971.**

A cross section of the 40 - 60 kW$_e$ thermionic reactor is shown in Fig. 4 with specific design details provided in Fig. 5. The drive actuators were located in a reduced radiation environment above the radiation shield. The core configuration included an inner region with both types of fuel elements (i.e., thermionic and U-ZrH) and an outer region with only U-ZrH fuel elements. Long lifetimes are predicted for the U-ZrH fuel elements as they were to operate under less demanding conditions than in the SNAP-8 system. Their cladding temperatures are about 110 K lower and the surface heat flux a factor of 2 to 4 lower in comparison with the U-ZrH fueled SNAP reactors.

The path of the NaK coolant flow through the reactor will be described. The core and reflector regions were separated by a shroud that directed the incoming NaK coolant through the reflector region before it flowed back through the core.

The coolant entered the side of the vessel near the bottom, was distributed in the small plenum area, and flowed up through the reflector region to remove neutron and gamma radiation heating. The coolant then entered the core region through openings in the shroud, was distributed in the region between the vessel head and the upper end of the fuel elements, and flowed down through the core. It passed through holes in the grid plate and through the shield block to the reactor outlet pipes. Materials used for the coolant and vessel (NaK-78 and 316 SS) were the same as those in the reference SNAP-8 U-ZrH thermoelectric reactor system.

Significant performance parameters are provided in Table 1. Only 46 thermionic fuel elements were needed to meet end-of-life conditions, including an allowance for power degradation or failure. One significant advantage of this system is that the fluid temperature exiting the core is the waste heat temperature. Notice that the temperature is between 827 (BOL) and 866 K (EOL)--this leads to a relatively small radiator for this type of system

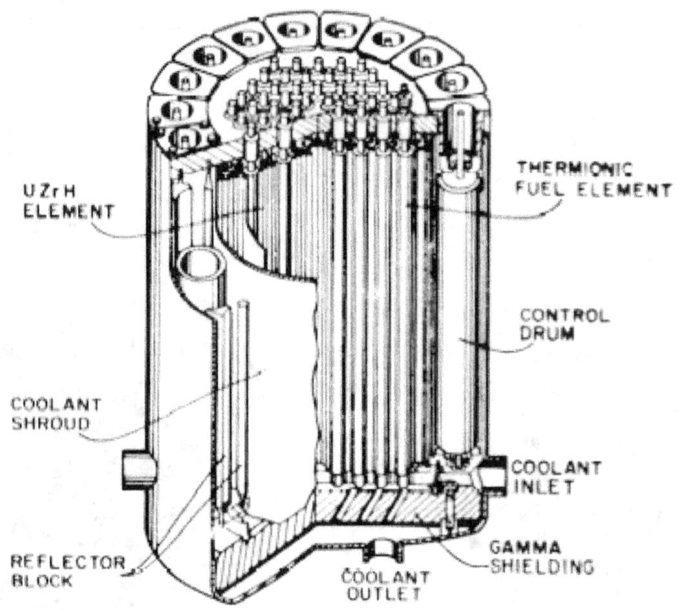

Fig. 4. Thermionic reactor schematic. *From* J. W. Holland et al.,1971

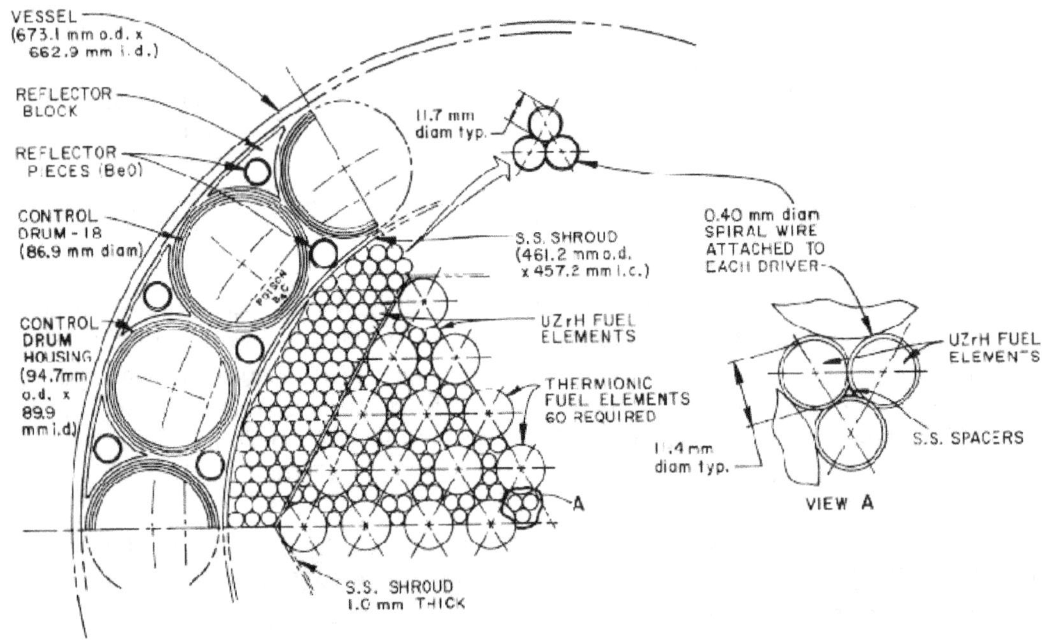

Fig. 5. Details of in-core thermionic reactor core configuration. *From* J. W. Holland et al., 1971.

Table 1. Thermionic reactor performance parameters.

Reactor outside diameter (cm)	67.3
Reactor length (cm)	96.5
Reactor mass (kg)	987
Number of thermionic fuel elements	60
Number of U-ZrH fuel elements	813
Number of control drums	18
Control drum diameter (cm)	82.3

	BOL	EOL
Gross electrical output (kWe)	53	58
TFE surface heat flux (W/cm^2)	24	29
Thermal power output (kW)	1,180	1,450
Coolant inlet temperature (K)	750	789
Coolant outlet temperature (K)	827	866

The 100-250 kW$_e$ power plant design used a fast reactor design contained 162 thermionic fuel elements. Criticality limitations required using uranium-235 as the fissile fuel. Though the reactor produced 164 kW$_e$ at 23 volts, losses in power distribution and conditioning equipment and in reactor coolant pumping requirements reduced the net power output to 120 kW$_e$. The heat rejection system was designed to radiate 1,510 kW$_t$ at a temperature of 1,000 K. Waste thermal energy was carried to the radiator by NaK-78 coolant loop that included a linear induction pump. The radiator utilized 1,620 heat pipes, containing sodium as the working fluid and employing niobium alloy for tubing.

Fuel burnup in a 120 kW$_e$ configuration was computed to be 1.6 X 10^{20} fissions / cm^3 in 20,000 hours. The Thermionic fuel element experienced a fast neutron fluence (> 0.1 MeV) of 4.6 X 10^{21} nvt during the same period.[14]

In the development program, the major objectives of the thermionic fuel element were to achieve:
1) temperature and current capability of 2,000 K and 12 amperes / cm^2 of emitter surface;
2) lifetime of 2 years at 2,000 K and 12 amperes / cm^2 and 5 years at 1,800 K and 6 amperes / cm^2;
3) demonstration of reproducible thermionic converter performance levels in-pile;
4) demonstration that any electrical discharges which might occur were not destructive to the fuel element integrity and would not result in excessive power losses; and
5) qualification to anticipated launch vibration and shock environments.

Key to thermionic fuel element development are high temperature materials. Fuels being considered included both carbide and oxide. Carbide fuels are attractive as fuel materials in the form of both 90 UC-I0 ZrC and UC because of their high melting points (2,723 K for UC and 3,023 K for 90 UC-10 ZrC), high uranium concentrations (approximately 12 g / cm^3), and high thermal conductivities (approximately 0.2 W / m-K). Difficulties existed, however, in their dimensional stability and thermionic converter performance. Oxide fuels indicated that a stoichiometry of 2.002-2.007 was desired. This was based on the requirements that the uranium chemical activity should not be so high as to affect the structural integrity of the tungsten cladding during irradiation and the oxygen chemical activity should not be so high as to cause reaction with the tungsten cladding.

Tungsten was selected as the cladding material, mainly because of its compatibility with both the carbide and oxide fuels. In addition, its low vapor pressure, high melting point, and superior creep strength at high temperatures were distinctive advantages.

Niobium was selected for the collector material because its thermal expansion coefficient closely matched that of the Al$_2$O$_3$ layer used to electrically insulate the collector from the sheath. In addition, Niobium is an excellent getter for oxygen and nitrogen and helps maintain a clean environment in the inter-electrode gap.

Copper was originally selected as the cesium reservoir material, but this choice was later changed to niobium. This change eliminated the need for a weld between the copper reservoir and the niobium collector. This weld had caused several converter failures.

Thermionic devices and fuel-clad capsules were tested in the Thermionic Test Reactor (TITR) at Gulf General Atomics.[15] Table 2 summarizes the results of thirty-eight in-core tested devices run between 1962 and 1973. Fuels used in the in-pile thermionic devices were either carbides or oxides. Progress in converter lifetimes increased in the longest tests from the 1,000 - 2,000 hour range in 1965 to 10,000 hours by 1970. The number of test hours in a full-length fuel element (called 6F3) reached 8,062 hours at an average power output of 647 watts and 11,084 hours on a partial length fuel element (called 2E2). Fuel performance found that the oxide-fueled tests have consistently exhibited good initial thermionic performance and have maintained this performance throughout the tests. However, carbide fuels have tended to exhibit less stable performance during such tests. The longest UO_2 fueled tests, namely: IC-14 (9,754 hours), 2E2 (11,084 hours), and 2El (12,535 hours), had stable performance within $\pm$ 10 percent. (In the case of the 2El test, a short occurred at 7,035 hours and its performance decreased approximately 15 percent over its lifetime.)

Table 2. Summary of General Atomic thermionic in-pile tests (as of 22 January 1973).

	Experimental Cells	Prototype Cells		Thermionic Fuel Elements			
	Mark VI	Mark VIIA	Mark VIIB	E Series (Two Cell)	F Series (One Cell)	F Series (Two Cell)	F Series (Six Cell)
Number of Tests	15	3	4	3	4	3	5
Total Test Hours	45449	4604	15915	27280	22152	16551	23972
Average Test Hours	3030	1535	3979	9093	5538	5517	4794
W/cm^2	6.3	5.2	3.7	4.5	3.8	4.0	2.6
Longest Test Hours	9754	2006	7881	12535	8560	8021	8062
W/cm^2	6.0	6.2	3.4	4.0	4.0	3.4	3.0
Failure Modes							
Carbide Fuel							
None	1	—	1	—	1	2	2
Short Circuit	1	—	—	—	2	—	1
Envelope Leak	10	2	2	1	—	—	1
Oxide Fuel							
None	1	—	—	1	—	—	1
Short Circuit	1	1	1	1	1	1	—
Envelope Leak	1	—	—	—	—	—	—

Additional experiments were performed using other materials. For example, tests revealed that $W(WCl_6)$ emitters performed with a relative power some 8 percent better that those with $W(WF_6)$ emitters; while oxide-fueled devices performed approximately 4 percent better than carbide-fueled devices. It was found that cells with molybdenum collectors produced about 7 percent more power than those with niobium collectors.

Factors that were effecting performance were determined to be: (1) leaks between the cesium or fission gas spaces and the containment gas (envelope failures); (2) a leak between fission gas and cesium spaces resulting in fission gas and fission products in the inter electrode space: (3) failure of the bonds in the collector-sheath tube structure, resulting in high collector temperatures; (4) contamination of the electrodes during fabrication; (5) electrode contamination due to fuel component diffusion through the emitter clad; and (6) inter-electrode spacing changes. Envelope failures were a major problem associated with braze joints. A solution to this problem was going to an all-welded design. Much of the fuel diffusion problem appears to have been solved by control of fuel stoichiometry and cladding structure. Further improvements were needed in dimensional stability to meet a 20,000 hour goal with a 1 millimeter cladding thickness.

To summarize the advantages of the in-core "flashlight" thermionic reactor concept include:
- The fuel element is similar to most other reactors;
- The high temperature components are thermally isolated from the core structure;
- Heat removal requirements are at conventional levels;
- Insulators are at a relatively low temperature; and
- The concept is scaleable over a wide range of power levels.

However, major difficulties encountered with this concept were found to:
- Fuel swelling that shortened diode fuel life;
- Fast neutron damage at higher power and flux levels;
- Output voltage limitations imposed by insulator arcing considerations;
- The need for larger diameter emitters; and
- Small values for the reactivity feedback coefficients.

This flashlight thermionic concept could be used in either a fast unmoderated reactor or a moderated thermal reactor. Selection of a moderated thermal reactor concept reduces both the minimum critical size of the core and the electric power output.[16]

Another thermionic reactor concept involved an externally fueled diode, as illustrated in Fig. 6.[17] Heat rejected from the collectors is removed by a liquid metal coolant loop or by heat pipes. Thermionic elements are connected in a series-parallel arrangement to build up the reactor output voltage and to provide greater system reliability. The high fuel volume fraction results in smaller core sizes than encountered with internally fueled thermionic reactor designs. Other advantages of this concept include the following:

- The insulator can be removed from the core.
- Fuel swelling may occur away from the cesium gap.
- Thermionic converter performance can be checked prior to reactor operation.
- The series-parallel electrical network provides flexibility optimizing system reliability.
- The system has a prompt expansion coefficient.
- Redundant cooling is possible.

There are disadvantages to the externally fueled diode concept:
- The concept has a growth limitation of a few hundred kilowatts-electric.
- It is difficult to add a moderator.
- The fuel element structure operates at emitter temperatures.
- Inter-electrode spacing must be maintained over a long distance.
- The fuel elements must be electrically isolated from each other, while the temperature is too high to use an insulator.
- Arcing paths may exist between fuel elements.
- The large size associated with dioxide fuels could give rise to very high currents.

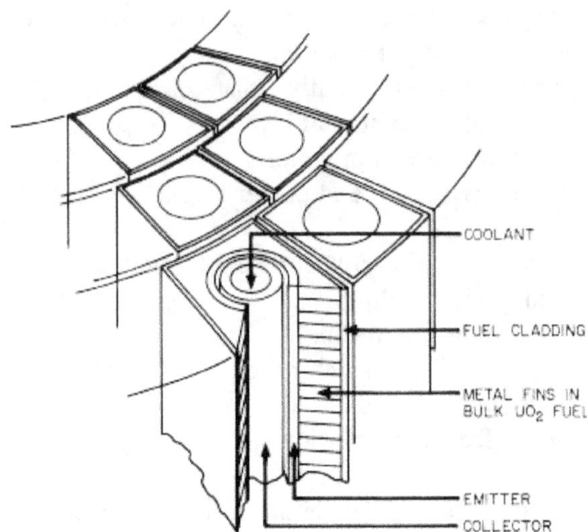

Fig. 6. Externally fueled, full-length thermionic diode modules. *From D. S. Beard, 1969.*

To summarize, in-core thermionic concepts suffer from the fact that the reactor and converters are physically combined or intertwined--a design relationship that results in heavier and larger reactors. Because of the proximity of the converters to the nuclear fuel, fuel swelling can result in an electrical short circuit in the converter should the emitter distort by as little as 0.5 mm.[18] In-core neutron damage to the alumina insulator also tends to limit thermionic diode lifetimes at high power, and therefore high neutron flux levels. The development of insulators more resistant to neutron radiation damage would offer improved performance and lifetimes.

From 1973 until 1982, the emphasize on thermionic power plants shifted to out-of-core heat pipe concepts (see Fig. 7).[19]   These concepts separate the heat producing and power conversion functions by placing the energy conversion devices outside the reactor core and transporting core thermal energy to them. Most of this out-of-core thermionic effort has been focused on a configuration in which the heat is transported by heat pipes and the converters are bonded directly to the high temperature heat pipes. Thermionic devices are more efficient at high temperatures. Therefore, the heat pipes were being designed for a temperature of 1,675 K. This results in a maximum fuel temperature of close to 2,000 K. Fuel swelling than becomes even more significant than with many other reactor designs. To reduce fuel swelling, hexagonally shaped fuel elements are used with the fuel in the form of UO_2 pellets in a molybdenum matrix that is brazed onto the heat pipe.

The nominal composition of the core fuel region is 60 percent UO_2 and 40 percent Mo. The operating characteristics for a 400 kW$_e$ out-of-core thermionic reactor are summarized in Table 3. Fuel swelling is estimated to be 5 percent.

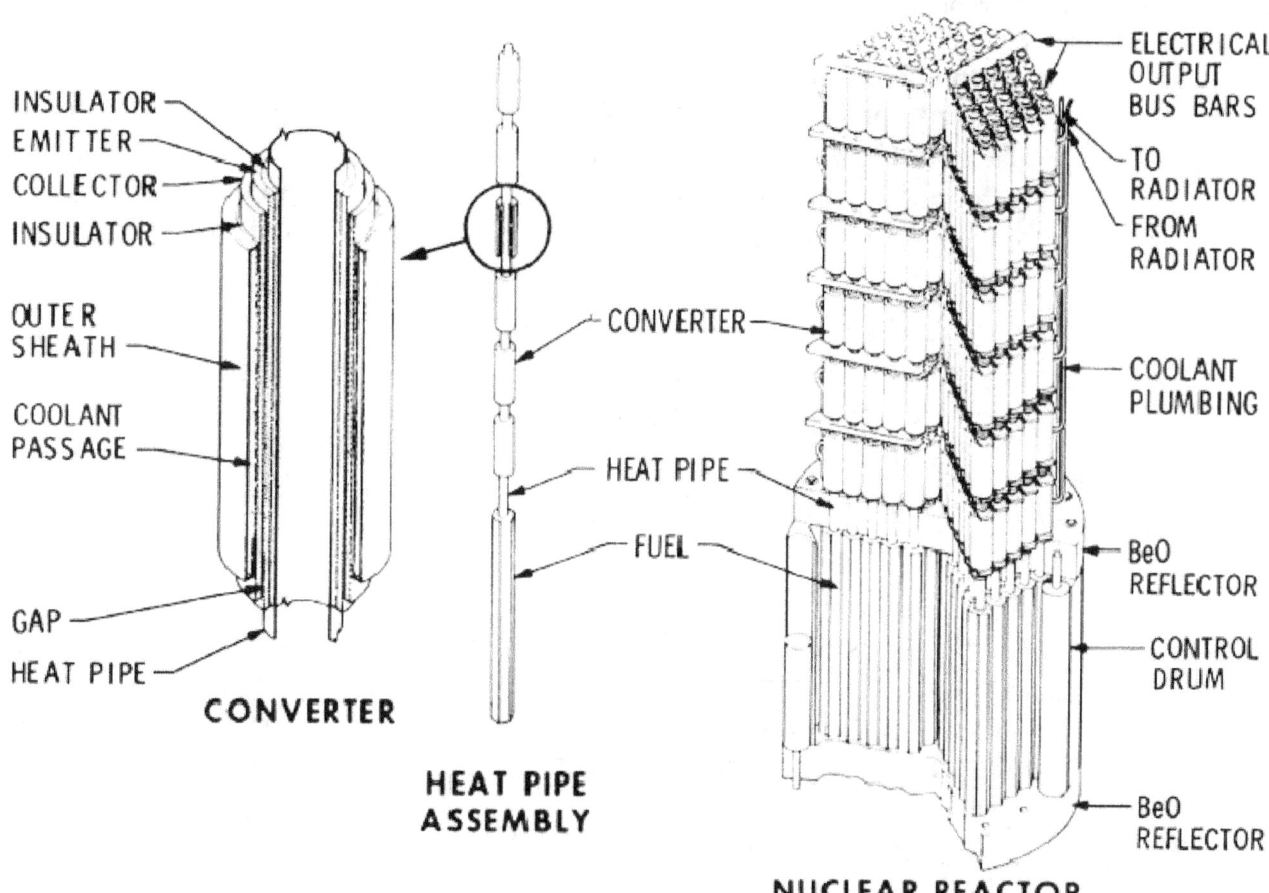

Fig. 7. Out-of-core thermionic reactor concept. *Courtesy of Los Alamos National Laboratory and Jet Propulsion Laboratory.*

Several significant problems exist with the out-of-core thermionic concept. Developing and demonstrating a long life, 1675 K heat pipe will require much more research. A high temperature electrical insulator is used around the heat pipe, since it must not electrically short the system while transporting thermal energy. This high temperature insulating material must match the thermal expansion coefficients of the heat pipe. "Seal on" (a ceramic made from silicon, aluminum, oxygen, and nitrogen) was tested but failed to provide the necessary lifetimes. It is difficult to avoid degradation of such an insulator. Another technical problem was the inability to obtain the same thermionic performance at 1,650 K as at 1,850 K, despite a considerable research effort.

Table 3. Out-of-core thermionic reactor operating characteristics.

Nominal electrical power output (kWe)	400
Thermal power level (MWth)	3
Thermionic converter efficiency (%)	15
Lifetime at full power (h)	75,000
Number of heat pipes	162
Nominal UO_2 content of fuel (vol %)	60
Average power density in Mo-UO_2 (MW/m^3 or W/cm^3)	78
Average power density in UO_2 (MW/m^3 or W/cm^3)	130
Power per heat pipe (kW)	18.8
Heat pipe axial heat flux (MW/m^2)	100
Heat pipe radial heat flux (MW/m^2)	0.91
Heat pipe temperature (K)	1675
Average fuel temperature (K)	1738
Maximum fuel ΔT (K*)	131
Maximum fuel temperature (K*)	1865
^{235}U burn-up (%)	5.8
Fission density in Mo-UO_2 (10^{20} fissions/cm^3)	6.3
Fuel volume swelling (%)	5
Reactivity change Δk due to fuel burn-up (%)	3.5
Reactivity change Δk due to thermal expansion (%)	1.5
Reactor Mass (kg)	
Mo-UO_2 fuel, total weight	372
^{235}U only	189
Mo heat pipes, 110 kg/m (to outer edge of reactor only)	60
BeO reflector	397
Control system	75
Reactor support structure	46
TOTAL	950

A modification of the out-of-core heat pipe thermionic reactor concept is shown in Fig. 8.[20, 21] This design approach eliminates the high temperature insulator and uses radiative coupling. A large number (500 - 5,000) of small heat pipes helps to minimize temperatures in the reactor core, thereby reducing fuel swelling. Heat pipes are clustered around each thermionic converter. The cold inside electrodes are connected through a sheath insulator to heat rejection heat pipes that radiatively couple to the waste heat radiator. However, this does not resolve the other problems associated with out-of-core thermionic power systems.

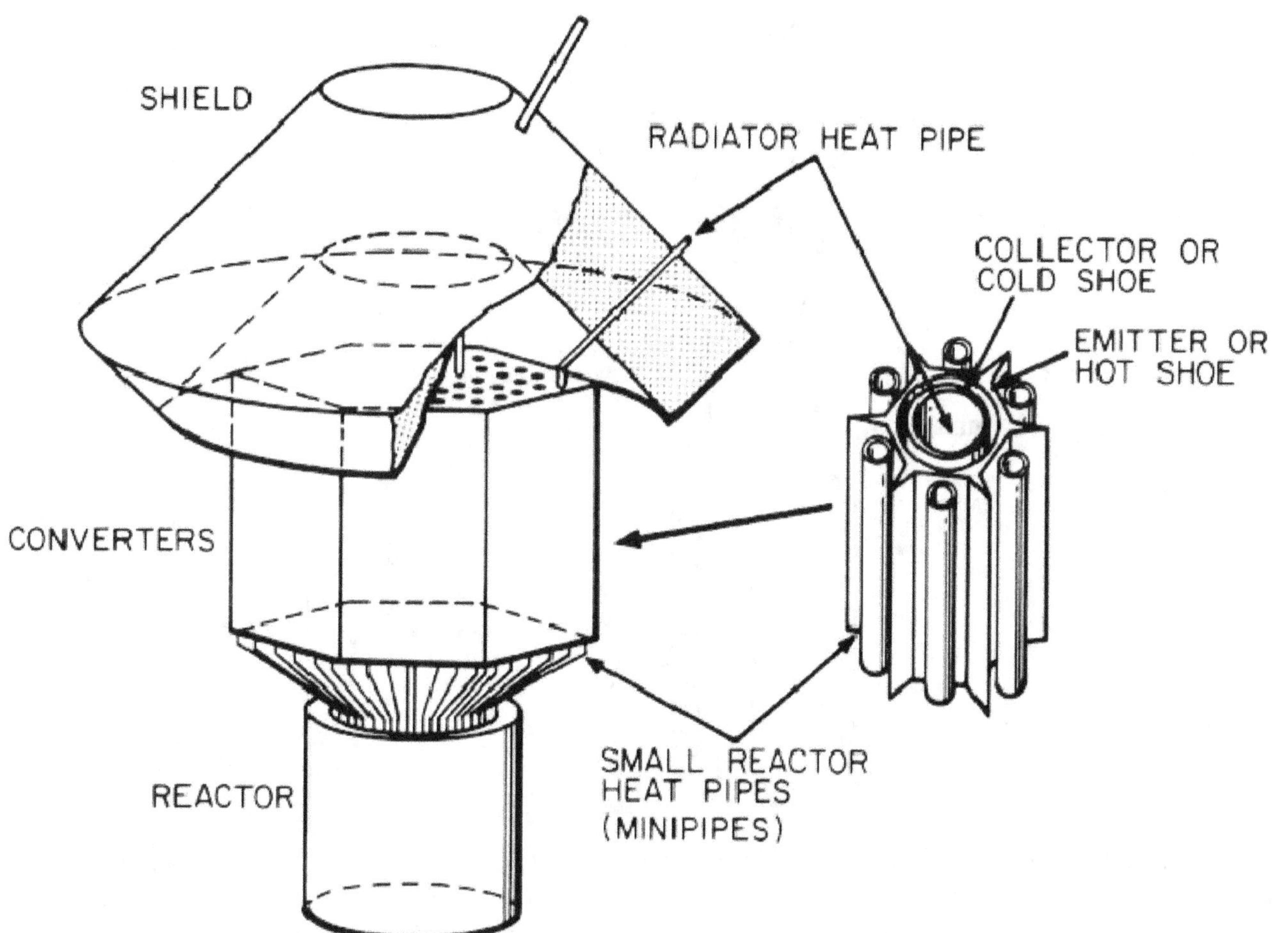

Fig. 8. Out-of-core radiation-coupled thermionic mini-pipe design. *From* E. J. Britt and G. O. Fitzpatrick, 1982 Proceedings.

Thermionic Developments From 1983-1994, Including The Fuel Element (TFE) Verification Program

During the first phase of the SP-100 program from 1983 to 1985, a 100 kW$_e$ thermionic space power system was designed having a fast spectrum reactor core with 180 6-cell thermionic fuel cells (TFE's). In 1986, the Thermionic Fuel Element Verification Program was initiated to resolve the technical issues identified with thermionic reactors during the SP-100 Phase I concept selection. The program's objectives were to resolve technical issues for a multi-cell TFE suitable for use in thermionic space reactors with electric power output in the 0.5 to 5 MW$_e$ range and a full-power lifetime of 7 to 10 y. The main concerns were fuel/clad swelling, insulator integrity for long irradiation times, and demonstration of performance and lifetime of TFEs and TFE components.[22] The program was later restructured to better address the performance and lifetime requirements of interest to the DoD (i.e., 5 to 40 kW$_e$, 1.5 to 5 y lifetime).

System Concept

Thermionic power systems are an attractive option for providing electric power in space because their radiators tend to be smaller than for other concepts (higher heat rejection temperatures and/or efficiencies); outside of the fuel and emitter, the temperatures are sufficiently low that refractory metals are not needed; and scaleability is from a few kilowatts to megawatts.[23, 24, 25] The heart of these systems is the thermionic fuel element (TFE). A single thermionic fuel cell is shown schematically in Fig. 8a[26] and 8b.[27] The TFE is a building block, as seen in Fig. 9.[28]

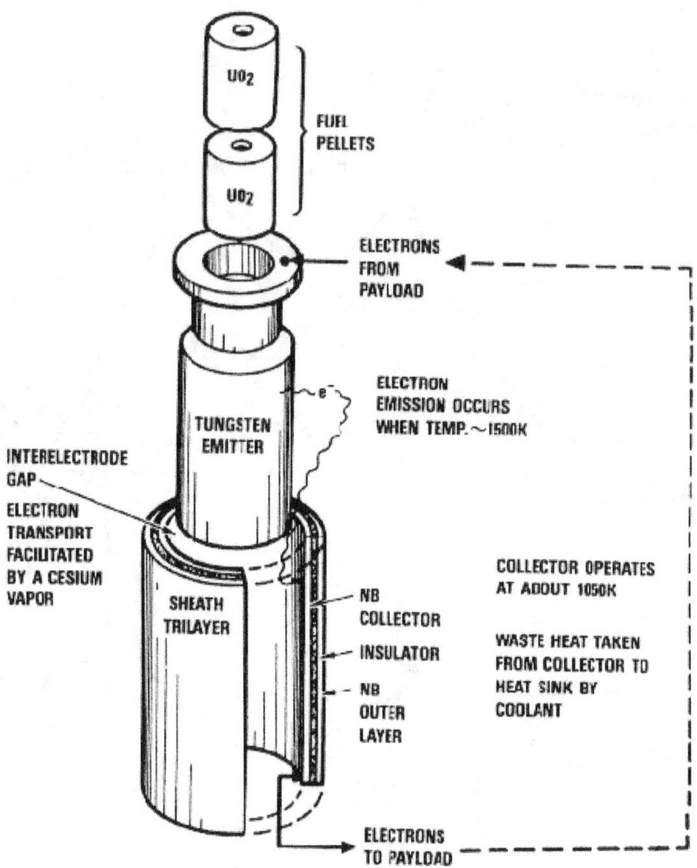

Fig. 8a. Schematic of a Thermionic Fuel Cell *From Strohmayer, W. H. and T. H. Van Hagan, 1985.*

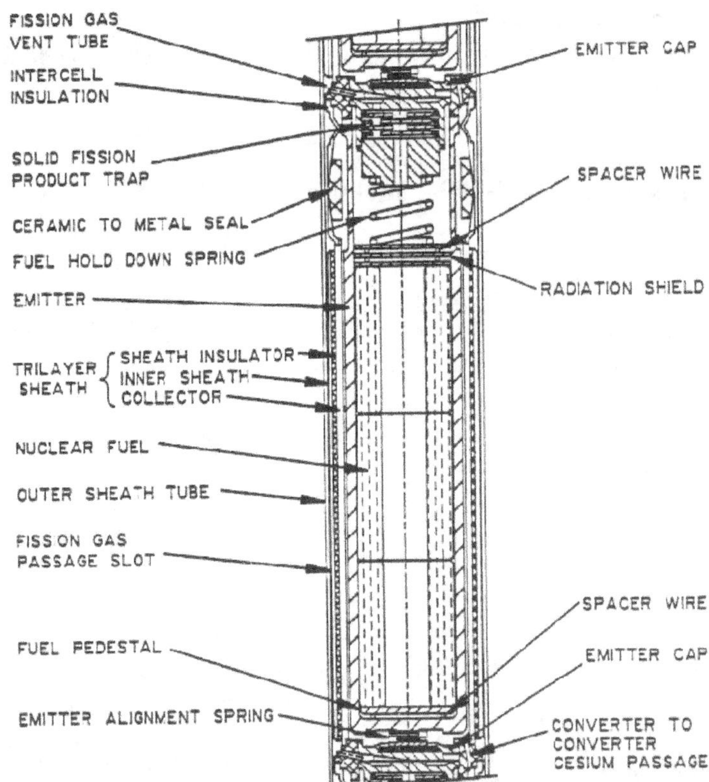

Fig. 8b.   More detailed schematic of a typical TFE, showing fuel emitter, ceramic-to-metal (insulator) seal, and other components. *From Michael G. Houts, et al., 1994.*

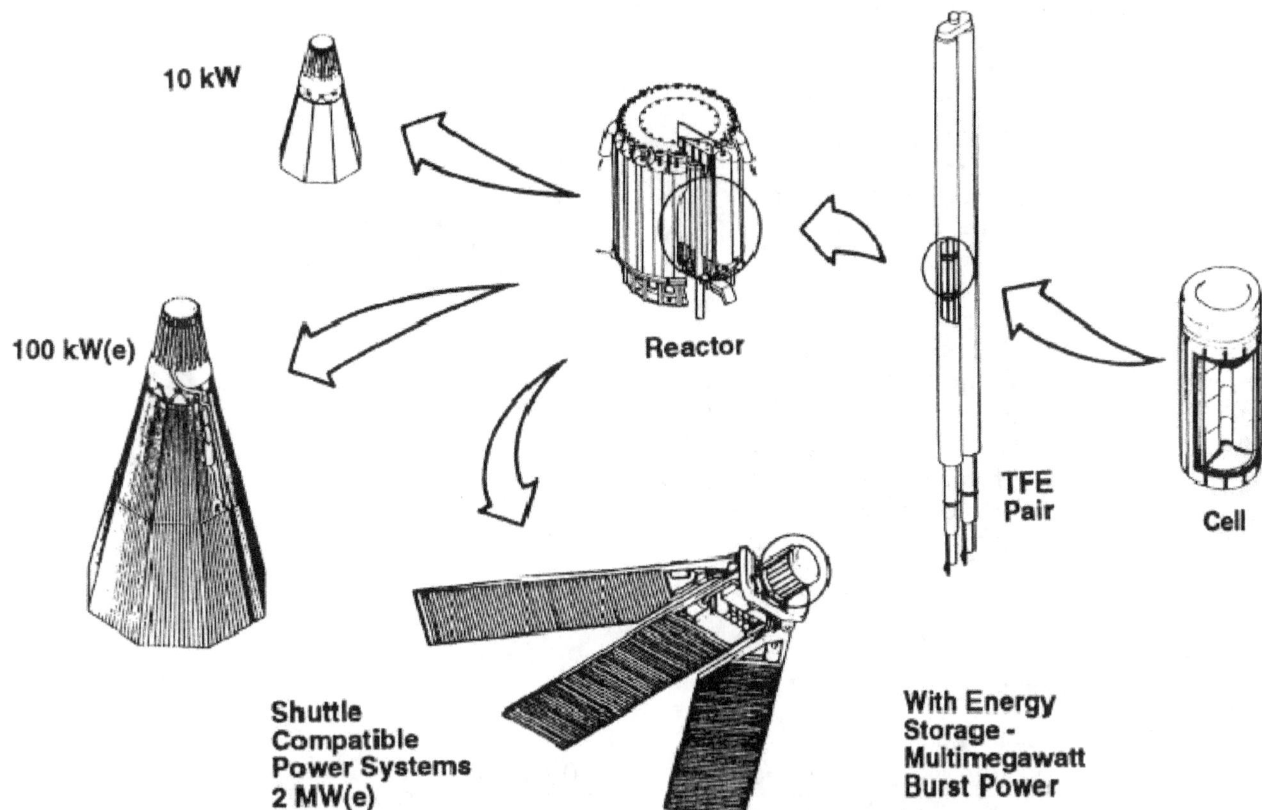

Fig. 9. Thermionic cell is the building block for space nuclear power systems to all levels. *From Samulelson, S., L. and R. C. Dahlberg, 1990.*

One of the objectives of the TFE Verification Program was to size the cell for a 2 MW_e space nuclear power system with a 7-y operating life. Another objective was to demonstrate scaleability from 500 kW_e to 5 MW_e. Table 4 summarizes systems parameters for the 2 MW_e system; Table 5 defines the TFE design parameters.[29]

Table 4. Two megawatt-electric system parameters. *From General Atomics 1988.*

Net electrical power (MWe)	2	**System Mass (kg)**	
System efficiency (percentage)	8.9	**Reactor**	8,720
Lifetime at full power (y)	7	**Shield**	2,803
Shield cone half-angle (deg)	12	**Coolant loop**	1,788
Reactor thermal power (WMt)	22.5	**Main radiator**	6,390
Coolant	Lithium	**Power conditioning and shielding**	1,230
Heat rejection (MWt)	20.3	**Power conditioning radiator**	262
Radiator temperature (K)	1,020	**Cables**	506
Radiator area (m^2)	426	**Structure and miscellaneous**	791
		Total	22,490

Table 5. TFE design definition. *General Atomics 1988.*

Performance	Values
Overall TFE:	
Output electrical power (We)	705
Efficiency (%)	8.9
Maximum voltage (v)	5.9
U-235 burnup (a/o)	4.1 average, 5.3 peak
Fluence (n/cm^2)	2.7×10^{22} average, 3.5×10^{22} peak
Converter:	
Converter power (Wt/We)	658/58.8
Thermal power/length (Wt/cm)	137.8
Emitter power flux (We/cm^2)	2.9
Diode current density (A/cm^2)	7.0
Thermionic work function (eV)	4.9
Emitter temperature (K)	1,800
Collector temperature (K)	1,070
Cesium pressure (Pa)	2.7
Converter output voltage (V)	0.49
Converter current (A)	140

Configuration	
Overall TFE:	
TFE length (active core) (cm)	100.6
TFE length (overall)	TBD
Sheath tube O.D. (cm)	1.8
Lead O.D. (cm)	2.2
Lead length (cm)	10.2
Converters per TFE	12
Converter	
Emitter O.D. x L x t (cm)	1.3 x 5.1 x 0.1
Emitter stem L x t (cm)	1.1 x 0.05
Diode gap (cm)	0.025
Trilayer thickness:	
collector(cm)	0.07
insulator (cm)	0.04
outer cylinder(cm)	0.07
Fuel specification	93% enriched UO$_2$; variable volume fraction
Intercell axial space (cm)	1.88

Technology Developments

From 1986 to 1994, component and TFE cell verification tests were performed through accelerated and real-time irradiation testing and analytical modeling.[30] These included six integrated TFE test. The major issues with developing a 7-y TFE were fuel/clad dimensional stability, seal insulator integrity, and sheath insulator integrity under temperature, irradiation, and applied voltage.

Specifically, the TFE tests were to:[31]

- Produce a TFE engineering design and specification.
- Develop required TFE assembly processes and process specifications, and to demonstrate manufacturing capability.
- Fabricate and test TFEs to demonstrate the assembly processes and the integrated performance of the components.
- Develop and verify a TFE model that can predict TFE performance and lifetime.

The first TFE test, 1H1, was to demonstrate the recovery of earlier converter technology, verify the performance of the reference cell, verify the viability of cell-fabrication processes, and demonstrate the ability to integrate a single-cell TFE into a TFE sheath tube with appropriate end fittings. The converter used a Y_2O_3 converter trilayer and an Al_2O_3 lead trilayer, and had a liquid cesium reservoir. The seal insulator was Al_2O_3 and Nb in the Litton taper configuration. Two anomalous events during 1H1 operation occurred. The first was apparently caused by an improper grounding that was corrected by a special ground attachment. The second was a loss of cesium in the interelectrode gap during the initial startup. Though corrected, the operating conditions of the cell had changed. Despite the incidents, 1H1 performed well for about 12,000 hours before its performance began to degrade seriously. It demonstrated the ability to fabricate and operate a TFE for a significant period of time.

The second TFE test, 1H2, was to verify the design and performance of an integral graphite cesium reservoir. The unit operated for 8,826 hours until a short was inadvertently created when a top-cap heater failed.

The third TFE test, 1H3, was designed to study the effects of allowing fission products to mix with the cesium in the interelectrode gap and to measure any degradation in performance. A hole was fabricated in the the emitter transition piece to allow communication between the fission gas plenum and the interrelectrode space, enabling the mixing of the fission products and the cesium. The unit operated well for about 12,000 hours. The I-V curve began to rotate counterclockwise and translate to the left and the tests ended after 13,541 hours. The failure was likely a helium leak from the test article containment resulting in a severe lowering of emitter temperature. 1H3 demonstrated 18 months of operation with a cracked emitter, failed insulator seal, or other flaw that would allow fission-product gas to enter the interelectrode space.

The fourth TFE test, 3H1, was a three-cell device operated from 3 May 1991 until June 1992. For a brief time there was a reduction in power because of improper cesium loading. Increasing the temperature of the cesium reservoir solved this problem. Then, I-V curves indicate that 3H1's cesium pressure was extremely low, causing the loss of power output. Further analysis of the cause of the failure requires post-irradiation examination.

The fifth test, 3H5, three-cell unit produced power for 8,042 hours before the TRIGA test reactor was shutdown. No observable performance degradation was noted during the testing.

The final test, 6H1, demonstrated the fabricability and performance of a six-cell TFE. It incorporated an improved cesium-reservoir temperature control and had an increased temperature control/safety margin in the cesium reservoir. The unit ran for 4,318 hours of irradiation prior to the shutdown of the TRIGA reactor.

Significant progress had been made in developing components capable of withstanding the neutron fluence (4 x 10^{22} n / cm^2, E > 0.1 MeV) and the burnup (5.3%) that would be experienced in a 2 MW_e, 7 year lifetime thermionic space nuclear reactor. TFE testing has demonstrated 18 months operation with no mechanism identified that would limit the TFE lifetime to less than 7 years. Also, demonstrated were higher burn up for fuel emitters to 3% at 1,800 K and 4% at 1,700 K--equivalent to over 5 y operation. Sheath insulators were tested under 10 - 12 V for 8 mo. (the test were stopped by a water leak with the test samples still in good condition) Insulator seals in converters were still functional after testing for 39,000 h. Graphite reservoirs were radiation tested to the equivalent of over 10 y operation.

However, a significant amount of research and development still needs to be performed on such items as: seals that reduce interrcell spacing to increase the effective fuel density and ease criticality limits; tungsten-to niobium joints that don't use tantalum to reduce parasitic neutron-absorption and ease reactor criticality limits; and on life testing.

Forty Kilowatt Thermionic Power Systems Program

In 1992, a 40 kW_e thermionic power plant program was initiated. The design life goal was 10 y; however, the initial lifetime requirement was 1.5 y. Two concepts were selected. One, called S-Prime Thermionic Nuclear

Power System, builds on the multi-cell TFE Verification Program as well as Russian technology from Topaz I (to be discussed in the next chapter). The second concept, called the SPACE-R Thermionic System, builds on the single cell Russian Topaz II technology (also covered in the next chapter). Initial mass calculations at 40 kW_e indicate both systems have a specific power of 18 W_e / kg and growth capabilities above 100 kW_e. The program did not proceed beyond the initial concept stage.

S-Prime Thermionic Nuclear Power System[32]

The S-PRIME concept is a scalable 5 to 40 kW_e using multicell Thermionic Fuel Elements in a zirconium hydride moderated reactor to achieve a specific mass of 18.2 W_e/kg The reactor is cooled with a single NaK-78 pumped loop, which rejects heat through a 24 m^2 heat pipe space radiator. Within the system, 95% of the components are based on the SNAP program space qualified components from SNAP 2, SNAP-10A, and SNAP 8. The remaining 5% of the components are based on the TFE Verification Program.

The moderated reactor converter assembly is shown in Fig. 10. The design uses 84 TFEs arranged into a 12-series by 7 parallel circuits electrical network to provide 30.7 V at 1,430 amps. The core thermal power is less than 500 kW_t with the use of zirconium hydride moderator in the form of 158 Hastelloy-N cladded pins. To limit hydrogen permeation through the barrier into the NaK-78 coolant passages, an internal surface of the cladding has a 2 - 3 mil SCB glass-enamel coating. Reactor reactivity control is provided by 18 poison back control drums located in the radial reflector assembly. Nine B_4C safety rods are used to assure water subcriticality and provide a secondary shutdown mechanism for the reactor. The reactor pressure vessel and its internal support is constructed from 316 stainless steel. Reactor design parameters are summarized in Table 6 and system level performance in Table 7.[33]

The status of key TFE components is given in Table 8. Though the S-PRIME system used conventional materials in a known operating environment and the TFE technology database, critical component testing listed in Table 9 was still needed to validate the design.

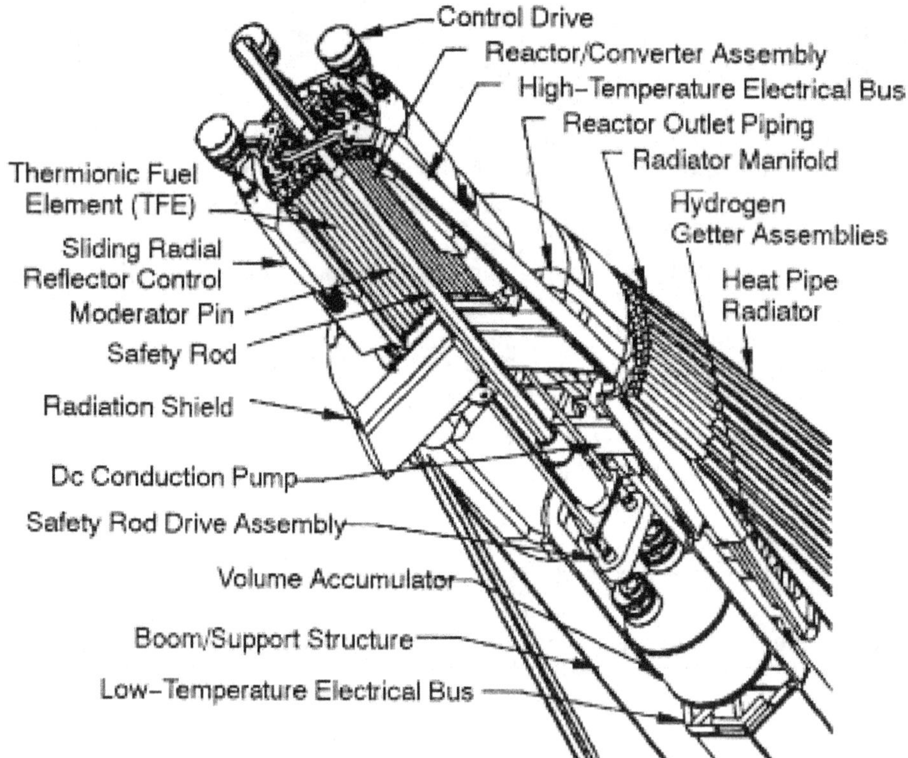

Fig. 10. S-PRIME reactor converter and shield assemblies. *From Joseph C. Mills, William R. Detterman, and Tom H. Van Hagan, 1994.*

Table 6. S-PRIME reactor design summary.

Description	Parameters
Fuel pellet OD (mm)	15.7
Emitter OD (mm)	18.0
Collector thickness (mm)	1.14
TFE OD (mm)	23.64
Core lattice pitch (mm)	24.02
TFE pitch (mm)	41.6
ZrH1.8 OD (mm)	22.26
Mod clad thickness (mm)	.05
Core height (mm)	614
Core OD (mm)	415
Reactor height (mm)	1115
Reactor OD (mm)	590
Reactor coolant	NaK-78
Coolant temperature range (K)	870-960
Fuel (enrichment)	UO2 (93%)
Moderator	Be/ZrH1.8
Radial reflector	Be
Axial reflector	BeO
Vessel	Stainless
Cladding	Stainless/Hastelloy

Table 7. S-PRIME system level performance.

Description	Parameters
EOM net user (kWe)	40
BOM gross power (kWe)	48.0
EOM reactor (kWt)	570
Reactor outlet temperature (k)	960
Loop temperature drop (K)	90
One-sided radiator area (m^2)	27.0
Manifold maximum dia (m)	1.11
Manifold height (m)	0.35
Maximum radiator dia (m)	2.63
Overall length (m)	6.78
Radiator half-angle (°)	10
Radiator slant height (m) (heat pipe condenser length)	4.42
Net system efficiency (%)	7.0
Specific power (We/kg)	15.0
<u>Subsystem/Assembly</u>	<u>Mass</u>
Reactor converter	770
Control and safety drives	50
Shield	587
Heat rejection	483
Power conditioning and control	563
Structure and boom	170
Full reentry shield	42
Total system (kg)	2665

Table 8. Status of key components.[34]

Key Components	Tested To	Life in 40 kWe Design (y)
Sheath insulators	4×10^{20} n/cm^2 (in-core with voltage)	0.75
Insulator seals	2.3×10^{21} n/cm^2	4.3
Graphite reservoir	3×10^{22} n/cm^2	>>10
Fueled emitters	>3% at 1800 K	>5
TFEs	13,500 h	1 to 1.5

Table 9. Critical component demonstrations required for S-PRIME.

Demonstration	Objectives	TI–SNPS Impacted Requirement
1. TFE Refractory Metal to Stainless Transitio Joint	Demonstrate a stable, long–lived transition zone between the riobium collector and the stainless reactor vessel.	Lifetime
2. TFE Intercell Transition Joint Fabrication (without tantalum)	Demonstrate fabricability of joint and thermal stability.	Mass performance power density
3. TFE Cesium/Graphite Reservoir/Cell Performance	Demonstrate chemical stability of Cesium–graphite reservoir, thermal coupling to reservoir can, and verify transient performance.	Startup and shut-down survivability
4. TFE Testability Demonstration	Demonstrate the method for functional test of individual cells, a complete multicell TFE, and a system of TFEs.	Testability
5. Hydrogen Getter – NaK Compatibility and Hydrogen Getter Performance	Demonstrate compatibility of yttrium and yttrium hydride with NaK at 1000°F. Demonstrate capability of yttrium getter bed to maintain hydrogen partial pressure in NaK below .01 Torr.	Lifetime and mass performance
6. Radiator Heat Pipe Performance	Demonstrate heat pipe start–up, shutdown, and long–term operation at reduced power.	Lifetime, survivability and mass performance

Space Power Advanced Core-Length Element-Reactor (SPACE-R) Power System[35]

The SPACE-R reactor power system is shown in Fig. 11. This is an incore thermionic reactor power system with UO$_2$ fuel, pumped liquid metal (NaK) cooling and a moderated core. The baseline SPACE-R design generates 40 kWe (EOL) with 10 year life. The radiator configuration and the shield core angle are chosen to match a 10 m separation between the reactor core and the payload. The system allows a full power operational test of the entire assembly by electric heating before fuel is loaded. The design is based on a single cell type similar to those proven in the Topaz II flight qualified reactors. The 46 cm diameter by 35 cm long reactor core is composed of 150 Thermionic Fuel Elements, Be metal clad yttrium hydride moderator elements, sodium-potassium eutectic alloy coolant, and appropriate stainless steel and Hastelloy containment. The core layout is shown in Fig. 12.

Single cell TFE's are similar to the Topaz II flight reactors (see Fig. 13). The claim is that these are simpler to fabricate than multicells, have lower cost, have a lower potential for fission gas contamination of the electrode space, higher reliability, provide higher fuel volume fraction and a smaller critical core size. Also, they allow for total system full power thermal and electrical checkout before introducing nuclear fuel. The reflector is of 10 cm thick beryllium metal that operates around 800 K. In it are 12 rotating control drums that experience a maximum fast fluence of 3×10^{20} nvt. The moderator is a zirconium-hydride to provide a positive temperature coefficient. Only one scram drum is needed to shutdown the reactor. Key characteristics are given in Table 10.

Key lifetime issue are related to nuclear ceramic fuel swelling, fission product leakage into the electrode spaces, hydrogen containment, beryllium swelling, coolant corrosion of containment or mass transport, and meteorite or debris damage to the coolant system or radiator. The low temperature operation of 900 K in the primary loop using NaK coolant and 815 K in the radiator using potassium coolant should result in minimum mass transport loops and long life concerns. Beryllium at 800 - 900 K will maintain dimensional stability over the 10-year life time. Since only one pair of interelectode seals is required and these are located in the cooler and lower fluence ends of the TFE. This helps prevent fission product leakage into the electrode space.

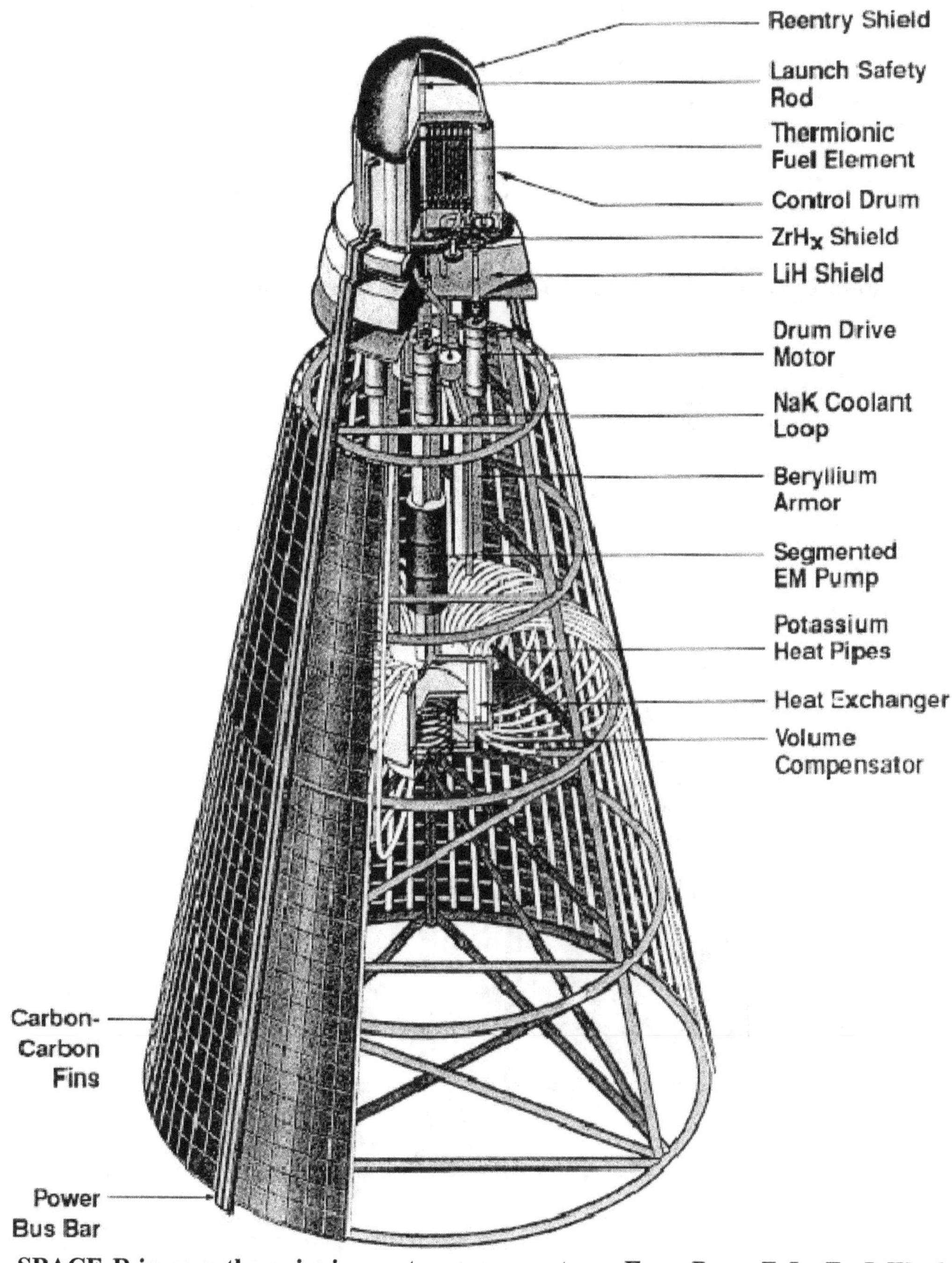

Fig. 11. SPACE-R in-core thermionic reactor power system. *From Begg, T. L., T,. J. Wuchte, and W. D. Otting, 1992.*

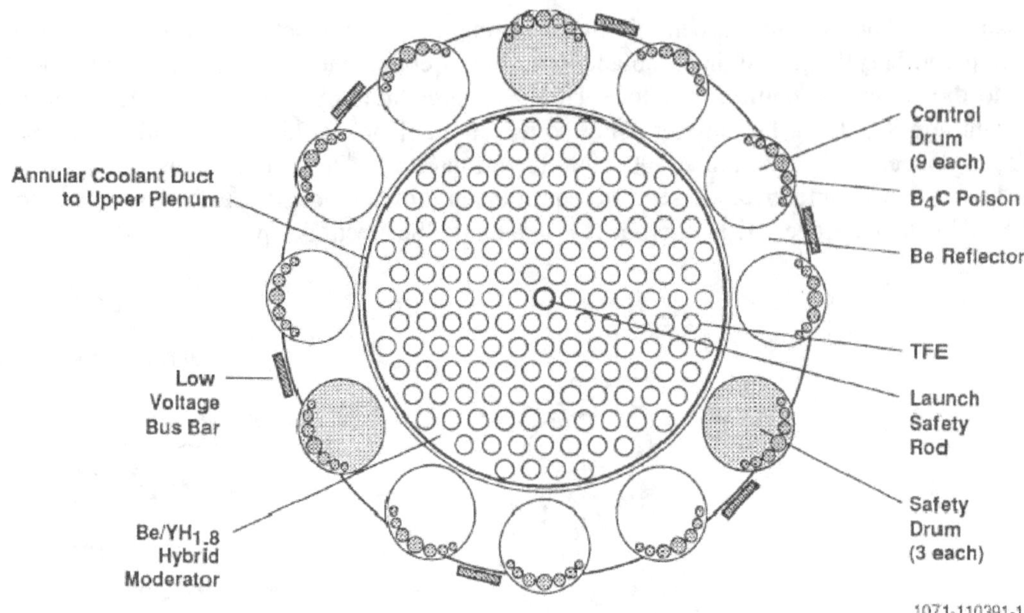

Fig. 12. Moderated core cross-section. *From Begg, T. L., T,. J. Wuchte, and W. D. Otting, 1992.*

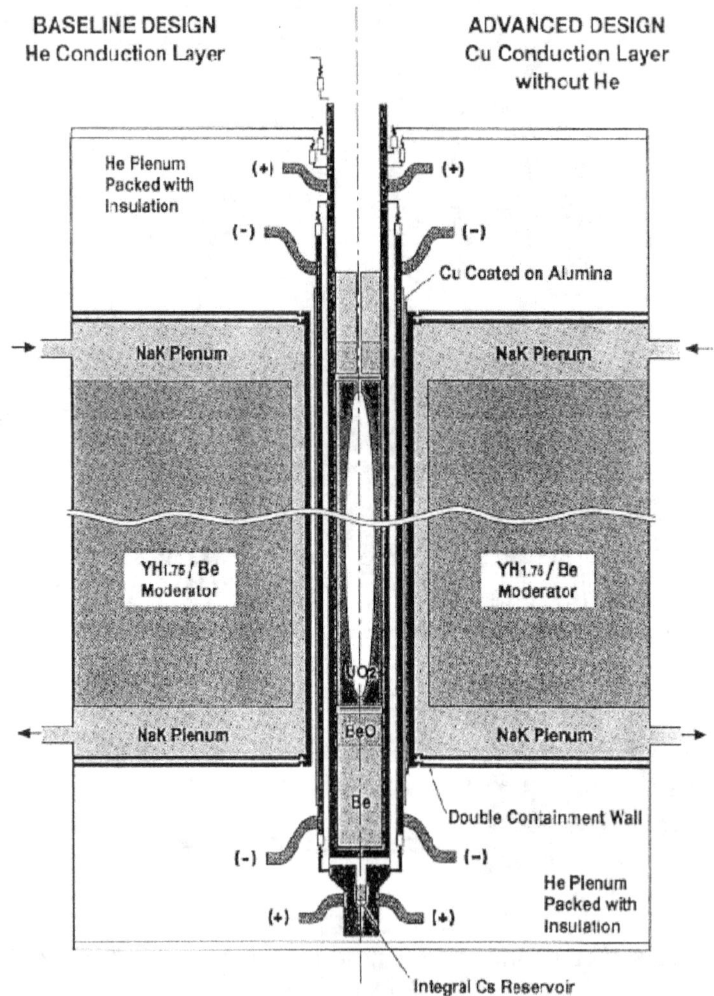

Fig. 13. SPACE-R TFE and core structure base line and advanced designs. *From Begg, T. L., T,. J. Wuchte, and W. D. Otting, 1992.*

Table 10. SPACE-R 40 kW$_e$ baseline system characteristics.

Reactor Thermal Power	611 kWt
Power Output (BOL)	44 kWe
System Efficiency	7.2%
System Degradation Allowed	10% in 10 yrs
Core Diameter/Length	46/35 cm
Power Conversion	Single Cell TFE's in Moderated Core
Number of TFE's	150
Number of Drivers	0
Avg Emitter/Collector Temp.	1823/900 K
Axial Thermal Power Ratio	0.7 (end/midplane)
Radial Thermal Power Ratio	0.7 (min/max)
Reactor Output V/A	24V/1900 A DC
Fuel	UO_2
U^{235} Loading	81 kg (90% Enriched)
Burnup	3.5% in 10 yrs
Smear Density	0.7
Moderator	Hybrid Be/$YH_{1.75}$
Reflector	Be
Axial	8 cm
Radial	10 cm
Number of Control Drums	9 (10.4 cm Dia. Be + B_4C)
Number of Safety Drums	3 (10.4 cm Dia. Be + B_4C)
Number of Incore Safety Rods	1 (B_4C with Be/YHx Follower)
Heat Rejection	
Primary Loop	EM Pumped NaK (6.45 kg/s)
Secondary Loop (Radiator)	K Heat Pipes
Avg. Radiator Temperature	815 K
Radiator Area	28 M^2
Shield	LiH + ZrHx
Reactor Mass	621 kg
Shield Mass:	
5M*	1270 kg
10 M	512 kg
System Mass:	
5 M	2781 kg
10 M	2201 kg

* Payload separation Distance

Space Thermionic Advanced Reactor-Compact (STAR·C)
Another thermionic concept investigated used an external location for the thermionic converters. Heat was transported from the reactor to the converters. In STAR-C,[36, 37] illustrated in Fig. 14, the solid core is composed of annular plates of UC_2 fuel supported in graphite trays. Heat is radiated from the radial core surface to a surrounding array of thermionic converters where electric output is generated. Planar thermionic converters operate in the ignited, high-pressure cesium mode with an inter-electrode gap of 0.1 mm. The emitter is tungsten, collector niobium, and insulator-seal alumina tri-layer. Reject heat is taken from the thermionic converter collectors through the radial core reflector by heat pipes to an extended surface immediately on the outside of the reactor, from which heat is radiated directly to space. Key parameters are given in Table 11. Design studies are the extent of the funded activities on this concept. STAR-C has the advantages of relatively simple thermionic

converters located outside the core, a simple fuel design, no pumped loops, and being a compact power plant overall. The major concerns relate to the development of high temperature, long-life heat pipes, quantity and containment of fuel swelling, sublimation of graphite, venting of fission gases without loss of fuel, and limited power system growth.

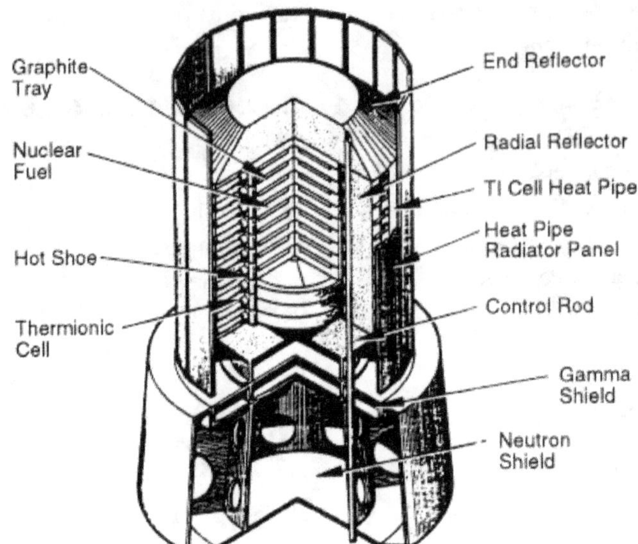

Fig. 14. STAR-C configuration. *Allen et al. 1991.*

Table 11. STAR-C key performance features.

Thermal Power (kWt)	340
Reactor Output (kWe)	42.8
Net Electrical Power (kWe)	40.9
Net System Efficiency (%)	12.0
Peak Fuel Temperature (K)	2,150
Core Surface Temperature (K)	2,000
Emitter Temperature (K)	1,854
Collector Temperature (K)	1,031
Main Radiator Area (m^2)	5.9
Mass (kg)	2,502

Summary

Significant progress has been achieved in the development of thermionic space reactors. For in-core thermionic systems, the good design feature is that all of the components outside of the thermionic fuel element are at temperatures that are straight forward for design purposes and experience. Temperature levels are 800 K or less. Difficulties with the TFE design result from very high temperature operation with emitter temperatures of around 1,800 K, the integral nature of the fuel and thermionic converter resulting in radiological and fission product interactions design problems, very tight tolerances needed in the converter designs, and the limitations on performing accelerated testing because materials are operating near their limits. The latter limits the rate that in-core thermionic fuel elements can be developed and demonstrated for long life systems. Out-of-core thermionic systems separate the radiological and fission products problem from the thermionic element development. Now, however. other components in the reactor become subjected to high temperatures and very high temperature heat transport is needed.

Chapter 6

Russian Space Nuclear Power Systems

Overview

The former USSR made extensive use of space reactors as a power source for radar ocean reconnaissance satellites. Fig. 1 summarizes the frequencies of USSR launches and Fig. 2 the duration of operation.[1] These satellites are more commonly known as RORSATs (Radar Ocean Reconnaissance Satellite). The reactor, entitled Romashka, (several designations are found in the literature) used thermoelectric power conversion as the primary electrical source for missions lasting up to 135 days in operational orbits near 250 km.[2] At completion of the mission, the nuclear power system was designed to be boosted to a long-lived storage orbit in the vicinity of 900 - 1,000 km altitude using solid-rocket propulsion. In 31 missions (see Table 1) [3] from 1970 - 1988, this disposal procedure failed twice, resulting in uncontrolled re-entries of the nuclear power system. The next-to-last mission experienced a partial malfunction with a disposal orbit altitude of about 700 km.

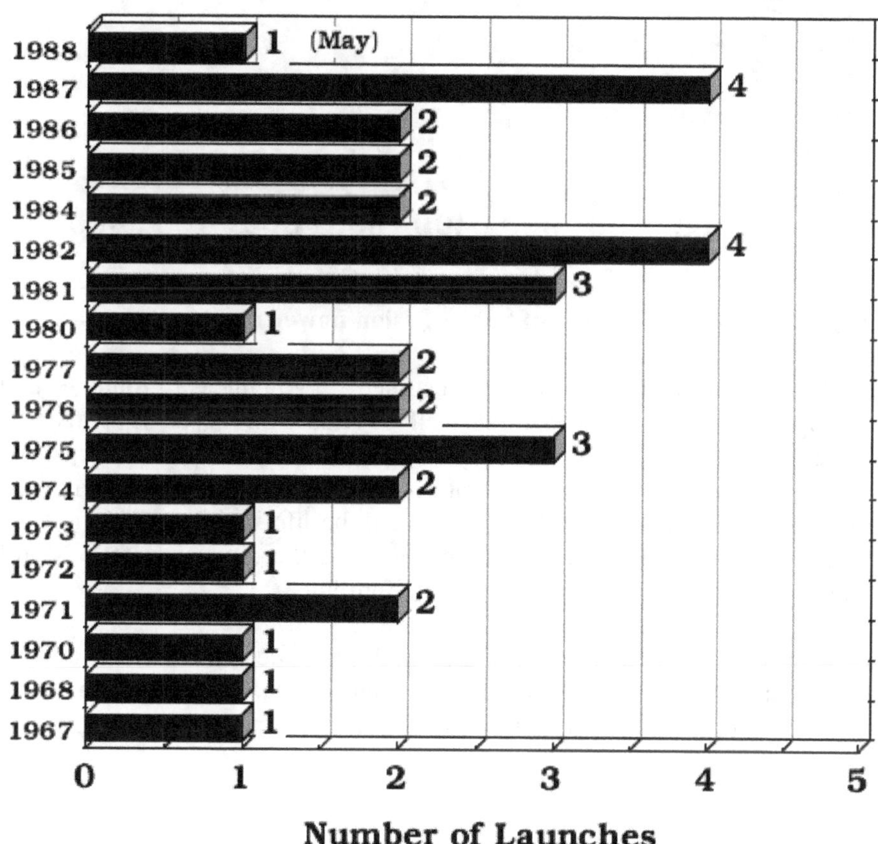

Fig. 1. Frequency of Soviet fission reactor launches.

In 1987 - 1988, the U.S,S,R, tested a different type of reactor power system using thermionic power conversion. Two space tests were performed, with one operating six months (Cosmos 1818) and the other operating 346 days (Cosmos 1867). These power plants are designated in the U.S. as Topaz I. Topaz I design output is 5 - 10 kW_e. The flight-tested units used a multi-cell thermionic fuel element with an output power of approximately 5 kW_e, one with a molybdenum emitter and the other with a tungsten emitter. The power system with the tungsten emitter operated for the longer period of time; degradation of performance occurred, with the thermal power increased to compensate for this degradation. Topaz I demonstrated 1 y operation in space. Experience with the multicell TFE used in Topaz I indicates that swelling and intercell leakage are significant life-limiting problems.

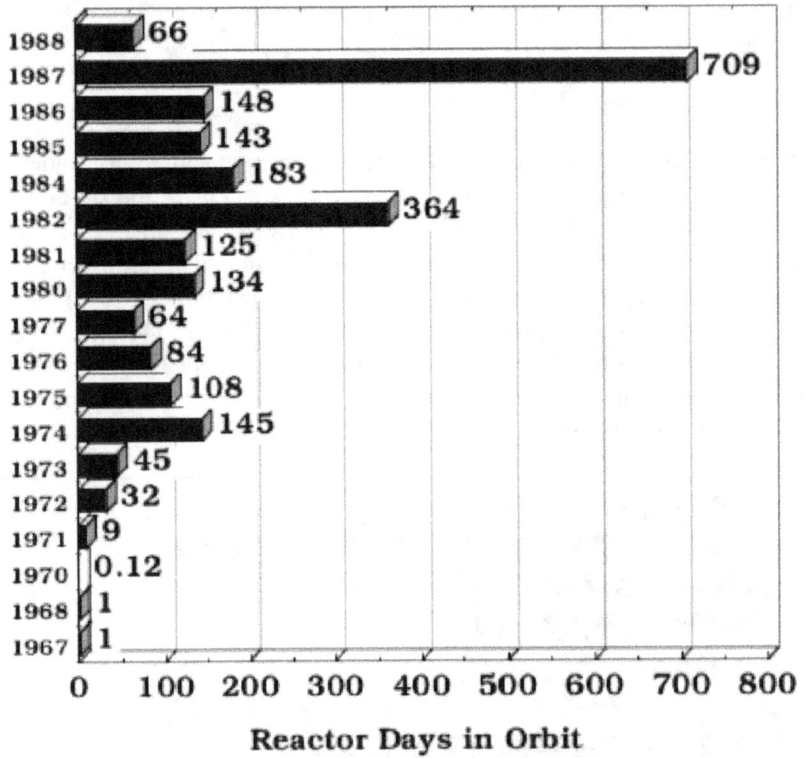

Fig. 2 Operation duration of Soviet fission power reactors in space.

Topaz II had a single TFE cell configuration. It has demonstrated 1.5 y in nuclear ground testing. A claim of 3 y lifetime is based on component life data. The single cell TFE used in Topaz II tends to correct problems experienced in the Topaz I design, partially by using a high void fraction. A primary life limiting element appears to be the loss of hydrogen from the ZrH moderator. The rate is about one percent per year. Also, with only 65 cents of excess reactivity at design the fuel burnup can be life limiting. This is especially true if the reactor cools down before a restart is achieved.[4] Another issue is the oxygen getter. Modifications may be needed in the cesium supply, but these do not appear to be life limiting.

In 1990, the U.S. and Russia started a joint program to design, develop, demonstrate, and advance the technology of thermionics. As part of this endeavor, the Russians delivered a Topaz II unfueled reactor to the U.S. and a joint U.S. and Russian test program was performed.[5] Planning at one time included a flight program; however, this never came about

Table 1. Soviet orbital reactor program history.

Number	Name	Launch Date	Termination Date	Lifetime
1	Cosmos 198	27 Dec 67	28 Dec 67	1 da
2	Cosmos 209	22 Mar 68	23 Mar 68	1 da
3	Cosmos 367	3 Oct 70	3 Oct 70	<3 h
4	Cosmos 402	1 Apr 71	1 Apr 71	<3 h
5	Cosmos 469	25 Dec 71	3 Jan 72	9 da
6	Cosmos 516	21 Aug 72	22 Sep 72	32 da
7	Cosmos 626	27 Dec 73	9 Feb 74	45 da
8	Cosmos 651	15 May 74	25 Jul 74	71 da
9	Cosmos 654	17 May 74	30 Jul 74	74 da
10	Cosmos 723	2 Apr 75	15 May 75	43 da
11	Cosmos 724	7 Apr 75	11 Jun 75	65 da
12	Cosmos 785	12 Dec 75	12 Dec 75	<3 h
13	Cosmos 860	17 Oct 76	10 Nov 76	24 da
14	Cosmos 861	21 Oct 76	20 Dec 76	60 da
15	Cosmos 952	16 Sep 77	7 Oct 77	21 da
16	Cosmos 954	18 Sep 77	~31 Oct 77	~43 da
17	Cosmos 1176	29 Apr 80	10 Sep 80	134 da
18	Cosmos 1249	5 Mar 81	18 Jun 81	105 da
19	Cosmos 1266	21 Apr 81	28 Apr 81	8 da
20	Cosmos 1299	24 Aug 81	5 Sep 81	12 da
21	Cosmos 1365	14 May 82	26 Sep 82	135 da
22	Cosmos 1372	1 Jun 82	10 Aug 82	70 da
23	Cosmos 1402	30 Aug 82	28 Dec 82	120 da
24	Cosmos 1412	2 Oct 82	10 Nov 82	39 da
25	Cosmos 1579	29 Jun 84	26 Sep 84	90 da
26	Cosmos 1607	31 Oct 84	1 Feb 85	93 da
27	Cosmos 1670	1 Aug 85	22 Oct 85	83 da
28	Cosmos 1677	23 Aug 85	23 Oct 85	60 da
29	Cosmos 1736	21 Mar 86	21 Jun 86	92 da
30	Cosmos 1771	20 Aug 86	15 Oct 86	56 da
31	Cosmos 1818	1 Feb 87	~ Jul 87	~6 mo
32	Cosmos 1860	18 Jun 87	28 Jul 87	40 da
33	Cosmos 1867	10 Jul 87	~ Jul 88	~1 yr
34	Cosmos 1900	12 Dec 87	~14 Apr 87	~124 da
35	Cosmos 1932	14 Mar 88	19 May 88	66 da

Romashka Thermoelectric Power Reactors

Design Features

Fig. 3 is a sketch of the RORSAT satellite configuration that incorporated the Romashka reactor. Fig. 4 is an engineer's sketch of the power plant configuration.[6] (These are based on information from the COSMOS 954 reentry over Canada to be discussed shortly). The general RORSAT power system characteristics are summarized in Table 2.[7]

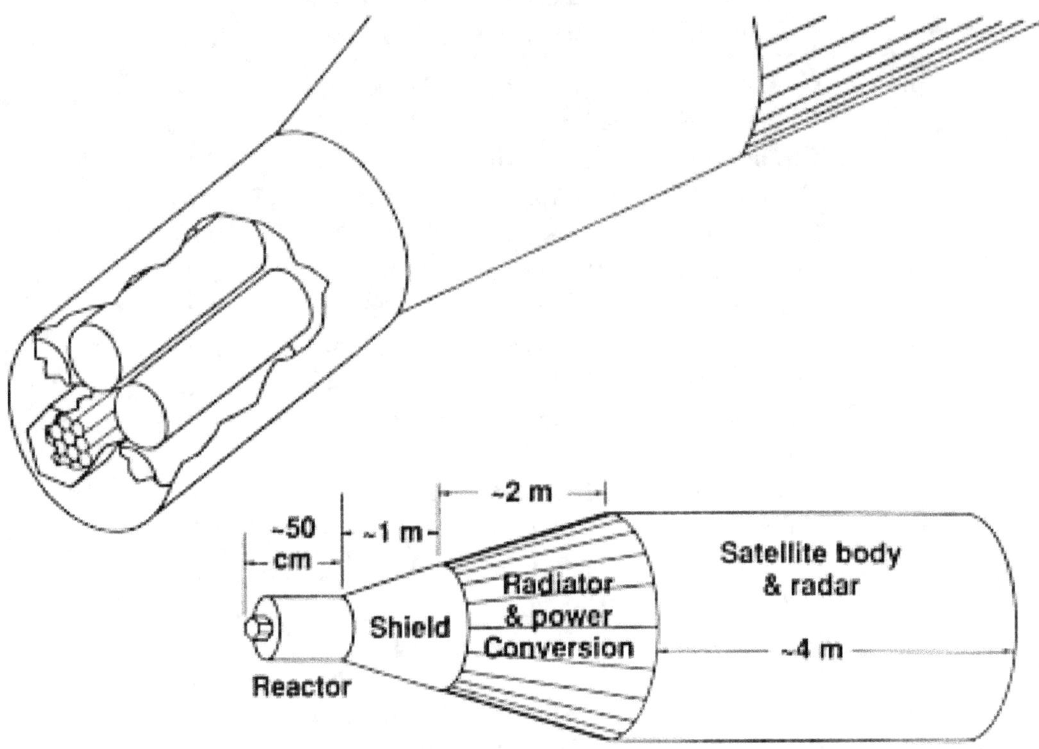

Fig. 3. Artist concept of COSMOS 954 satellite showing the reactor. *From Gary L. Bennett, 1992.*

The Romashka (BES-5) reactor had a fast neutron spectrum.[8] It contained 37 steel clad rods about 0.6 m long with uranium-molybdenum alloy as fuel. The fuel core of the reactor was 0.2 m in diameter, 0.6 m long and weighed, as an assembly, 53 kg. The 30 kg of uranium was more than 90% enriched ^{235}U. Thermal power was on the order of 100 KW_t and the power from the thermoelectric converters was approximately 3 KW_e. The power conversion subsystem used silicon-germanium semiconductor material with an upper working temperature of 1273 K.[9] The reactor weighed 385 kg, including the radiation shielding. Shielding consisted of LiH and stainless steel supplemented by tungsten and uranium alloys.

The reactor was controlled by six cylindrical control elements made of beryllium that could be reciprocated along the reactor axis. The control cylinders are filled with BC_2 to act as a reactor poison against leakage of radioactivity through the radiation shield. Fig. 5 shows a cutaway view of the reactor and Fig. 6 a view of the Romashka fuel disk.

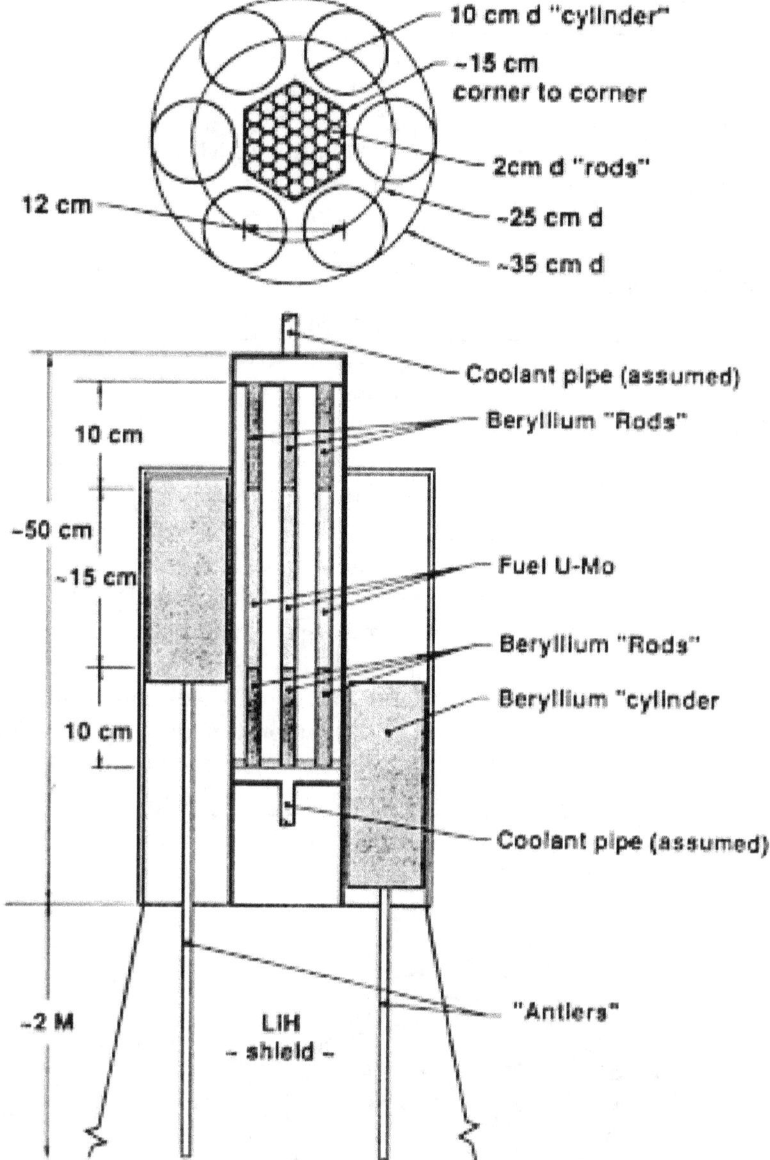

Fig. 4. Engineer's sketch of the COSMOS 954 reactor. *From Gary L. Bennett, 1992.*

Flight Safety Considerations For RORSAT Satellites

The Soviet Union (U.S.S.R.) routinely used to fly spacecraft in low Earth orbits that included a nuclear reactor as a power source. When their low orbit mission was completed, the Soviets used a boost system to increase the orbital altitude of the reactor. By increasing the orbital altitude and, therefore, the orbital lifetime of a shutdown reactor, sufficient time is permitted before eventual reentry for the accumulated radioactivity in the core to decrease to acceptable levels. In the event that the nuclear power system falls from its operational (working) orbit after a failure of the boost system, the Soviet aerospace nuclear safety philosophy includes a core dispersal mode. Soviet space reactors are reported to be equipped with a backup safety system that will disperse the reactor core in such a way so that the radiation dose to people living in any potentially contaminated area will not exceed 0.5 rem (5 mSv) during the first year after reentry and surface contamination.

Table 2. Romashka power system.

Thermal Power (kWt)	≤100
Conversion System	Thermoelectric
Electrical Power Output (kWe)	≤5 (-1.3 to 2)
Fuel Material	U-Mo (≥3wt% Mo)
Uranium-235 Enrichment (%)	90
Uranium-235 Mass (kg)	≤ 31 (-20 to 25)
Burnup (fissions/g of U)	≤2 x 10^{18}
Specific Fuel Thermal Power (W/g of U)	-5
Core Arrangement	37 cylindrical elements (probably 20 mm dia)
Cladding	Possibly Nb or stainless steel
Coolant	NaK
Coolant Temperature Outlet (K)	≥970
Core Structural Material	Steel
Reflector Material	Be (6 cylindrical rods)
Reflector Thickness (m)	0.1
Neutron Spectrum	Fast (- 1 MeV)
Shield	LiH (W and depleted U)
Core Diameter (m)	≤0.24
Core Length (m)	≤0.64
Control Elements	6 in/out control rods composted of B$_4$C with LiH inserts so prevent neutron streaming and Be followers to serve as the radial reflector
Overall Reactor Mass (kg)	<390

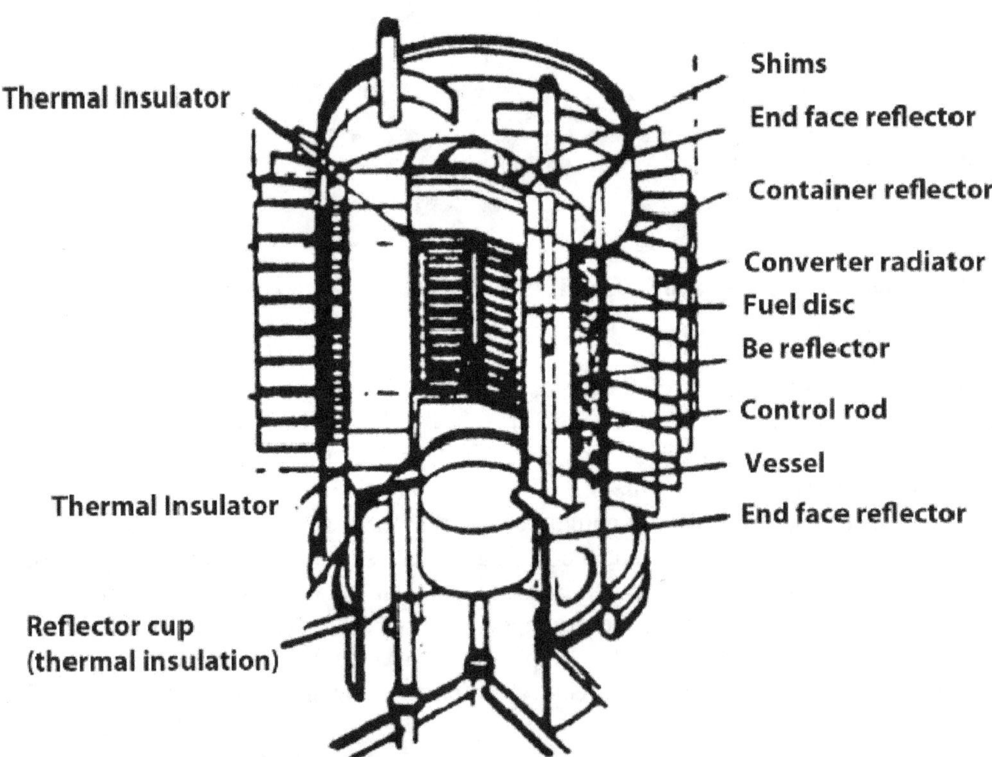

Fig. 5. Cutaway view of an early ground-based Romashka reactor showing 11 fuel disks. *From G. Bennett, 1989.*

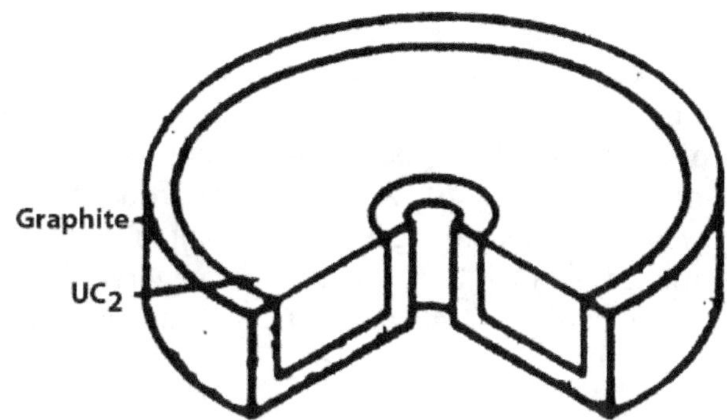

Fig. 6. Cutaway view of a Romashka fuel disk. *From G. Bennett, 1989.*

Soviet aerospace nuclear safety design was put to the test in a major accident in 1978--an international incident that focused world attention on the use of space nuclear power sources.[10, 11, 12] On 24 January 1978, Cosmos 954 entered the Earth's atmosphere over Canada's Northwest Territories. A massive airborne and ground search and recovery program was undertaken under the direction of the Canadian Atomic Energy Control Board. This effort was named Operation Morning Light. A 124,000 square kilometer area of the Northwest Territories as well as selected portions of northern Alberta and Saskatchewan were identified for search. Debris was found in a 600 km tract stretching southward from Great Slave Lake. The Soviets stated that the reactor was designed to burn up on atmospheric reentry. Table 3 summarizes the Cosmos 954 debris that was recovered. Although no large fuel particles were found, several large fragments with high radioactivity levels were discovered. For example, the four steel plate fragments appeared to be part of the periphery of the reactor core and may have been cylindrical in shape. Radioactivity levels in the beryllium pieces, believed to be part of the reflector, were mainly attributed to the activation products associated with impurities in the material. Some recovered fission products were believed to be the remnants of a cladding material.

The largest fragment of Cosmos 954, actually two pieces, was a complex structure nicknamed the "antlers" (see Fig. 7). This structure consisted of six legs or tubes, three still attached to a circular base plate and three others that apparently broke loose on impact. The complex structure of the entire assembly appeared consistent with that of a reactor control function. The boron-rich link on the movable rods suggested that these were reactor control or shutoff rods and a large amount of lithium hydride would provide a neutron shield for the payload.

Table 3. Summary of COSMOS 954 debris recovered in Canada.

Material	Size, Weight	Radiation Level When Found
Large fragments		
4 steel plate fragments	Largest 225 mm × 75 mm × 6.5 mm weighing 272 g	To 200 R/h on contact
41 beryllium rods (some with niobium sheaths)	100 mm × 20 mm diam. About 51 g each	600 mR/h to 100 R/h on contact; 30 to 150 mR/h at 1 meter
6 beryllium cylinders	250 mm × 100 mm diam. 3.6 kg each	5 to 15 R/h on contact 40 to 800 mR/h at 1 meter
Tubes, rods, plate (the "antlers")	6 rods about 1 meter long attached to plate. 20 kg.	To 15 R/h on contact Variable over the debris
Steel tube (the "stovepipe")	About 500 mm long, 360 mm diam. 2.5 mm wall. 18.2 kg	Non-radioactive
Other chunks, flakes, slivers	Variable	Most 10 to 30 R/h on contact one to 500 R/h
Small particles		
About 4000 recovered	a) spherical. 1 mm to less than 0.1 mm diam. Mass 5 mg to less than 100 μg	Most in the range 10 to 100 mR/h at 1 meter
	b) irregular, or flaky. Some several mm on a side and very thin and friable	

The small particles found appeared to represent the reactor fuel. These were scattered over 100,000 square kilometers. Their mean particle diameter as a function of distance is shown in Fig. 8. Near the reentry trajectory, these particles were about 1 mm in diameter, while those found 200 km to the south were typically 0.2 mm. There appeared to be two types of small particles. The majority were fuel particles that had a crystalline outer layer, which was analyzed as being uranium dioxide. A metallic core beneath this surface was apparently made up of a variable uranium/molybdenum alloy containing 5 to 15 percent or more molybdenum. The uranium was found to be nearly 90 percent enriched uranium-235. A second category of particles were low density, often flakey, thin, curved, or twisted. These particles contained some uranium alloyed with stainless steel or in some cases, niobium and tantalum.

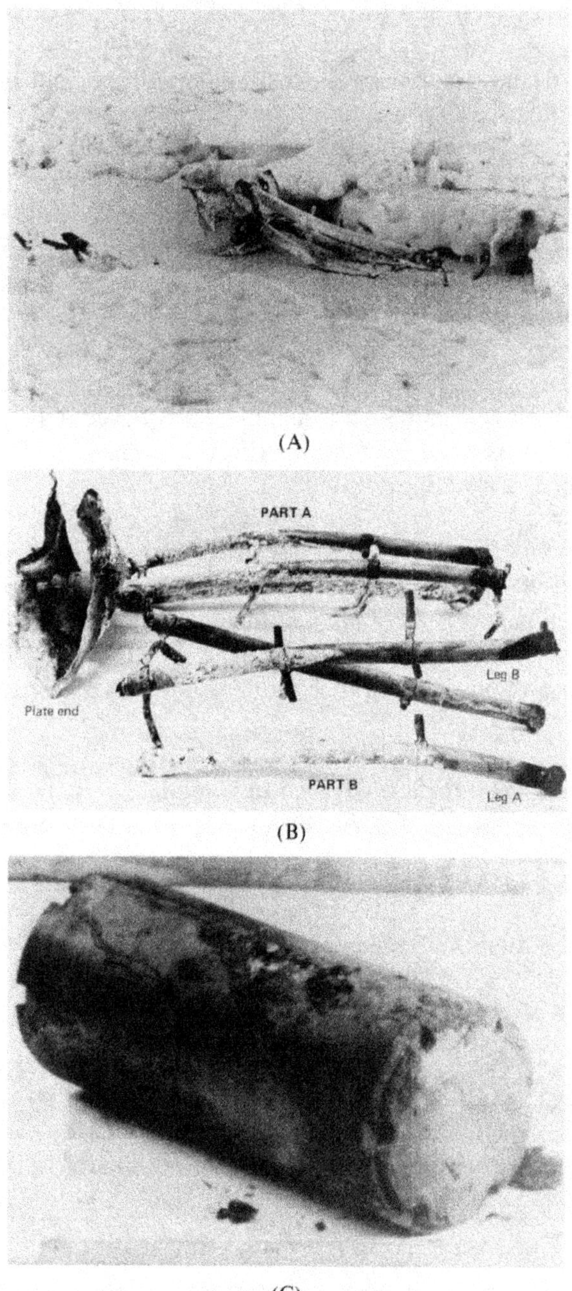

(A)

(B)

(C)

Fig. 7. COSMOS 954 debris. (A) Shows the "antlers" as discovered in the ice and snow on the Thelon River, Canada (29 Jan. 1978); (B) shows the "antlers" as photographs at Canada's Whitehall Nuclear Research Establishment (the white coating is hydrated lithium hydroxide; (C) is a cross section of one of the legs of the "antlers." *From W. R. Gummer, et al., 1980.*

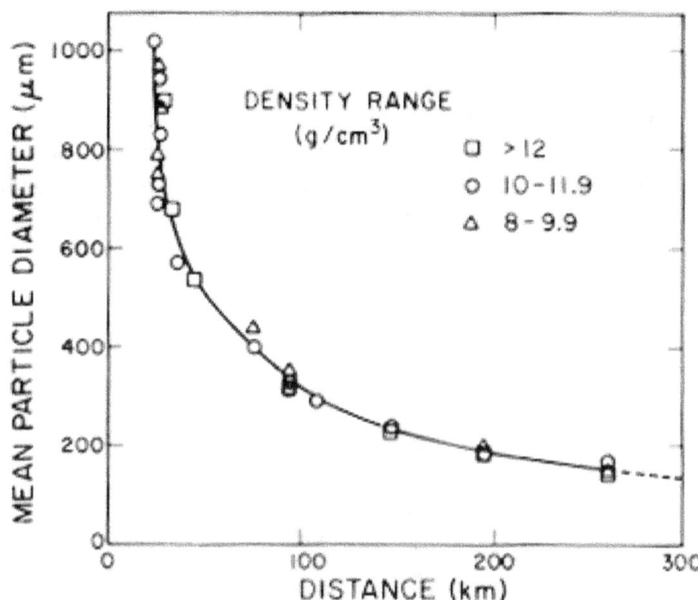

Fig. 8. Mean particle diameter of debris from COSMOS 954 versus distance from reentry trajectory. *From W. R. Gummer, et al., 1980.*

Except for those small particles that drifted southward, it appears that most debris fell on Great Slave Lake or on unpopulated and little frequented areas to the northeast. There is reasonable assurance that all accessible, significantly sized radioactive fragments from Cosmos 954 were located and removed in the cleanup operations. The radioactivity inventory indicated that about 20 percent (or some 4 kg) of fuel had come back to Earth as of 1 April 1978. Various estimates made on the actual particles recovered in specific sections of the search area indicate that distribution of these particles may be as high as several hundred per square kilometer. The average was one particle every 4,000 square meters. The effects of the debris identified or observed on any part of the natural environment was considered insignificant.

In a second incident, Cosmos 1402 was launched by the Soviet Union on 30 August in 1982 into an orbit with an apogee of 279 km, perigee 354 km and inclination of 89.6 degrees.[13] When it failed to boost to a higher orbit for disposal, the safety system split the satellite into three sections. Before the satellite was split into fragments, the reactor was shutdown by a command from Earth. One segment burned up upon reentry on 30 December 1982.[14] The main structure entered the dense layers of the terrestrial atmosphere over the central area of the Indian Ocean at 1:10 a.m. (local Moscow time) on 24 January 1983.[15] The reactor core appeared to have burned up over the southern part of the Atlantic Ocean on 7 February 1983 at 1:56 p.m. (Moscow time).[16] Because burnup occurred over ocean areas, no fragments of Cosmos 1402 were located. Air sampling indicated that the reactor did indeed burnup in the upper atmosphere.[17]

Another, almost incident occurred with Cosmos 1900.[18,19] It was launched into orbit on 12 December 1987. Radio contact to the satellite was disrupted on April 9, 1988 and the satellite failed to boost on command on 13 April 1988. The conditions that existed at the time were: (1) failure to correct the operating orbit on command from Earth; (2) normal operation of the orientation system and stabilization according to the readings of the flight trajectory parameters; (3) the reactor seem normal, but with no electrical power output; and (4) a sufficient supply of fuel for engine stabilization in conjunction with a lack of orbital correction. The satellite was designed to separate the reactor part of the spacecraft and boost it to a higher orbit if any anomalies were detected. If there was a failure to boost the reactor, a second safety system would operate to eject the core. The reported probability of failure to boost or separate and disperse was 10^{-4}. On the night of 30 Sept. - 1 Oct. 1988 Moscow time during the first part of rotation number 4,727 from an altitude of approximately 180 kilometers, the higher altitude dispersion system caused the separation of the reactor from the remainder of the satellite and boosted it to an orbit of 773 km at the apogee and 725 km at the perigee. The rotation period is 99.4 minutes and the life span is approximately 200 years.

Topaz I Reactor Power Plants

Design Features

Topaz I (pictured in Fig. 9) power systems were flown on Cosmos 1818 and Cosmos 1867. Topaz I was a direct conversion nuclear power source, using a multi-cell thermionic power conversion devices located within the reactor. These systems were approximately 6 kW$_e$, thermal power between 110 and 150 kW$_t$, operated at an emitter temperature of ~ 1,865 K. These flights were to prove the working capacity of thermionic Space Nuclear Power Systems (SNPS) in a flight environment. Long life was not an objective of the flight program.

Fig. 9. Topaz reactor[20]

A cutaway of the Topaz I reactor showing the principal subsystems and design features is shown in Fig 10.[21] The reactor core is composed of a set of thermionic converters arranged in a moderator. The reactor control is effected by means of rotatable cylindrical control drums of boron carbide in a beryllium reflector. The fuel and moderator arrangement helps attain a high radial uniformity of energy generation. The power plant includes a cesium delivery system. NaK coolant is used for heat removal. Also, included is the commutation section and current leadouts. All of this is housed in a single body.

A schematic of the fuel element, Fig. 11, portrays the fuel element as divided into five fuel cells. The thermionic converter was combined with the fuel element to produce a single power generating channel. This included the urania fuel, cathodes made from a tungsten alloy or the molybdenum alloy VM-1, anodes made from the niobium alloy VN-2, beryllia insulators, stainless steel outer casings and cesium vapor in the interelectrode gap.

The Topaz SNPS included:[22]
- An incore thermionic reactor with 12 rotating control drums;
- A control drive actuator for four groups of drums;
- A cesium vapor supply system;
- A heat rejection system;
- A starting system; and
- An automatic control system with storage battery.

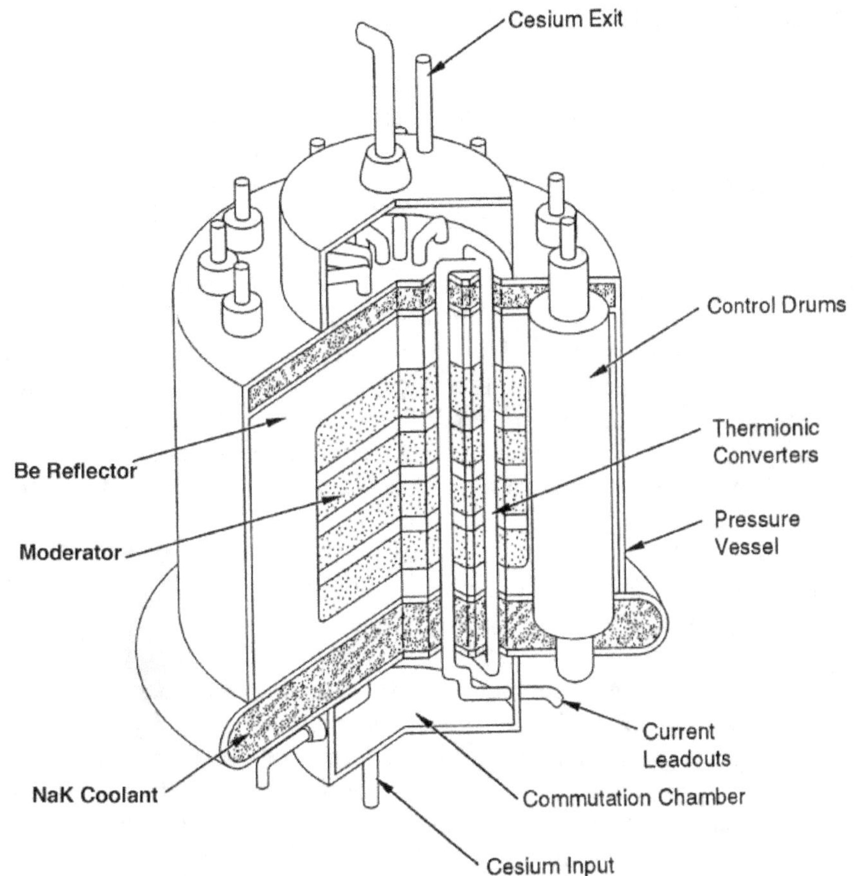

Fig. 10. Configuration of the TOPAZ I reactor. *From G. Bennett, 1989.*

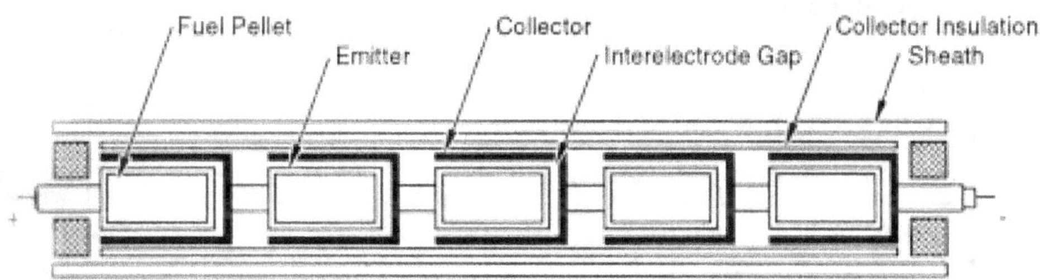

Fig. 11. Basic arrangement of the Multicell TOPAZ Thermionic Fuel Element (TFE). *From G. Bennett, 1989.*

The reactor included a zirconium hydride moderator, beryllium reflector, and the thermionic fuel elements (TFEs). The TFEs are connected into two independent sections. These sections are the working segments for spacecraft equipment and the pump section of the electromagnetic pump (EMP). Urania (UO$_2$) fuel is used.

The TFE emitter envelop material on the first flight was monocrystalline molybdenum; on the second flight it was monocrystalline tungsten. Gaseous fission products are removed from nonhermetic fueled emitters into the interelectrode spacing directly. A three-layer design is used in the collector assembly. Niobium alloy is used for the collector, beryllium oxide for insulation, and stainless steel for the TFE outer sheath. The TFEs are electrically connected into working and pump sections in the upper and bottom commutation chambers in cesium vapor. The spacecraft equipment feed section has 6 parallel branches incorporating from 8 to 12 TFEs connected in series. A conductive electro-magnetic-pump (EMP) is used for coolant pumping. The pump section has all TFEs connected in parallel. Thirty per cent redundancy is provided in the EMP energy supply by the number of

TFEs in the pump section. Hermetic ceramic-metal seals are used for current conductors from the commutation chambers.

The 12 rotating control drums in the radial reflector incorporate neutron poison surface segments of boron carbide (B_4C). Two groups of control drums are used for regulation of reactor thermal power, while the other two are used for reactivity compensation.

The cesium vapor supply system passes through the interelectrode gaps of the TFEs to remove gaseous impurities by absorption. This uses a pyrolitic graphite cesium trap, while noncondensed gases are removed to outer space. The cesium vapor supply system includes a cesium thermostat with electric heaters to maintain the desired temperature in the cold zone of the thermostat. The cesium vapor generator includes devices to keep the liquid phase of cesium in a certain position and prevent its penetration into vapor ducts under microgravity conditions. The SNPS lifetime is limited by the amount of liquid cesium in the vapor generator (2.5 kg). In the cesium vapor generator, a temperature regulator provides the capability to operate at three different settings to maintain temperature in the range of about 30 K. These can be controlled from earth using radio commands.

A single-loop heat rejection system is used with the eutectic alloy NaK as a coolant. The radiator is a tube-screen-type design with a rigid structure and the ability to withstand mechanical loads in all operating regimes. The emissivity of the radiator surface coating is 0.85 or greater.

An automatic control system and storage battery are located in the spacecraft equipment section. The controller functions include: start-up to bring the SNPS to power; maintaining nominal operating parameters; maintain output spacecraft bus voltage; and shut down on spacecraft control system signals. At nominal operating conditions, the required electric power is provided by maintenance of working section current by means of proper correction of the reactor thermal power. The outlet coolant is controlled not to exceed 880 K. The bus voltage is regulated using a ballast load whose heat is absorbed by the coolant loop.

Details of the power system are given in Tables 4 and 5.

Table 4. Summary of characteristics of TOPAZ I reactors flown on Cosmos 1818 and Cosmos 1867.

Parameter	Value
Working section electric power (kWe)	about 6
Output voltage of working section (V)	32
Working section current (A)	180
Output voltage of pump section (V)	1.1
Pump section current (A)	1200
Coolant maximum temperature (°C)	610
Coolant temperature rise in the core (°C)	80
Cesium consumption rate (g/day)	6 - 20
Cesium vapor pressure (Pa)	266 - 730
Radiator surface (m²)	7
Mass of SNPS (without storage battery) (kg)	1200
Size (m)	
length	4.7
diameter	1.3

Table 5. Design features of the TOPAZ I thermionic reactor system.

Characteristics	Value
Diameter of fueled emitter (mm)	10
Outer diameter TFE (mm)	14.6
Total number of TFEs	79
Number of TFEs in the working section	62
Number of TFEs in the pump section	17
Average moderator fraction	0.72
Core diameter (mm)	280
Core height (mm)	364
Uranium-235 fuel load (kg)	12
Outer diameter of reactor (mm)	460
Efficiency of 12 control drums (% K_{eff})	7
Subcriticality of the reactor in "cold" condition (% K_{eff})	6.2
Thermal reactivity effect (% K_{eff})	1.8

For the thermionic fuel elements in Topaz I, the chloride-deposited tungsten layers on Mo alloy monocrystalline substrates have been subjected to in-pile tests of up to 17,340 h duration and high temperature (1,900 K) tests of over 13,000 h. For the collector, after 1.5 y of in-pile testing opposite a tungsten emitter, a condensed mass transfer layer has reached a thickness of 300 nm. The layer consists primarily of tungsten, but also contains some oxygen, carbon, and cesium. The UO_2 fuel behavior under irradiation is very complex. The high temperature at the outer edge (~ 1,900 K), combined with the relatively large diameter of the fuel pin (~ 10 to 15 mm), implies high temperatures at the center of the fuel pin. Within a matter of hours after startup, the fuel column restructures so that the original stack of pellets has become a single fused structure. The interior of this structure is an isothermal void, which extends the length of the TFE. The 25 to 40 mm gap between the fuel pellets and the interior of the emitter closes due to sublimation and redistribution of the fuel. This void and the micropores originally present in the UO_2 fuel pellet are swept to the interior void. As burn up proceeds, xenon, krypton, and a few volatile fission products are similarly swept from the fuel into the interior. This central void is vented by a passage in a screw hold-down plug fitting in the end of the TFE.

Extensive testing on components has been performed in Russia. Coated zirconium hydride material to reduce hydrogen loss has been tested in a reactor with losses of hydrogen being less than 1% of the initial inventory. Zirconium hydride swelling data from neutron irradiation show a volume change of 2% at 823 K for a fluence of 1.5×10^{21} n / cm^2. The temperature in the system is actually 50 to 75 K less during design operation. Cladding the hydride improves performance by six orders of magnitude, and the use of a cover gas was found to provide another factor of five improvement.

Flight Operations

Flight operations of Topaz I were carried out to demonstrate the working capacity of thermionic systems in a space environment. The flight operations were successful with Cosmos 1818 operating for 143 days and Cosmos 1867 for 342 days. For safety, the working orbit altitude guaranteed a spacecraft orbital lifetime of not less than

350 years--which is sufficient for nuclear fission products to decay to a safe level. Space testing ended in both cases as planned, with the exhaustion of cesium in the vapor generator. There was a short circuit experienced about 70 minutes after startup in the first case and 55 minutes in the second. It took about 90 s for both SNPS systems to transfer from the short circuit current regime to the nominal regime and continue to operate. Some differences were seen in Cosmos 1867 than with ground testing. This was attributed to differences in hydrogen leakage.

Topaz II Reactor Power Plant

Design Features
A second form of thermionic reactor, entitled Topaz II, has been purchased by the U.S. for testing and evaluation. Topaz II also was designed to provide ~ 6 kW_e output, but a single thermionic cell fuel element has replaced the multicell fuel element used in Topaz I. A significant advantage of Topaz II is the ability to test the entire system at full temperature in an electrically heated configuration.[23] The single-cell fuel element makes this possible. The characteristics of the Topaz series of space power systems are summarized in Table 6 .[24, 25]

A schematic of Topaz II is shown in Fig. 12.[26] At the beginning-of-life, the reactor produces approximately 115 kW_t with a conversion efficiency of 5.2%; and at the end-of-life, the reactor produces 135 kW_t with a conversion efficiency of 4.4%. The Topaz II is cooled by a liquid metal of eutectic sodium-79% potassium-21% (NaK) that remains liquid during all phases of the Topaz II lifetime, excluding the end-of-mission shutdown. The NaK coolant removes the waste heat from the reactor and transports it to the radiator, where it is rejected to space.

Table 6. TOPAZ reactor characteristics.

	TOPAZ I	TOPAZ II
Electrical Power (kWe)	5 to 6	6
Voltage at Lead (V)	5 to 30	28 to 30
Thermal Power (BOL/EOL)(kWt)		115/135
Number of TFEs	79	37
Cells/TFE	5	1
System Mass (kg)		1,061
Emitter Diameter (cm)	1.0	1.73
Core Length (cm)	30	37.5
Core Diameter (cm)	26	26
Reactor Mass (kg)		290
Moderator	$ZrH_{1.8}$	$ZrH_{1.85}$
Emitter	Mo/W	Mo/W
Emitter Temperature (K)	1,773	1,800 to 2,100
Coolant	Pumped NaK	Pumped NaK
Number of Pumps	1	1
Type of Pump	Conduction	Conduction
Pump Power	19 TFEs	3 TFEs
Reactor Outlet Temperature (BOL/EOL)(K)	773/873	560/600
Coolant Flow Rate (kg/s)		1.5
Cesium Supply	Flow Through	Flow Through
Axial Reflector	Be Metal	Be Metal
Radial Reflector	Be Metal	Be Metal
Number Control Drums	12	12
Shield Mass (kg)		390
Radiator		Tube and Fin
Radiator Area (m^2)		7.2
Radiator mass (kg)		50

The Topaz II core is a right circular cylinder 260 mm in diameter and 375 mm high.[27] Thirty-seven cylindrical thermionic fuel elements (TFEs) are arranged in an approximately triangular pitch within a block of moderator. The moderator is epsilon phase zirconium hydride (ZrH_{2-x}) with hydrogen stoichiometry in excess of 1.8. The moderator is canned in a stainless steel calandria that has 37 circular channels in it to accommodate the TFEs and NaK coolant. The coolant channel gap between the TFE outer sheath and the calandria wall is a grooved surface.

The fuel pellets are contained within an emitter tube of single crystal Mo-3%Nb with a tungsten coating on the outer surface to enhance thermionic emission. Each TFE has approximately 40 fuel pellets. The fuel is high purity and high density (96% theoretical density) UO_2.[28]

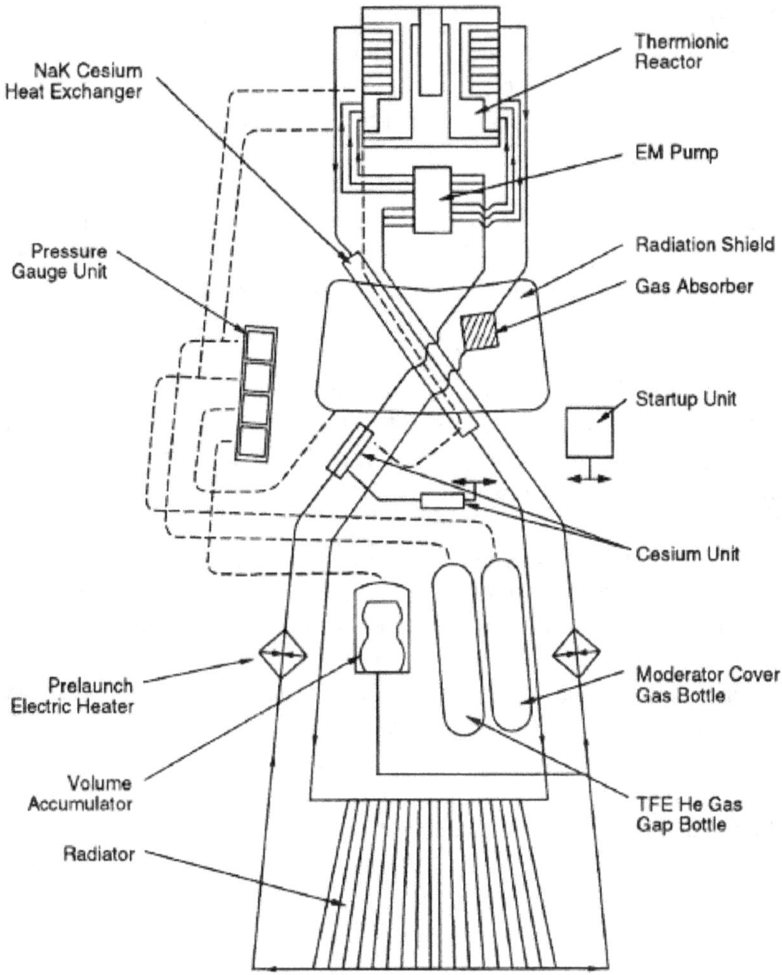

Fig. 12. Topaz II schematic. *From Topaz CoDR, 1992.*

A gas mixture of approximately 50% CO_2, 50% helium, and other trace gases is maintained within the moderator/axial reflector region to help inhibit the release of hydrogen from the $ZrH_{1.85}$, and to increase the heat transfer from the $ZrH_{1.85}$ to the outer surface of the vessel and to the coolant channels.

The thin-walled, stainless steel cylinder reactor vessel encloses the TFEs, moderator calandria, and axial beryllium metal neutron reflectors. It supports the core and TFEs and provides plena for the cesium vapor, helium gas, and NaK coolant.

Outside the reactor vessel within the radial reflector are 12 rotating control drums. Three of these drums are used in the safety system. The remaining nine are for control, and are driven by a common mechanism. The radial reflector assembly is held together with two fused tension bands. The control drum bearings and drive trains are lubricated with MoS_2.

Fig. 13 is a cross section of the Topaz II TFE. The fuel contained within each thermionic fuel element is highly enriched (96%) UO_2 in the form of pellets stacked within the cavity of the emitter. Each fuel pellet is ~ 8 mm high with an outer diameter of 17 mm. The fuel pellets possess a central hole with a diameter of 4.5 mm or 8 mm, depending upon the position within the core, to help flatten the power profile in the radial direction. The fuel height is 355 to 375 mm, where the height of the fuel can be varied at the time of loading to compensate for variations in fabrication that can affect core excess reactivity. The maximum fuel temperature is ~ 1,775 to 1,925 K, and the end temperatures are ~ 1,575 K.

Topaz II has regenerating cesium supplies and vents fission gases outside the reactor directly to space. The emitter contains the fuel and fission products and serves as the source of thermal electrons. Emitter strain limits system life, so the mechanical properties of the emitter are extremely important. The emitter is made from monocrystal molybdenum alloy substrate with a chemical vapor deposition (CVD) coating of tungsten. The tungsten coating is for improved thermionic performance. This coating is deposited from chloride vapor, and is also monocrystalline. The outer tungsten layer is enriched in the isotope ^{184}W in order to limit adverse neutronic effect. The emitter temperature is 1,873 K.

Beryllium oxide (BeO) pellets on both ends of the fuel stack provide axial reflection. The BeO pellets have central holes that match up with the holes in the fuel. The pellets are stacked to a height of 55 mm. They are used to compensate for variations in the fuel loading. The total height of the core, including the BeO end reflector, is 485 mm.

The monocrystal molybdenum collector tube is coaxial with the fueled emitter. High collector temperatures are desirable to reduce radiator size, while lower temperatures reduce thermionic back-emission and keep the dissociation pressure of hydrogen in the moderator within bounds. A temperature of 925 K is used at the outlet of the reactor; the temperature is about 100 K lower at the inlet.

Between the collector and emitter is the interelectrode thermionic gap. This gap is 0.5 mm.[29] There are scandium-oxide (Sc_2O_3) spacers between the emitter and collector surfaces in the interelectrode gap. These spacers prevent the emitter from shorting to the collector as a result of emitter distortion caused by fuel swelling.

External to the collector is a helium gap between the collector insulator and inside diameter of the inner coolant tube. The helium provides a good thermal bond for the transfer of heat to the coolant, while maintaining electric insulation of the thermionic fuel element. A helium bottle is located in the radiator region that maintains helium in the gap over the lifetime of the system. The stainless steel sheath completes the TFE structure. The sheath supports the TFE and provides a heat transfer surface to the NaK coolant.

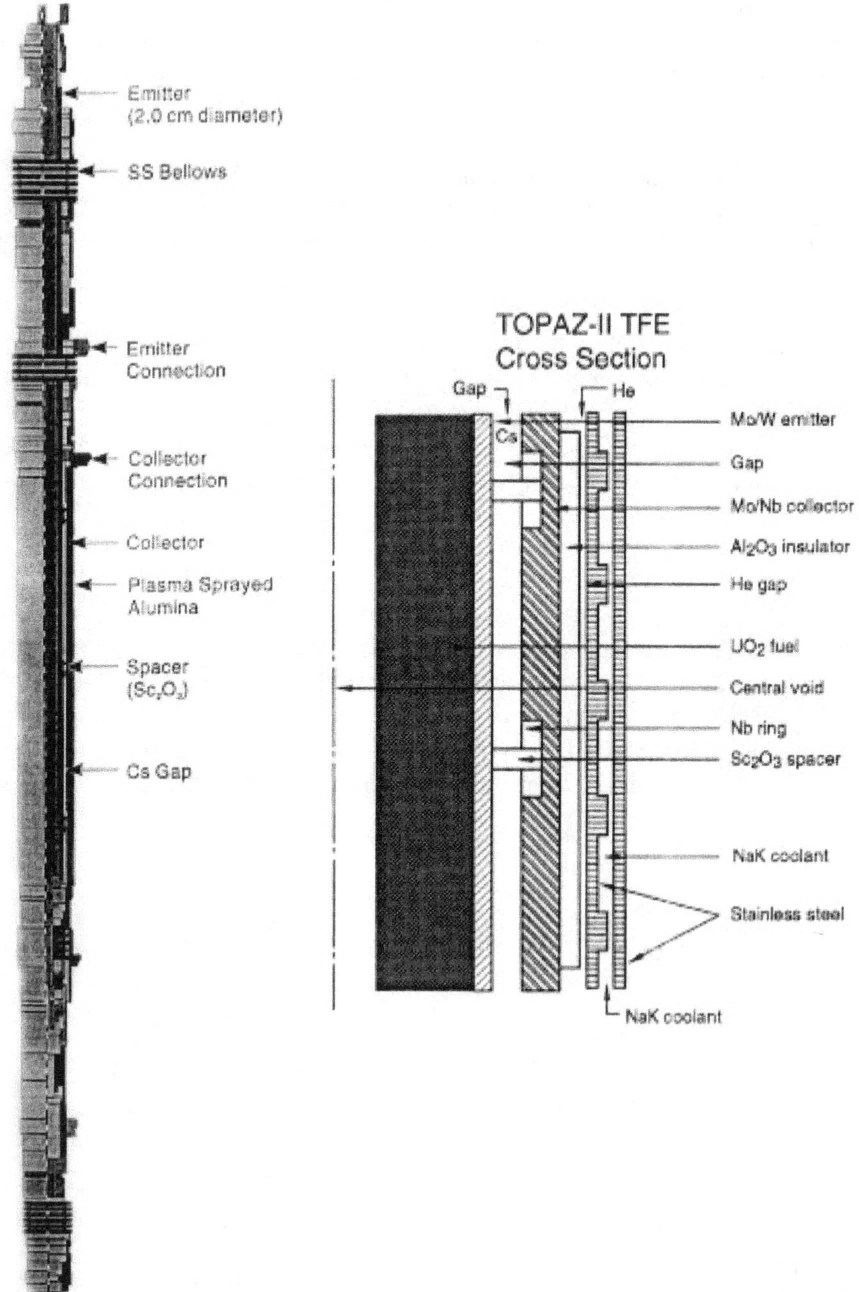

Fig. 13. Topaz II thermionic fuel element cross section. *From Topaz CoDR, 1992.*

The gas and coolant systems within the Topaz II include the NaK coolant, cesium supply system, CO_2/He cover gas, the He thermionic fuel element gas gap, the argon/He gas in the volume accumulator, and the helium gas in the radiation shield. The last four systems are fed from pressurized bottles.

The Topaz II uses eutectic NaK to remove heat from the reactor. The NaK must be kept liquid during launch. A single DC conduction pump powered by a group of three dedicated TFEs connected in parallel are used. The coolant piping is divided into two groups of three channels as it passes through the pump. An on-board current source is used for startup.

The cesium supply is a still that feeds cesium vapor to the interelectrode gap and vents any gases to space. Cesium venting is 0.5 g / d for this system.

A tube-and-fin radiator is employed in the shape of a truncated cone. The surface of this cone is formed from steel tubes welded to circular manifolds at the top and bottom of the radiator. Copper fins are welded to these tubes. To improve emissivity of the fins, a glass coating is used that adheres well and has good thermal resistance.

A shadow shield in the shape of a truncated cone is used for radiation attenuation. Both end caps are concave downward, spherical, and thick walled. The sides of the shield are thin-walled steel. The space between the end caps is filled with LiH for neutron shielding. Four coolant pipes pass through the shield at angles designed to minimize radiation streaming. A stepped channel through the shield contains the control drum drive shaft.

For Topaz II, extensive component and systems testing has been performed. Table 9 provides summary information for some of the major components including: the number of components tested, the type of testing, and the time at test. During the development phase, component testing was performed in two stages. The first was to preliminarily assure functional sufficiency and identified potential problems in an informal manner. The second phase included more formal tests to ensure that the components met the defined acceptance criteria.

Tables 7 and 8 (prepared by Susan Voss, Los Alamos National Laboratory) summarizes the power plant testing data on Topaz II. The longest test was 14,000 h. Transient behavior, shown in Fig. 14, shows stable operation.

Fuel development for the Topaz II program was discontinued in Russia in 1988. The Russian researchers had concluded that the existing fuel and TFE components would provide an estimated operating life of 4 to 5 years at a reactor power of 6 kW_e. The basis for the lifetime estimate is 12,500 hours of reactor operation and parametric testing of key lifetime variables such as: fuel swelling, fission gas release, and the irradiation performance, thermal stability, mechanical strength, and compatibility of all TFE materials.[30]

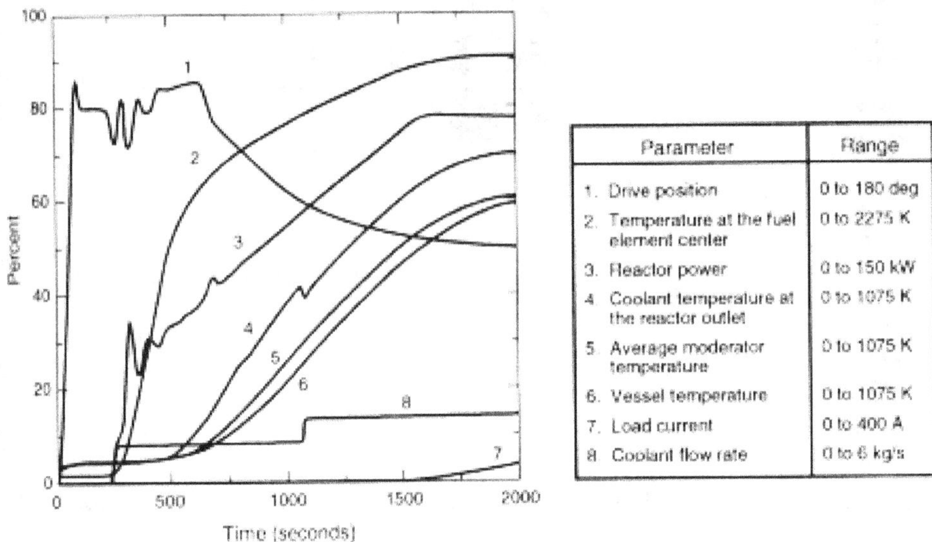

Parameter	Range
1. Drive position	0 to 180 deg
2. Temperature at the fuel element center	0 to 2275 K
3. Reactor power	0 to 150 kW
4. Coolant temperature at the reactor outlet	0 to 1075 K
5. Average moderator temperature	0 to 1075 K
6. Vessel temperature	0 to 1075 K
7. Load current	0 to 400 A
8. Coolant flow rate	0 to 6 kg/s

Fig. 14. Topaz II regular power plant startup *From Topaz CoDR 1992.*

Safety

Topaz II was originally designed for operation in geosynchronous orbit. Therefore, the Russians were not concerned with dispersal after operation. If reentry occurs, it is doubtful that complete break up at sufficiently high altitudes will achieve full reactor dispersal.

A number of design modifications were being considered to meet U.S. safety philosophy. These included the inclusion of a reentry thermal shield to avoid breakup if reentry occurs and the addition of removable poison in the annulus of a number of TFEs or removal of fuel during launch to ensure against reactor supercriticality when immersed and flooded with water. For the planned U.S. flight in 1996 -1997, the initial operational altitude was planned to be well above a "sufficiently high orbit" for fission product decay.

Another safety concern is the delayed positive temperature coefficient in the reactor core. This coefficient has been confirmed experimentally, and the Topaz reactors have been proven to be experimentally controllable. The delayed positive temperature coefficient effect is due primarily to moderator spectrum hardening when temperature is increased. This leads to fewer moderator captures and more fuel captures. The positive feedback time constant is very long (~ 330 s) relative to the control system. The long time constant is due to a large heat capacity and high thermal resistance. An effect of having a delayed positive feedback coefficient is to reduce the amount of excess reactivity required to very low levels (65 cents). This results from no negative temperature defect and a minimum of burnup reactivity loss to compensate for. The reactor has the unique feature that startup prompt disassembly accidents are not probable.

If the reactor has a loss of coolant accident and the control system does not shut down the reactor, Topaz II has a built-in safety feature to shut down the reactor. The ZrH moderator will heat up, releasing the hydrogen, shutting down the reactor.

Table 7. Component testing

	Single Component Tests			System Test		
Component	Test Description	Number of Tests	Time on Test (h)	Test Description	Number of Test	Time on Test (h)
1. Reactor	Thermo-Physical	3	12,000	Electrical Tests	7	12,500
	Mechanical (Dynamic)	2		Mechanical		
				• Ground Transport	4	
				• Static	2	
				• Dynamic	4	
				Cold Testing	4 & 1	
				Nuclear Tests	6	
2. Control Drum Drive	Operational	5	13,000	Electrical Tests	6	14,000
	Lifetime	6		Mechanical		
				• Ground Transport	1	
				• Static	1	
				• Dynamic	3	
				Cold Testing	3	
				Nuclear Tests	4	
3. Safety Drive	Mechanical	3		Mechanical		
	Thermo-Physical	3		• Ground Transport	12	
	Climatic	3		• Static	6	
	Operational	3		• Dynamic	12	
				Cold Testing	12 & 3	
4. Radiation Shield	Mechanical (Static)	2		Electrical Tests	7	14,000
	Lifetime	2		Mechanical		
	Characteristics and			• Ground Transport	4	
	Material Changes			• Static	2	
				• Dynamic	4	
				Cold Testing	4 & 1	
				Nuclear Tests	6	
5. Cesium Unit	Thermal Lifetime	7	26,400	Electrical Tests	7	14,000
	Operational Lifetime	3		Mechanical		
				• Ground Transport	4	
				• Static	2	
				• Dynamic	4	
				Cold Testing	4 & 1	
				Nuclear Tests	6	

Table 7. Component testing (continued)

	Single Component Tests			System Test		
Component	Test Description	Number of Tests	Time on Test (h)	Test Description	Number of Test	Time on Test (h)
6. Radiator	Thermal Lifetime	1	10,000	Electrical Tests	7	14,000
				Mechanical		
				• Ground Transport	4	
				• Static	2	
				• Dynamic	4	
				Cold Testing	4 & 1	
				Nuclear Tests	6	
7. Volume Compensator	Thermal Lifetime	7	40,000	Electrical Tests	7	14,000
	Operational	5		Mechanical		
				• Ground Transport	4	
				• Static	2	
				• Dynamic	4	
				Cold Testing	4 & 1	
				Nuclear Tests	6	
8. Ionization Chamber	Thermal Lifetime	3	26,300	Electrical Tests	7	14,000
				Mechanical		
				• Ground Transport	4	
				• Static	2	
				• Dynamic	4	
				Cold Testing	4 & 1	
				Nuclear Tests	6	
9. Pressure Sensor Unit (4)	Lifetime	1	13,500	Electrical Tests	6	2,000
	Operational	1		Mechanical		
				• Ground Transport	3	
				• Static	2	
				• Dynamic	3	
				Cold Testing	3 & 1	
				Nuclear Tests	6	
10. Start-up Unit	Operational	3		Electrical Tests	3	
	Mechanical			Mechanical		
	• Ground Transport	6		• Ground Transport	3	
	• Dynamic Launch	2		• Static	1	
	Thermal Lifetime	6		• Dynamic	3	
				Cold Testing	1 & 1	
				Nuclear Tests	1	
11. Thermal Cover	Mechanical (Static)	2		Mechanical		
	Mechanical (Dynamic)	1		• Ground Transport	4	
	Operational	2		• Static	2	
				• Dynamic	4	
				Cold Testing	4 & 1	

Table 8 TOPAZ II systems test data.

Test	Date	Time at Test	Findings
Plant 23	1975-76	2,500 h at 6 KWe 5,000 h total test	Ground Demonstration Unit. Significant degradation of power due to TFE.
Plant 31	1977-78	4,600 h	Flight Demonstration Unit. Same as TFEs as Plant 23 Startup with Automatic Control System (ACS). 4,600 h was planned test time.
Plant 24	1980	14,000 h	Ground Demonstration Unit. Same TFEs as Plant 23 Startup with ACS Provided life testing of many components.
Plant 81	1980	12,500 h at 4.5 to 5.5 kWe	Ground Demonstration Unit. New TFE design. Did not complete 1000-h electric testing prior to nuclear testing. NaK leaked 150 h into nuclear test. Fixed NaK leak and continued testing.
Plant 82	1983	8,300 h at 4.5 to 5.5 kWe	Ground Demonstration Unit. Loss of flow progressing to a loss-of-coolant accident. Final shutdown of the reactor due to loss of H from the $ZrH_{1.85}$. No major structural damage to the reactor.
Plant 38	1986	4,700 h at 4.5 to 5.5 kWe	Flight Demonstration Unit. Test ended due to a NaK leak in upper radiator collector First ground test with the temperature regulator as part of the ACS.

Flight Readiness

A joint launch by the U.S. and Russia of Topaz II was being planned in the 1996-1997 time period.[31] It would have used information from the two unfueled reactors that the U.S. purchased for electric heating testing in the Thermionic System Evaluation Test (TSET) program.[32] Goals of this program included: learning from the Russians, evaluating performance within the design limits of Topaz II, and training a cadre of U.S. experts on space nuclear power systems. By using purchased Topaz II power plants, the U.S. obtains insights into Russian technology, insights into complete non-nuclear satellite qualification and acceptance methodology, knowledge of Russian safety methods, and reduced cost to develop U.S. thermionic systems. It would also have demonstrated electric propulsion options and measured the nuclear electric propulsion self-induced environments.

If the U.S. had purchased and flown a Topaz II power system, a new automatic control system would be needed to meet U.S. qualification standards for space applications and that is compatible with U.S. launch systems. Also, it is uncertain whether the nuclear fuel would have been procured from Russia or fabricated in the U.S. If fabricated in the U.S., the fuel would have needed to meet both U.S. and Russian quality standards. Other modifications included an anti-criticality device, thermal cover and reentry shield. All of these modifications were considered relatively minor.[33]

Russian Flight System Approach

According to U.S.S.R. documents concerning regulation of the use of space nuclear power sources operations with regard to radiation safety, their systems meet the following radiation safety standards:[34]

- isolation of space nuclear power sources from the Earth population by keeping the power sources intact and undamaged for the sufficient period of time until their activity drops to safe level; and
- if the first principle cannot be realized, disperse the power source to the levels assuring the population safety in the radiological contaminated area.

Safe radiation levels for individuals recommended by the International Commission for Radiation Protection amounts to 5 mSv per year. Accordingly, the duration of isolation of a reactor with total power of 10^3 to 10^4 MW-h should be no less than 300 years and the size of the particles of the nuclear fuel at dispersion of the space nuclear power sources should not exceed 100 µm with their dispersion over the area of no less than several hundred square kilometers.

Additional requirements are:

- startup of the reactor on the orbit of the satellite;
- shutdown of the reactor at the end of useful life or in case of an emergency situation;
- preservation of subcriticality of the reactor until the object with the space nuclear power source is launched to the orbit and also during its reentry.

The main candidate methods of radiation safety used for realization of the first principle include:

1) The injection of the satellite into a "safe" orbit. The injection into the "safe" orbit requires a nominal orbit of 800 km or higher. The Topaz reactors, Cosmos 1818 and Cosmos 1867, were operated in a circular orbit with the approximate altitudes of 800 km. However, the standard altitude of U.S.S.R. satellites is 200 to 300 km. Therefore, to transfer the objects into a "safe" orbit, it is necessary either to provide for additional fuel in the last stage of the launch vehicle and ensure its restarting or introduce an additional transfer stage into the launch vehicle. The payload penalty is approximately 20%.
2) Transfer of only the reactor from the low operational orbit to the "safe" orbit. The mass loss in this case is no more than 8%. This method of transfer was to be used in Cosmos 954 type satellites where the operational orbit altitude is 265 km and the "safe" orbit is 900 km. At the end of useful life or in case of an emergency, the reactor is separated from the satellite and is transferred to the "safe" orbit by means of a solid-fueled rocket.
3) Transportation beyond the Earth's activity sphere (Earth escape). The current disposal range of 800 to 1000 km is a working orbit for many other satellites. Therefore, this disposal region may be banned for the transfer of the used space nuclear power systems. The cost of doing this could result in a payload reduction of more than 80%. That is why this method has not been used.
4) Controlled descent to a desired area. Another solution of the problem of protection of the near Earth space is a controlled descent of the space nuclear power system into a desired region of the Earth's surface keeping them intact with subsequent burial. This requires the addition of thermal protection while moving into the atmosphere. There is some penalty over moving to the higher orbit because of the addition of the thermal protection.

Backup methods of radiation safety include:

1) Aerodynamic breakup. This method of dispersion is used in satellites like Cosmos 954 where natural physical phenomenon are use to disperse the space nuclear power sources upon reentry. The method used is given in Fig. 15. During breakup of fuel elements with a uranium-molybdenum alloy core in optimal conditions, the ultimate size of the particles will not exceed 100 µm. The success of this method was demonstrated in the Cosmos 954 reentry. The breakup of the Topaz reactor with uranium dioxide core has been demonstrated in ground testing to also have particles that do not exceed 100 µm.
2) Destruction by chemical reagents. Experiments have demonstrated the possibility of fine-dispersion destruction of the material of the fuel elements by converting them into liquid or gaseous compounds. However, there seem to be certain limitations to this method: the need for a reaction space, preliminary cooldown of the reactor and neutralization of the reagents to prevent overheating and thermal destruction of

the flow channels of the reactor, the great length of the process (hours and even days), and the large stock of the regent required (up to 10 kg per kg of the material to dispersed).

3) Reactor self-destruction due to overheating. The reactor can be used to overheat and melt the fuel elements. Using a nozzle for escape, the particles formed can be micron size. This has the advantages of being a short duration process and requires no additional mass. However, it has the disadvantage of being difficulty to control the release of the reactor materials and it does not disperse other activated materials.

4) Destruction by explosion of chemical explosives. This method disperses not only the fuel elements, but also the construction materials. Using explosives behind the shadow shield shows 90 to 95% of the material of the destruction products are of the size less than 100 µm. The size of the charge is 0.15 to 0.25 of the mass of the reactor with the shield.

5) Aerodynamic breakup. The use of refractory fuel materials, such as uranium oxides, carbides, nitrides, are extremely difficult to destroy by aerodynamic heating. However, if the fuel is in the form of spherical particles with the size of 100 to 500 µm, than the particles are practically the size needed to meet the radiation safety requirements.

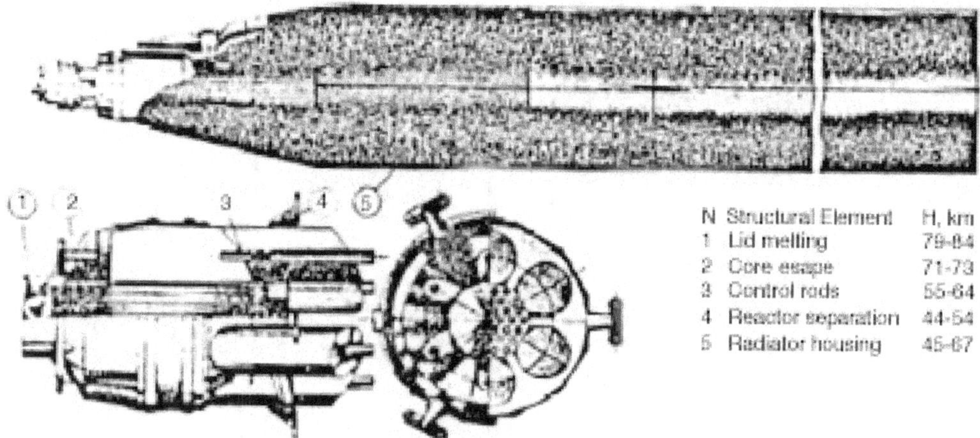

Fig. 15. Schematic of Cosmic 954 type satellite with reactor equipped space nuclear power system and its aerodynamic breakup sequence. Gafarov, Albert A., Boris I. Bakhtin, Alexey V. Kosov.

The Soviet satellites use a combination of the above methods to meet radiation safety standards. Boost systems are on board satellites deployed at very low orbits, like the RORSAT satellites, to transfer them to higher orbits at the end of life or in case of a malfunction. Back up boost systems are also on board to separate the reactor core and boost it to higher orbits using commands or sensors if the main boost system fails to function as planned. In addition, the power systems are designed for aerodynamic breakup, so material will be dispersed over a large area, with particle sizes less than 100 µm.

A significant difference in design philosophy between the U.S. and U.S.S.R. is the reactor temperature coefficient. The U.S. calls for a negative temperature coefficient in reactor power systems. The Russians have a positive temperature coefficient, claiming that this is an added safety feature.[35] The positive temperature coefficient reduces the built-in reactivity required in the reactor. It has a time constant of about 5 minutes; the thermal time constant is on the order of seconds. Thus, the reactor controller can easily compensate for the positive temperature coefficient and operate stably.

Summary

From 1967 - 1988, the Russians launched some 33 satellites that incorporated Romaskha reactors with thermoelectric power conversations units. Power levels ranged from several hundred watts to a few kilowatts. Missions lasted up to 135 days in operational orbits near 250 km. At completion of the missions, the satellites boosted the reactors to long-lived storage orbits in the vicinity of 900 - 1,000 km. This disposal procedure failed twice (Cosmos 954 and 1402) and once only boosted the satellite to an orbit of 700 km (Cosmos 1900).

Topaz I demonstrated 1 y operation in space, while Topaz II has demonstrated 1.5 y of nuclear ground testing. A claim of 3-y lifetime is based on component life data. Experience with the multicell TFE indicates that swelling and intercell leakage are significant life-limiting problems. The single cell TFE tends to correct these problems, partially by using a high void fraction. A primary life limiting element appears to be the loss of hydrogen from the ZrH moderator. The rate is about one percent per year. Also, with only 65 cents of excess reactivity in the design[36] the fuel burnup can be life limiting. This is especially of concern if the reactor cools down before a restart is achieved. Another issue is the oxygen getter. Modifications may be needed in the cesium supply, but these do not appear to be life limiting.

Chapter 7

Project Prometheus, Gas-Cooled Reactor Power System[1]*

Project Prometheus was established in 2003 by NASA to develop nuclear-powered systems for long-duration space missions using electric propulsion. Spacecraft exploring the outer planets are severely limited if they depend on solar energy as a source of electrical power. Hence, the interest in developing nuclear power as an energy source. The initial application was to be the Jupiter Icy Moons Orbiter (JIMO) to explore Callisto, Ganymede, and Europa. Other possible missions included: Saturn and its moons; Neptune and its moons; comet and multi-asteroid sample return; and Kuiper Belt rendezvous (see Fig. 1). The project was terminated in 2005 after having selected a gas-cooled reactor with a directly coupled Brayton energy conversion system for further development. Fig. 2 is a conceptual view of the spacecraft and module.

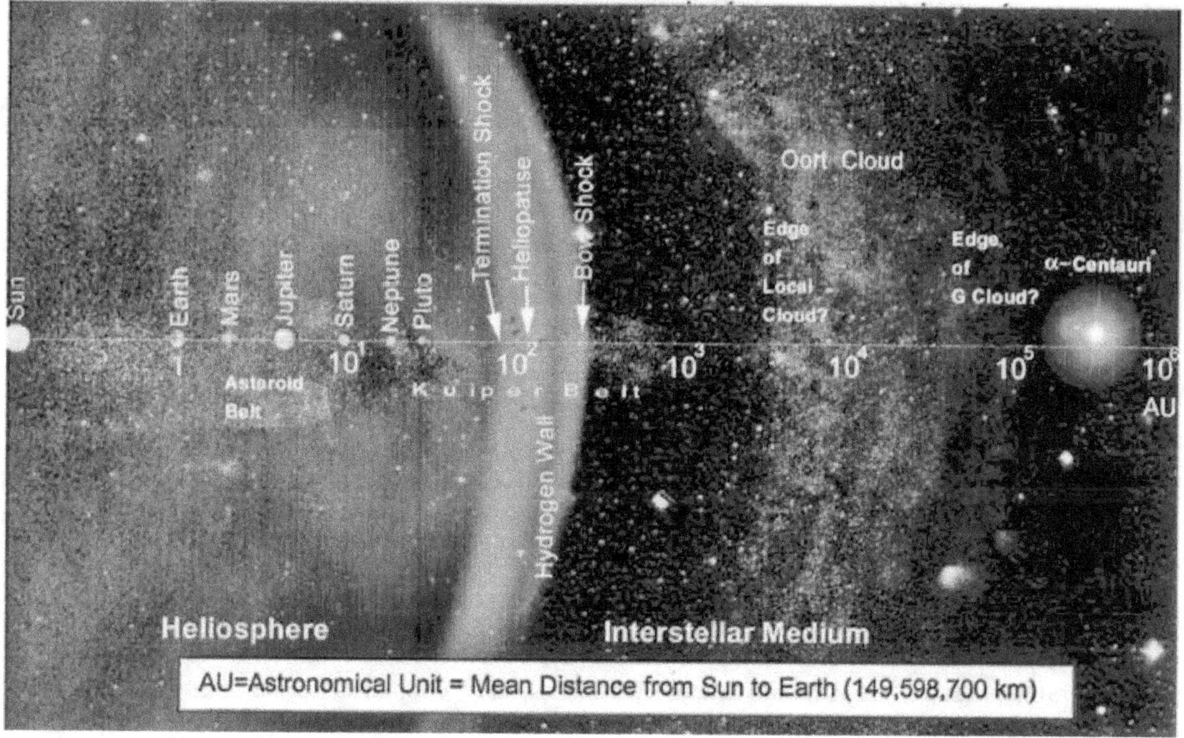

Fig. 1. Illustration of possible deep space missions. Courtesy *of NASA*.

The duration of the proposed missions ranged from 10 to 12 years for JIMO, to a maximum of 20 years for the Kuiper Belt rendezvous. Approximately 200 kW of electric power is needed for the envisioned missions. The Deep Space Vehicle was to have a payload accommodation envelope with a mass capability of no less than 1,500 kg. The original time schedule was a launch of JIMO in 2015. The controlling activities to meet the 2015 date were: (1) developing reactor module materials that will successfully meet the JIMO mission duration, and (2) establish critical nuclear testing facilities to support the design timeline.

----

* An extensive Bibliography is included in "Project Prometheus Reactor Module Final Report," prepared by Knolls Atomic Power Laboratory and Bettis Atomic Power Laboratory, SPP-67110-0008, April 24, 2006.

In addition to deep space missions, the reactor module technologies were to be developed for Lunar and Mars surface power reactors. These included the nuclear fuel, reactor core materials and coolants, and instrumentation and control.

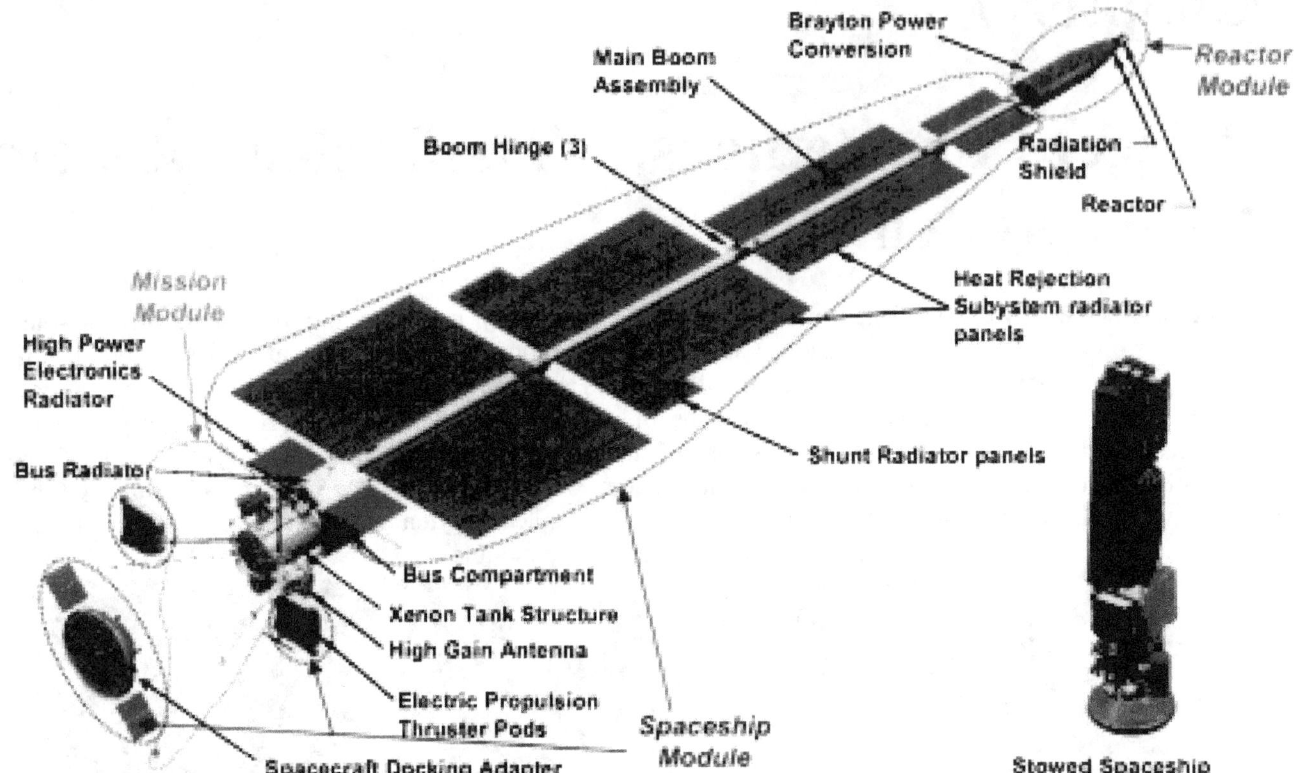

Fig. 2. Conceptual view of the spacecraft and modules. *From Knolls Atomic Power Laboratory and Bettis Atomic Power Laboratory*

Design Concept

The key requirements for the development of the reactor module are given in Table 1. The table also includes the impact of the requirements on the reactor module design and implementation approach to meet the requirements. Table 2 provides a listing of some key additional requirements derived from the JIMO space system and environmental requirements.

The selected gas-cooled reactor Brayton system is depicted in Fig. 3. The reactor consists of a core with cylindrical fuel pin elements arranged within the core structure and a reactor vessel to direct coolant flow and provide structural support for the core and reactivity controls. The coolant is an inert gas (He-Xe) to cool the core and transport energy around or through a shadow shield to the Brayton energy conversion system. The power plant schematic indicates two redundant Brayton power conversion loops to increase reliability. The thermal power level is $\sim 1,000$ kW$_t$, and operates at a reactor outlet temperature of $\sim 1,150$ K.

Table 1. Key requirements, impacts, and implementations.

Key Level 2 Requirement	Impact on Reactor Module	Implementation
The Space Nuclear Reactor design shall utilize technologies that facilitate extensibility to surface operations.	Consideration in the selection of design and materials compatible with Lunar and Mars missions.	Must consider compatibility of pressure boundaries and external surfaces with surface environments.
The Project shall use a Deep Space Vehicle that provides jet power greater than or equal to [130] kW of primary thrust during thrust periods.	200-kWe Reactor Module power output required to deliver net thruster power based on JPL orbital mechanics studies.	Plant Electrical Power ≈ 200 kWe Plant Thermal Power ~ 1 MWt
The Project shall design the Deep Space Vehicle to have an operating lifetime greater than or equal to [20] years.	20 year life is long term requirement for very deep space missions. The JIMO requirement is for 12 years.	Initial design efforts to support 15 year operational life. Long term design goal is to satisfy 20 year life requirement.
The Project shall use a Reactor Module that is capable of generating the maximum electrical power required by the Spaceship for cumulative minimum of [10] years, and is capable of generating the minimum required electrical power for the rest of the operating lifetime.	This requirement permits the option of reducing power in order to conserve reactor energy or reduce pressure and temperature during non-thrust phases. This may maximize Reactor Module life for the most demanding follow-on missions to the outer solar system.	Trade studies would be required to determine if reduction of power would improve Reactor Module longevity.
The Project shall comply with the Prometheus Single Point Failure Policy as documented in the Prometheus Project Policies Document 982-00057.	Single point failure locations shall be avoided. Where this is not practical (e.g., reactor), it must be demonstrated that alternatives to single point failure are not available and sufficient robustness must be shown to mitigate risk of failure.	Where practical, redundancy would be part of the Reactor Module design.
The Project shall be able to autonomously detect and correct any single fault that prevents thrusting in less than or equal to [1 hour]. (Note: Missing thrust during many of the mission phases severely jeopardizes mission success, and therefore should be prevented or minimized.)	This requirement must be considered in the design of instrumentation and control for a self-regulating plant and design for recovery from transients for which the module would be designed.	Robust and redundant system architecture for instrumentation and control. Automatic recovery from transients must be considered in system design.
The Spaceship shall survive without Ground System commanding for at least [50] days in the presence of a single failure.	Must consider this, with other autonomy and single point failure requirements, in design of the control system.	Design for redundancy and robustness wherever practical. The spaceship cannot survive in deep space for more than a short time without reactor power.
The Project shall assure that all Science System hardware in its deployed configuration, except approved science hardware, shall remain within the protected zone of the reactor radiation shield.	Coordination between the shield and spaceship designs is required to assure that maximum dose levels are not exceeded. Shielding of local electronics will also be required.	Shielding sufficient to reduce payload neutron flux to 5E10 n/cm2 and payload gamma flux to 25 kRad Si damage and cover roughly a 12° by 6° cone angle.
The Project shall obtain launch approval as specified in the Prometheus Launch Approval Plans.	To meet this requirement, satisfaction of various governing safety requirements would have to be demonstrated by NRPCT and NASA.	Design features will be required to assure safety. Safety assurance must be considered during design of certain Reactor Module elements.
The Spaceship total dry mass at launch shall not exceed [25,000] kg.	Minimum module mass is a goal and a selection criteria for design.	High temperature reactor is required to minimize overall mass

Note: Values in [brackets] were not firm and thus subject to review.

Table 2. Additional key requirements.

Requirement	Impact on Reactor Module
The Spaceship launch configuration shall be compatible with a [5-m] launch vehicle payload fairing (dynamic envelope dimensions 4.5m diameter, 26m height), or smaller (Ref (360))	Arrangements and overall sizing must fit within allocated space inside the fairing. This mainly constrains the radiator area, which drives the heat balance design space of the Reactor Module. Preliminary studies limited radiator area to less than 450m².
The Spaceship shall accommodate the solid particle mission environments defined in {ERD} with a probability of meeting end-of-mission (EOM) requirements greater than or equal to 0.99 ... (Ref (361))	Protection from orbital debris and micrometeoroids is required, especially of the crucial pressurized components and moving assemblies.
The Spaceship shall be designed to accommodate the radiation environment specified in the Environmental Requirements Document (982-00029). (Ref (360))	Solar, galactic and Jovian radiation sources, coupled with reactor radiation, must be considered during electronics selection, shielding trades and material evaluation.
The Spacecraft Module shall be capable of rejecting [682] kWt of heat from the Reactor Module. (Ref (360))	If radiator size is constrained, temperature and flow rate must be maximized to reject sufficient heat. This impacts the Reactor Module heat balance.
The Prometheus flight hardware shall be designed and verified to meet applicable functional, performance, operation, and other design requirements without damage or degradation when exposed to the design environments specified herein (in ERD). (Ref (361))	In addition to particle and radiation environments described above, the module must withstand launch loads and other space environments. Section 10 has more details about environments.
During no-thrust periods of Science Orbits, the Deep Space System shall continuously point a Spaceship-fixed vector to commanded directions in the target-centric reference frame to within 20, 20, and 20 mrad (3 sigma) about the reference frame X, Y, and Z axes respectively... (Ref (360))	This requirement drives the need for positional stability of the spaceship. Counter-rotating Braytons, or alternate localized means to offset angular momentum, might be necessary to provide the needed stability.

Note: Values in [brackets] were not firm and thus subject to review.

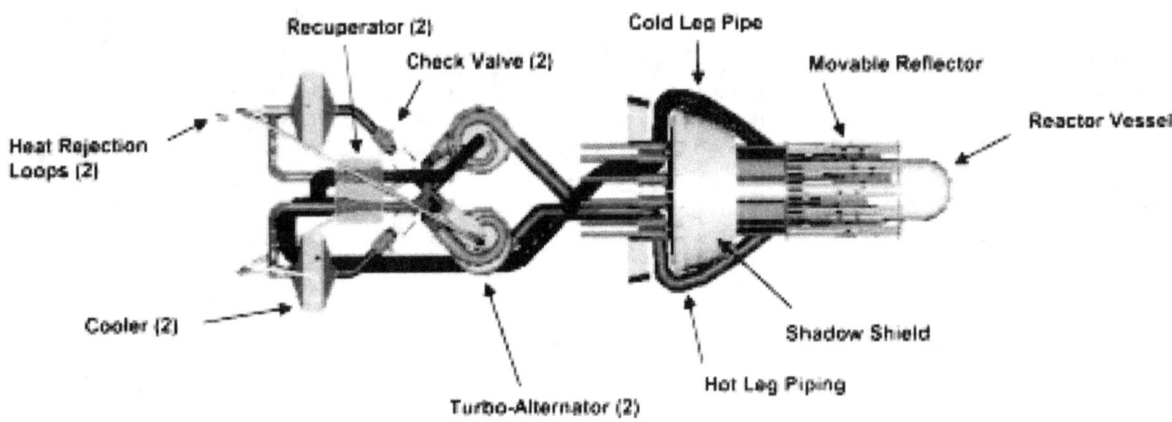

Fig. 3. Prometheus power plant schematic. *From Knolls Atomic Power Laboratory and Bettis Atomic Power Laboratory*

The gas-cooled reactor, Brayton conversion concept was selected after an intensive effort assessing: reactor types including liquid metal, heat pipe, and gas-cooled; and conversion systems including Brayton, Stirling, Rankine, thermoelectric, in-core thermionics, thermophotovoltaics, magnetohydrodynamics, and alkali metal thermal-to-electric conversion. Taken into account were the development status of such items as fuel materials, cladding and core structure materials, fuel configurations, reactor neutron energy spectrum, reactor safety and reactivity features, shielding and reflector materials, and heat rejection materials and systems.

Reactor Design

The fuel element concept and the assembly of fuel elements within the core are shown in Fig. 4.

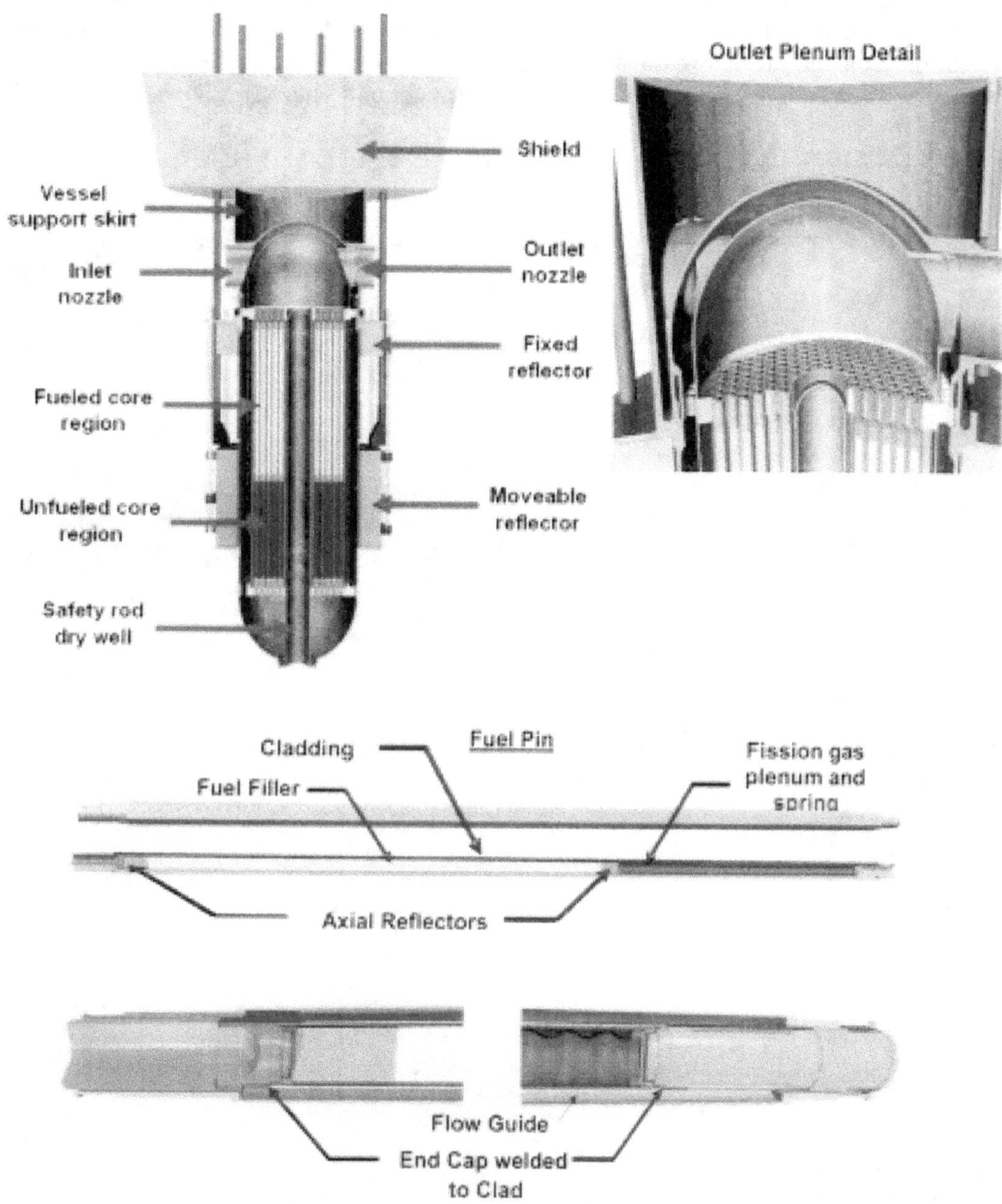

Fig. 4. Notational core and fuel element configuration. *From Knolls Atomic Power Laboratory and Bettis Atomic Power Laboratory*

The fuel element consist of ceramic fuel pellets, a gas gap to accommodate swelling, a cladding liner to improve material compatibility, and the cladding which prevents fission gas escape. The fuel is UO_2 with several refractory metals as well as silicon carbide under consideration for cladding. To accommodate the fission gas released from the fuel pellets without producing excessive clad strain from gas pressure accumulation, a fission gas plenum is situated at one end of each fuel element. Structural support for the fuel elements is provided at the support structure at only one end to allow for differential growth between the fuel elements and the structure. Several core geometries were under considerations at project termination. These are shown in Fig. 5.

Open Lattice Design

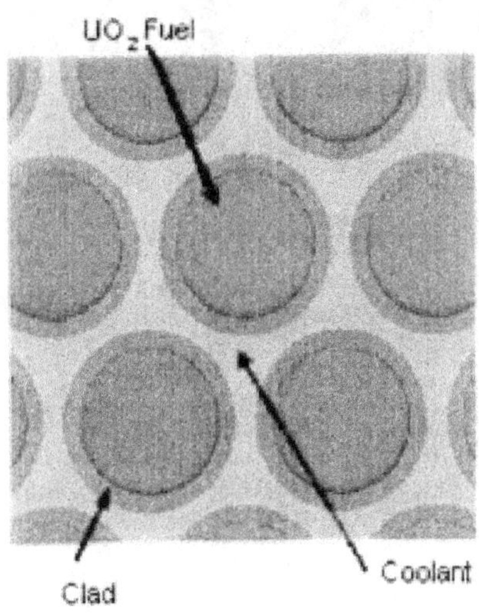

"Pin In Block" Design

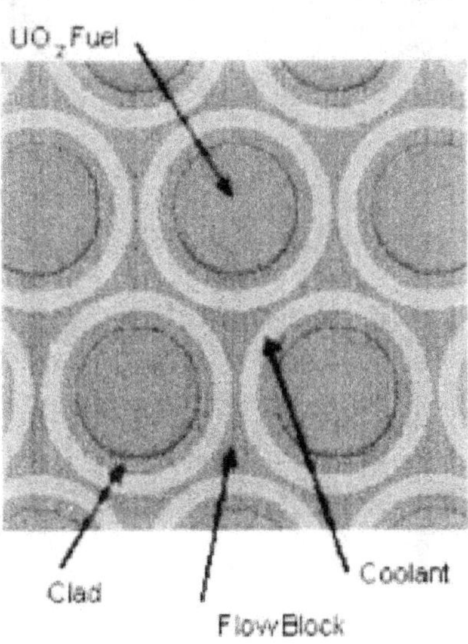

Modular Cermet Design

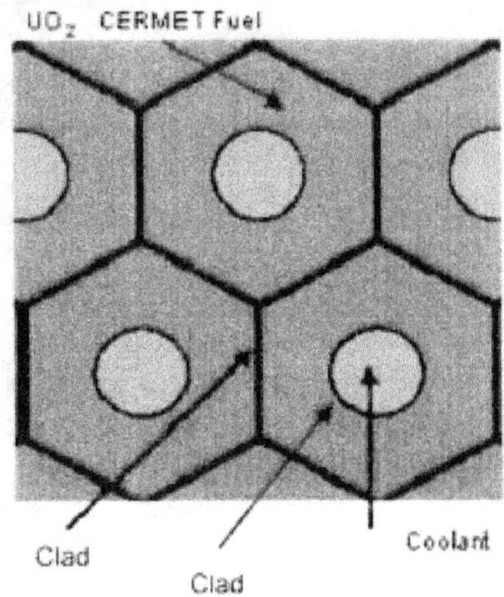

Significant Characteristics of the Open Lattice

- The design provides the lowest mass geometry.
- Rod mechanical distortion could damage the fuel. A wire wrap or other mechanical support is needed to protect the fuel pins, although, at the expense of an increased core pressure drop.
- Initial assessments indicate axial flow will not result in flow-induced vibration issues, but testing is needed.
- The lack of defined coolant channels limits the ability to control flow to individual channels.
- The lack of a core block structure provides a limited heat sink and will likely result in the highest fuel temperature excursion during a transient.

Significant Characteristics of the Annular Flow "Pin in Block"

- Flow is contained in each distinct channel, forcing coolant flow across each pin.
- Flow can be metered to each channel to achieve a uniform temperature rise up each channel.
- The geometry is heavy due to the mass needed for the block.
- The extra mass is separated from the fuel and is less of an effective heat sink in reducing transient fuel temperatures than designs where the fuel pin is in contact with the core block. The magnitude of this effect was not quantified at the time of project termination.
- The fuel pins need to be positioned in the channel with sufficient surrounding space to prevent overheating.
- Inlet orificing is used to distribute flow and balance the coolant temperature rise.

Significant Characteristics of the Modular Cermet Design

- Dedicated flow paths force coolant flow in each channel.
- Designed in modular sections for manufacturability.
- Most of the temperature drop is from the clad to coolant with no internal gap and high metal conductivity. Design is more sensitive to large uncertainties in the gas convective heat transfer coefficient.
- Concept requires optimistic fabrication and performance design assumptions to be mass competitive.
- The design considered has a very high ceramic volume fraction of 60%.
- The design provides the best use of structure to reduce transient fuel temperatures.

Fig. 5. Comparison of primary core geometry options. *From Knolls Atomic Power Laboratory and Bettis Atomic Power Laboratory*

Reactor core parameters for a number of the material and arrangement options are shown in Fig. 6 for 1 megawatt thermal subsystems. Table 3 provides a more detailed breakdown of certain reactor core options. Shown are the effects of core design changes with changes in cladding material, core configurations, coolant pressure, and required reactor power to meet the nuclear and mechanical design requirements.

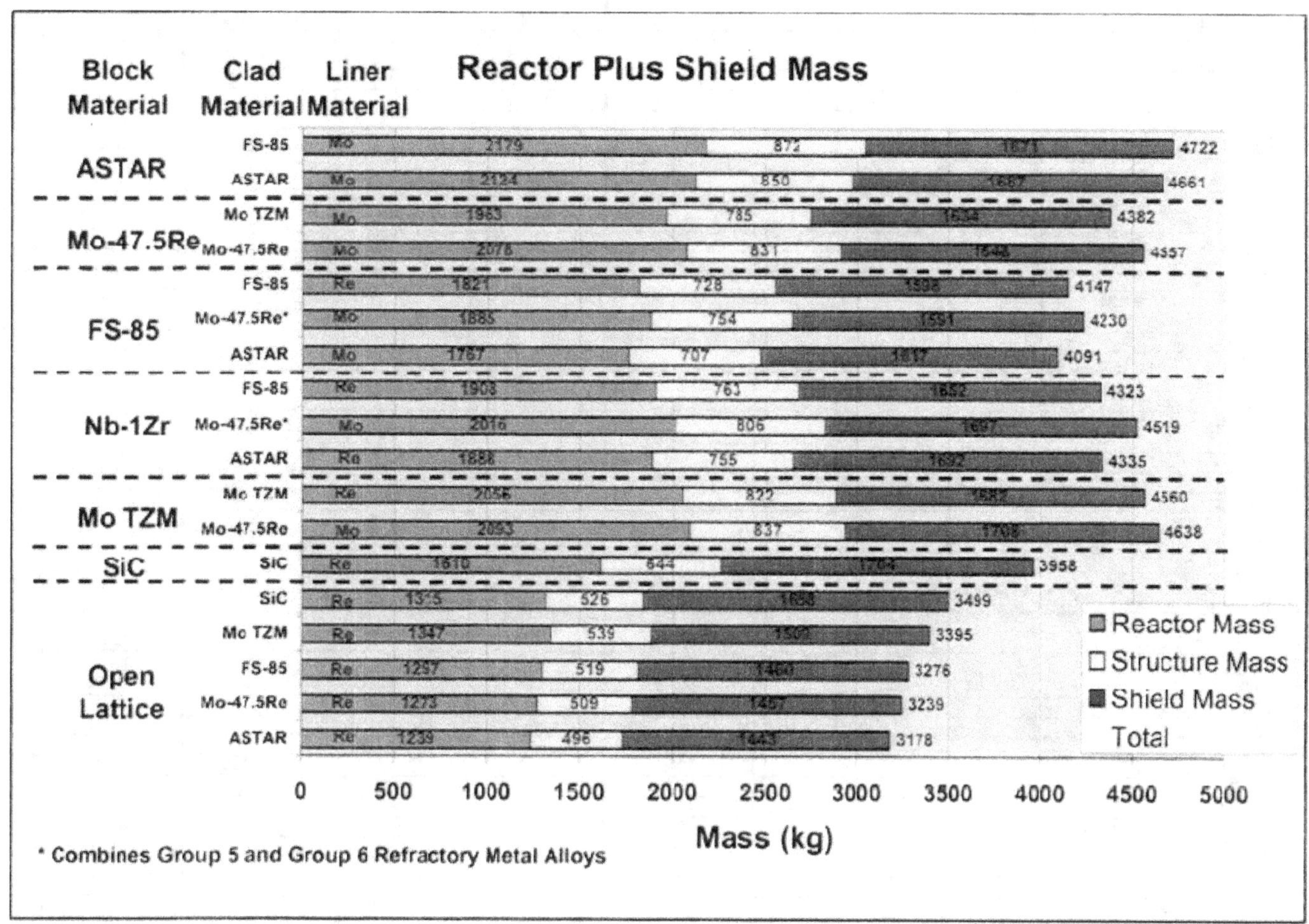

Fig. 6   Reactor plus shield mass for various material options. *From Knolls Atomic Power Laboratory and Bettis Atomic Power Laboratory*

The core is enclosed in a pressure vessels and surrounding the pressure vessel is a combination of fixed and movable reflector segments. A control system is used to position the movable reflector segments to startup, maintain the desired operation temperature, compensate for fuel burnup, and shutdown the reactor. At least one safety rod is included in the reactor for use during core assembly, transport and launch. This would be withdraw prior to initial criticality, once a stable orbit was achieved.

Core exit temperature is limited to 1,150 K to permit the use of more conventional materials in the reactor and energy conversion system. The hot gas from the core is expanded through a turbine. The turbine is connected on a common shaft to the compressor and an alternator. The alternator converts part of the turbine power to electric power for the spacecraft. This alternator power is high frequency, three-phase and is conditioned and distributed in the PCAD (Power Conditioning and Distribution) sub-system to provide high voltage to the electric propulsion units and low voltage to the computers and instruments. A parasitic load radiator is used to shed excess electrical power.

From the turbine, the gas passes through a regenerative heat exchanger (a recuperator) and a gas cooler. The cooled gas is then pumped back through the recuperator and to the core by the compressor. In the cooler, heat is transferred to a pumped liquid loop (water or NaK) and radiated to space through radiator panels.

Table 3. Representative reactor core parameters.

General Parameters	Base Case	Low Power	High Pressure PCEH-129	Open Lattice PCFD-474	TZM	SIC
Power (MWth)	1	0.5	1	1	1	1
Full Power Years	15	15	15	15	15	15
Fuel Type	UO2	UO2	UO2	UO2	UO2	UO2
Fuel Form	Ceramic	Ceramic	Ceramic	Ceramic	Ceramic w/Eu03 Poison	Ceramic w/Eu03 Poison
Geometry	Modular Annular Flow Block	Modular Annular Flow Block	Modular Annular Flow Block	Open Lattice	Modular Annular Flow Block	Modular Annular Flow Block
Clad Material	Mo-47.5Re	Mo-47.5Re	Mo-47.5Re	Mo-47.5Re	TZM	SIC
Control Device	Sliders	Sliders	Sliders	Sliders	Sliders	Sliders
System Pressure (MPa)	2	2	4	2	2	2
Gas Composition (%He/%Xe)	78/22	78/22	78/22	78/22	78/22	78/22
Gas Composition (g/mol)	31.5	31.5	31.5	31.5	31.5	31.5
Vessel Material	Alloy-617	Alloy-617	Alloy-617	Alloy-617	Alloy-617	Alloy-617
Block Material	Mo-47.5Re	Mo-47.5Re	Mo-47.5Re	N/A	TZM	SIC
Shield Material	Be-B4C-W	Be-B4C-W	Be-B4C-W	Be-B4C-W	Be-54C-W	Be-B4C-W
Tcold - Average @ nozzle B.E. (K)	880	880	880	880	880	880
Thot - Average @ nozzle B.E. (K)	1150	1150	1150	1150	1150	1150
Dimensions						
Vessel Outside Diameter (cm)	51.81	49.71	54.55	54.66	51.87	63.32
Vessel Thickness (cm)	0.46	0.36	0.64	0.43	0.46	0.49
Vessel Length (cm)	159.6	131.2	139.9	137.2	149.8	145.8
Reflector Outside Diameter (cm)	85.1	73.1	78.1	78.45	85.44	86.9
Fuel Pellet OD (cm)	1.819	2.236	1.72	1.475	1.852	2.067
Gap Thickness (cm)	0.022	0.024	0.022	0.021	0.022	0.032
Clad Thickness (cm)	0.051	0.051	0.051	0.135	0.051	0.102
Fuel Pin OD (cm)	1.955	2.386	1.866	1.788	2.008	2.325
Fuel Pellet U235 Loading Density (g U235/cc)	8.26	8.26	8.26	8.26	7.58	7.44
Channel Thickness (MAFB) (cm) or Distance Between Pins (OL) (cm)	0.216	0.188	0.135	0.145	0.213	0.234
Pitch (cm)	2.614	2.977	2.35	1.94	2.213	3.01
Core Volume (L)	135.6	79.48	95.53	86.0	137.1	145.9
Number of Pins	288	144	288	402	288	228
Core Fuel Height (cm)	60.8	55.5	53.3	49.8	59.9	61.0
Gas Plenum Height (cm)	31	20	26	26.5	22	15.5
Number of Control Elements	12	12	12	12	12	12
Number of Safety Rods	1	1	1	1	1	1
Safety Rod Diameter (cm)	12.72	9.52	11.00	11.98	12.95	15.17
Shield Thickness (cm)	66.03	66.71	67.39	68.07	66.17	65.91
Shield Leading Edge Diameter [D_{degree}] (cm)	102.66	89.54	94.64	93.32	103.26	104.78
Shield Cone Angle (degrees)	6/12	6/12	6/12	6/12	6/12	6/12
Masses						
U235 Fuel Load (kg)	375	256.6	294.9	263	355	344
Reactor (kg)	2076	1350	1731	1340	1793	1432
Additional Reactor Components (kg)	831	544	692	536	717	573
Shield (kg)	1648	1334	1520	1511	1665	1661
Total Mass (Rx with Shield) (kg)	4557	3238	3943	3387	4175	3666
Key Results						
Nuclear						
Peak Burnup-B.E. (% FIMA)	2.18	1.58	2.78	2.88	2.1	2.14
Max Local Peaking Factor	1.49	1.49	1.49	1.48	1.48	1.48
Slider/Drum Worth - Most Reactive/Least Reactive Rod Out (Δp)	0.11	0.14	0.1	0.13	0.13	0.16
Mechanical						
Peak EOL Volumetric Fuel Swelling (%)	4.9	4.1	4.8	5.1	4.8	3.6
Metal - EOL Primary Membrane Von-Mises Clad Stress (MPa)	21.8	25.2	32.4	13.7	44.4	N/A
Primary Membrane Clad Creep Strain (%)	1	1	1	1	1	1
Primary Membrane Vessel Hoop Stress (Mpa)	22.6	26.6	32.4	12.6	48.1	25.0
Thermal Hydraulic						
DPcore/Psystem (CORE ONLY) (%)	1.12	1.06	1.10	0.99	1.01	1.00
Max Surface Heat Flux (W/cm^2)	16.6	14.9	19.9	15.8	16.3	17.5
Max Linear Heat Generation Rate (W/cm)	102.2	111.9	116.6	88.8	101.9	127.8
In-Fuel Power Density Core Average (W/cc)	26.4	19.1	33.7	35.1	25.6	26.0
Peak Fuel Temp BOL - B.E. (K)	1636	1637	1638	1622	1636	1646
Peak Fuel Temp over Life - B.E. (K)	1773	1771	1775	1775	1769	1720
Peak Clad Temp over Life - B.E. (K)	1275	1262	1232	1311	1274	1264

Configuration Reliability

The Brayton configuration study included the feasibility of 200 kW$_e$ or more designs using existing approaches. Adding multiple turbines will necessitate the use of check valves and/or control valves which appear feasible, but are also developmental. The number of Brayton loops is a function of meeting the reliability requirements and mass goals. A single loop is the simplest design, but deviates from the single point failure tolerance criteria and requires a momentum compensation system because of the rotational forces of the turboalternator unit. Multiple Brayton units can meet the single failure tolerance criteria. However, these configurations increase system mass and lower overall thermal efficiency. Experience has shown that redundancy can correct for manufacturing defects, human error, or an unexpected event and thus save a mission. Some parts of the power system are impractical to make redundant--the reactor, the reactor coolant loop, the boom, and the xenon propellant tank.

Various configuration architectures were evaluated in order to assess the effects of better meeting the single failure tolerance criteria. The assessment included the changes in operating conditions, mass penalties, and variations on power plant reliability. Some possible configurations are illustrated in Fig. 7. Architecture variations included one to four Brayton units, some full sized units to deliver 200 kW$_e$ and others half size units used in combination to deliver the 200 kW$_e$. Examples of changes in operating parameters are illustrated in Fig. 8.

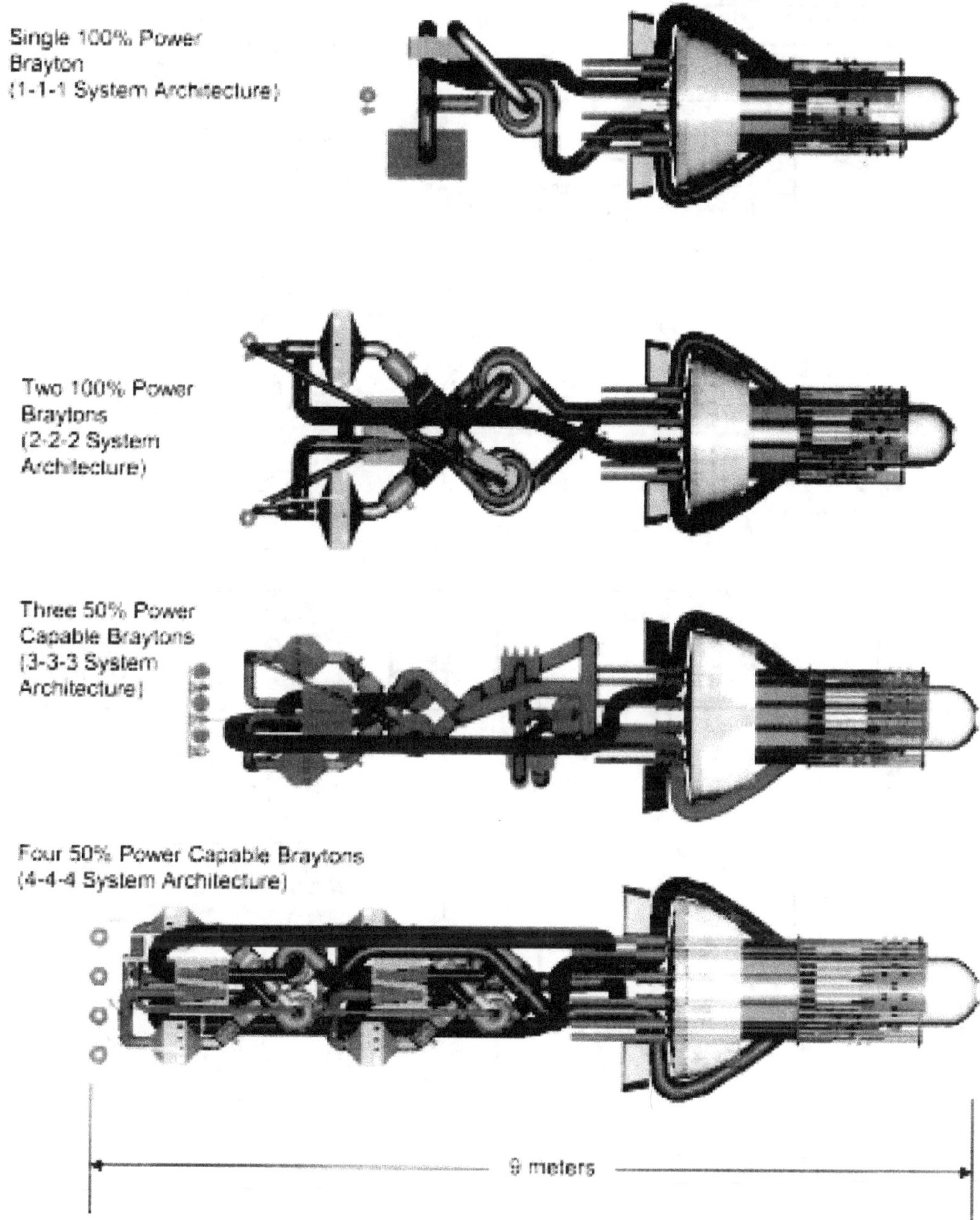

The designation of 1-1-1 or B-R-G is B is the number of Brayton units, R is the number of recuperators sized to support the operation of one Brayton unit at 100% of rated power and G is the number of gas coolers sized to support the operation of one Brayton unit at 100% of rated power

Fig. 7. Possible system arrangements with various numbers of Brayton conversion systems. *From Knolls Atomic Power Laboratory and Bettis Atomic Power Laboratory.*

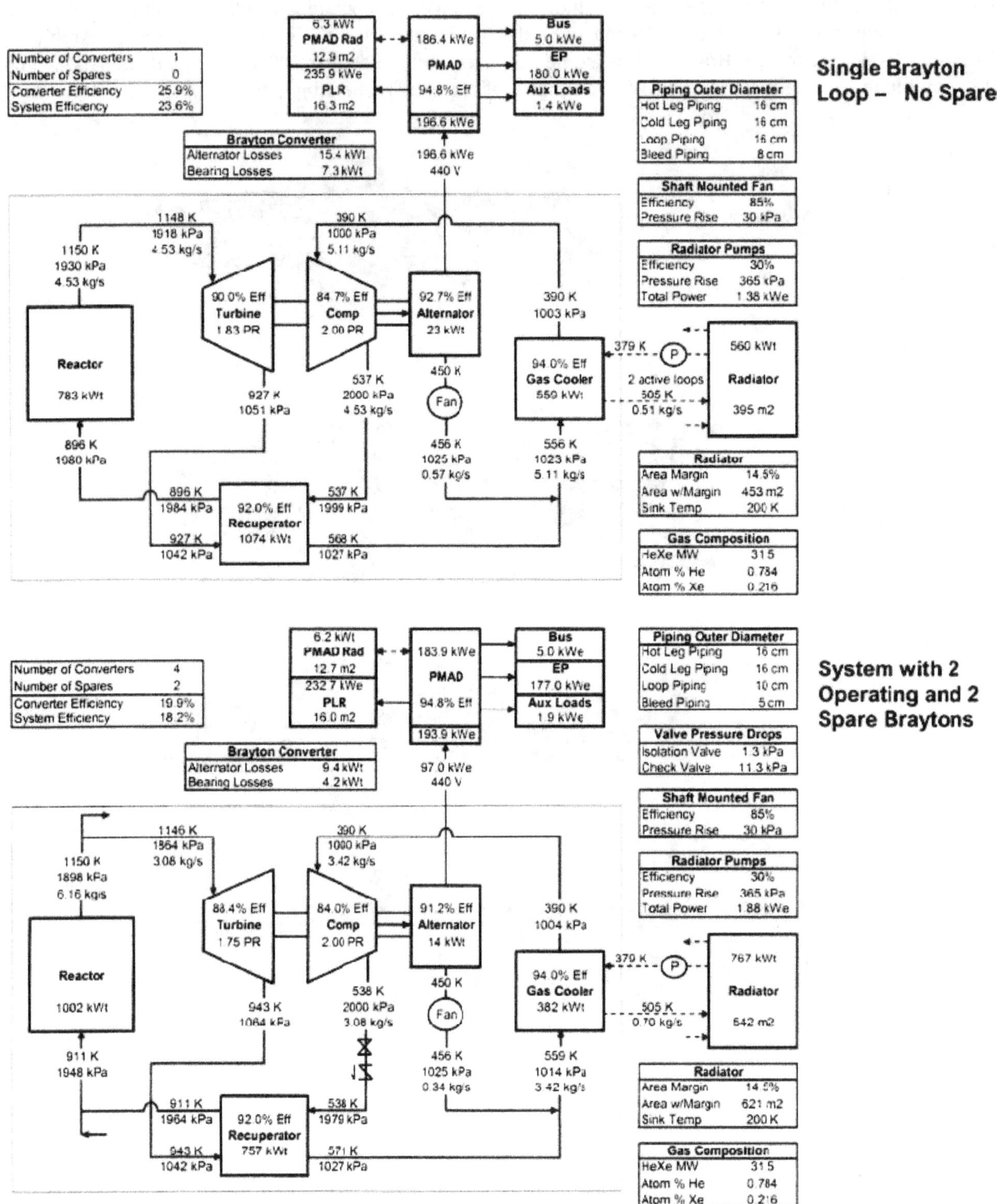

Fig. 8. Examples of heat balances for single and multi-Brayton architectures. *From Knolls Atomic Power Laboratory and Bettis Atomic Power Laboratory*

To assess configuration reliability, the reliability of each component was evaluated and these assessments combined into the various configuration architectures.[2] A range of values for the turbocompressor units was determined and so the configuration reliability assessments were made for: 0.95, the low estimate; 0.97, intermediate; and 0.99, the highest assessment. The component reliabilities assessments are shown in Table 4.

Table 4. Assumed component reliabilities and scaling sensitivities.

System Component	Reliability Approach	Failure Mode	Estimated Reliability per unit
Piping Area per loop	Reliability scaled to piping length and cross-section. Failures related to defects, fatigue or micrometeroid strikes.	Leak to space	.9996 per loop
Recuperator External Area	Reliability scaled to external area. Failures related to weld or braze defects, fatigue or micrometeroid strikes.	Leak to space	.9999 per HX (.9998 - 200 kW HX)
Recuperator Internal Area	Reliability scaled to internal area. Failures would need to be sufficiently large to impact heat transfer.	Internal leak (fails)	.9963 per HX (.9953 - 200 kW HX)
Cooler Area	Cooler has substantial internal area. Failure estimate derived from initial vendor development. Failure could be caused by defects, corrosion, overpressure or mircometeroid.	Leak to space	.9813 for all Coolers
Similar Material Welds	Weld failure rate based on expected high temperature material weld challenge. 2 pipe welds assumed for each component, including valves, and for tees.	Leak to space	.9996 per weld
Dissimilar Material Welds	Dissimilar material welds assumed for a transition from superalloy to titanium pressure boundary at cooler. Dissimilar welds assumed to be less reliable than normal welds due to potential brittle phases.	Crack/Leak to space	.9990 per weld
TCA Housing	Based on area and complexity of housing geometry.	Leak to space	.996
Turbine & Compressor	Research: Garrett data reflects 0.97 and Hamilton & Sundstrand's research reflects 0.99.	Leak to space, Mechanical failure	.95 - Low estimate .97 - HS estimate .99 – Garrett
Rotor & Shaft	Mechanical failures during start-up or as a result of overheating/overstressing.	Mechanical failure	.998 and .999
Bearing	Educated Decision based on consultation with bearing experts. Most well designed bearings fail as a result of other component performance degradation.	Seize	.997
Alternator Winding (electrical buss)	Estimate based on research on high temperature windings.	Loss of load/Cable short	.994
Electrical Penetration	Based on estimated Brayton conductor size. Scaled to penetration wire circumference. Failure due to fatigue, material degradation, or overheating.	Leak to space	.997 for 100 kW .994 for 200 kW
Isolation Valves	Educated Decision, Sticking in place. Failures only impact reliability if change in operating Brayton needed.	Sticking, electrical failure	.98
Check Valves	Educated Decision. Stuck open or closed.	Sticking	.99
Heat Rejection Loop	Long loop piping with many welds and flexible connections. Requires deployment. Failure modes include fatigue, overpressure, defects, corrosion, and micrometeriods.	Leak	.97
Heat Rejection Pump	Two heat rejection pumps per loop with check valves. No in-space failure noted in literature associated with pumped water systems.	Mechanical	.999 per loop

In order to compare reliabilities between various configurations, the values were appropriately weighted. Representative system reliability comparisons are shown in Fig. 9 for a single Brayton system and four Brayton system using two 100 kW$_e$ Brayton turbocompressors with two spares.

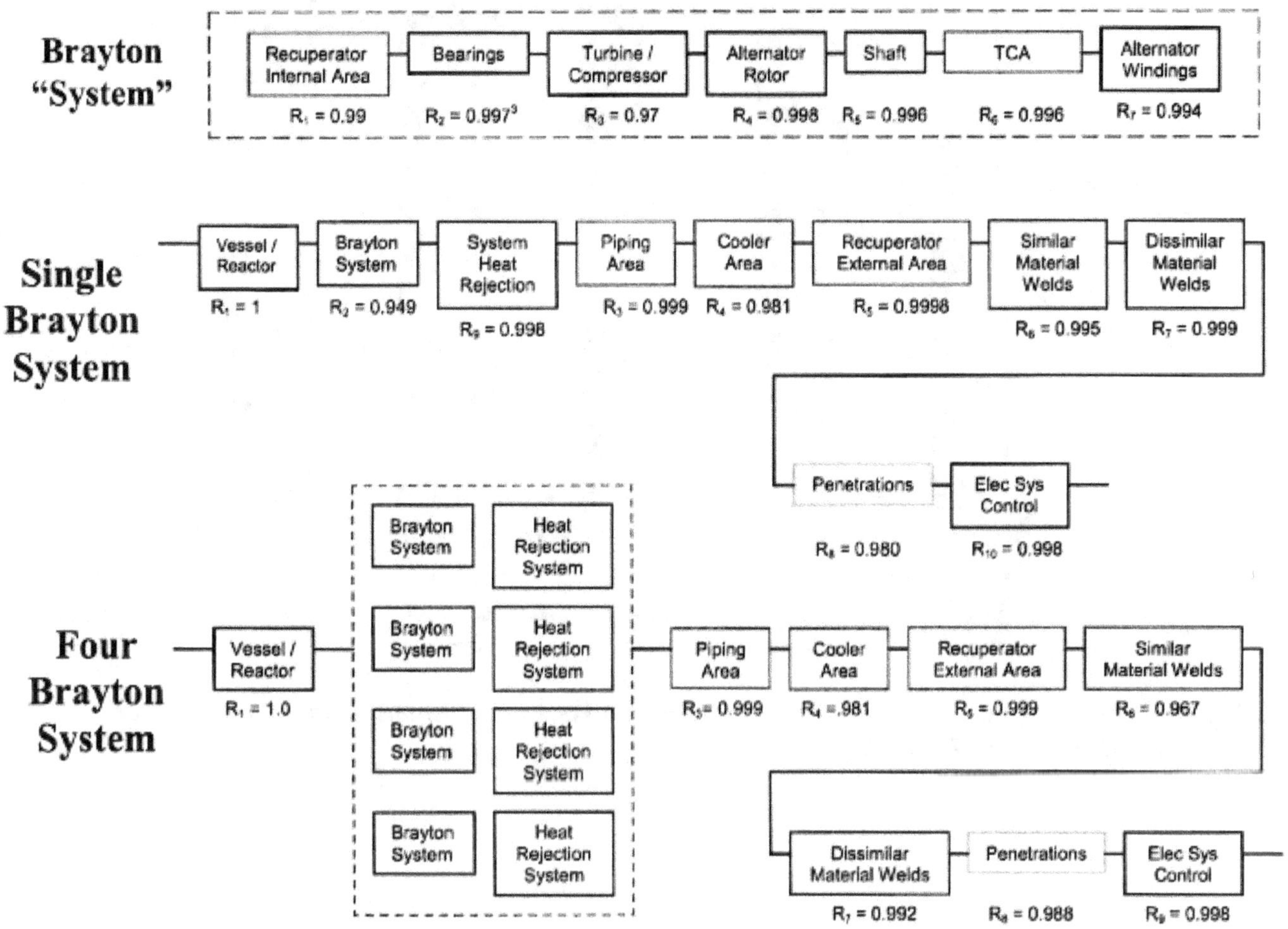

Fig. 9. **Brayton component reliability block diagram.** *From John Ashcroft, et al., 2007.*

The mass changes (Fig. 10) and the assessments on reliability (Fig. 11) are shown for representative configurations. A single Brayton 200 kW$_e$ power system has the lowest mass by over 1,000 kg. Two 100 kW$_e$ Braytons are heavier, and lack the desired redundancy. Three 100 kW$_e$ Braytons with two powered at a time and two 200 kW$_e$ Braytons with one operating and one on standby weigh about the same but have a significant mass penalty over a single 200 kW$_e$ Brayton unit. The configuration with two 200 kW$_e$ Brayton units shows improved reliability for turboalternators of 0.95 and 0.97 reliability and provides redundancy in case of single failure points. The three unit configuration with half power Brayton cycles has lower overall reliability. The four 100 kW$_e$ Braytons operating in pairs of two are much heavier and are much lower in reliability.

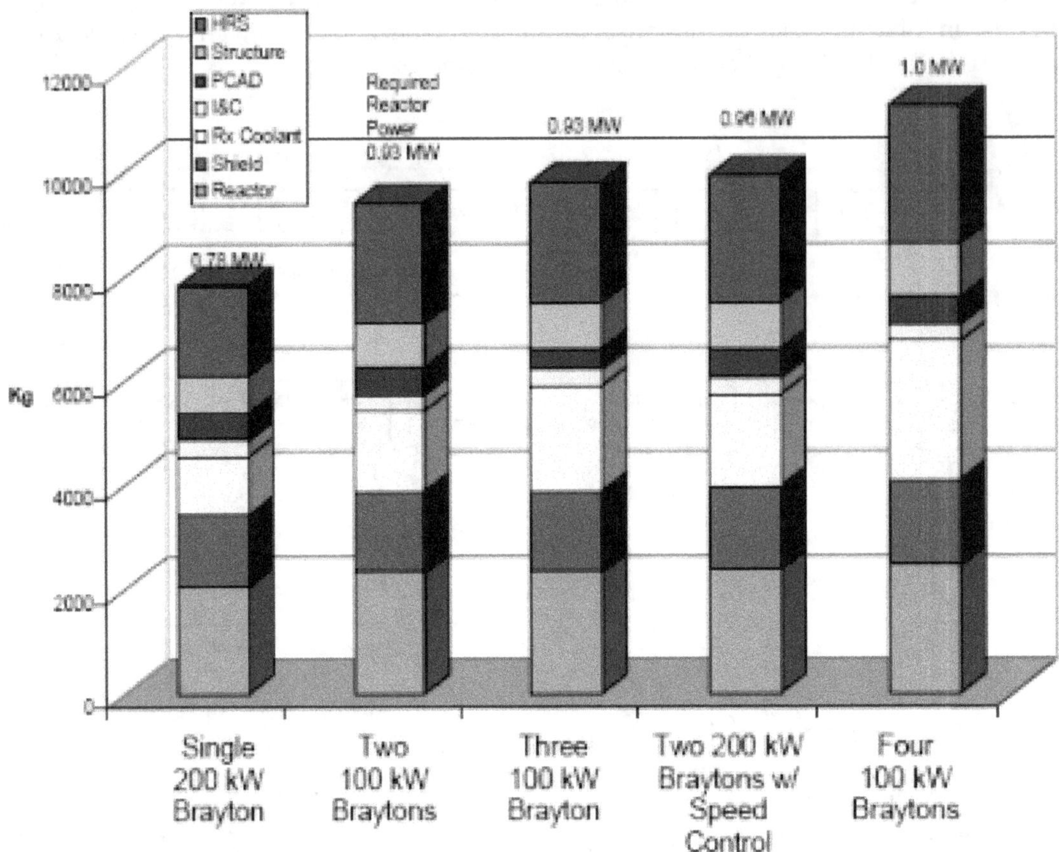

Fig. 10. Mass comparisons for key system architectures. *From Knolls Atomic Power Laboratory and Bettis Atomic Power Laboratory*

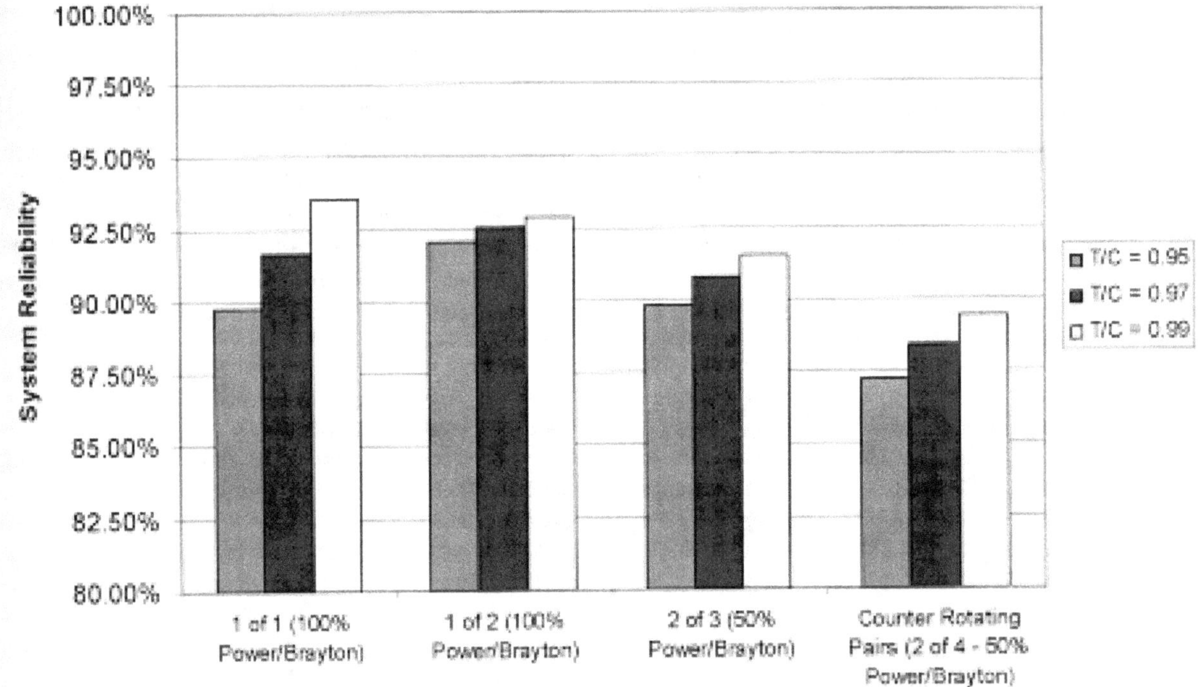

Fig. 11. Comparison of system reliability versus number of Brayton cycle converters. *From John Ashcroft, et al., 2007.*

Materials Assessment

Materials for such long life systems are a major concern. Various candidate materials for components and the developmental concerns are given in Table 5.

Table 5 list materials envisioned for use in the Prometheus program.

Component	Material Option	Operating Condition	Developmental Concern
Fuel	UO_2	900-1773 K $\sim10^{22}$ n/cm^2	Swelling/Cracking at Low Fluence/Burn-up/burn-up rate, fission gas release rate uncertainty
	UN		Fission Product Chemistry, fission gas release rate, porosity evolution
Fuel Cladding	Nb-1Zr	900-1300 K $\sim10^{22}$ n/cm^2	Creep Capability, Radiation-Induced and Interstitial Embrittlement
	FS-85		Phase Stability, Radiation-Induced and Interstitial Embrittlement
	T-111		Phase Stability, Radiation-Induced and Interstitial Embrittlement
	Ta-10W		Radiation-Induced and Interstitial Embrittlement
	ASTAR-811C		Interstitial Embrittlement, Phase Stability, Fabricability
	Mo TZM		Irradiation Embrittlement, Irradiated Creep Capability, Fabricability
	Mo-47Re		Radiation-Induced Embrittlement, Phase Instability
	SiC/SiC		Hermeticity, Fracture Toughness, Conductive Compliant Layer
Liner	Re, W, or W-Re	900-1500 K $\sim10^{22}$ n/cm^2	Embrittlement, Hermeticity, Reaction with fuel/cladding, Neutron Poison
	None		FP attack of cladding
Fuel Spring	W-25Re	800-1300 K $\sim10^{22}$ n/cm^2	Radiation-Induced Embrittlement, Relaxation
	Ta alloys		Radiation-Induced and Interstitial Embrittlement, Relaxation
In-Pin Axial Reflector	BeO	900-1300 K $\sim10^{22}$ n/cm^2	Irradiation swelling, He Gas Release, ^{6}Li Neutron Poisoning, BeO Handling Concerns
Core Block	Refractory Metal	900-1200 K $\sim10^{22}$ n/cm^2	Fabricability, Neutron Absorption
	Graphite		Fracture Toughness, C transport to refractory metal fuel
	Nickel Superalloy		Irradiation Damage, C/O transport to refractory metal fuel
In-Core Structure	Refractory Alloys	900-1200 K 10^{22} n/cm^2	Fabricability, Radiation-Induced and Interstitial Embrittlement
Reactor Vessel	Nimonic PE-16	Up to 900 K 10^{21} n/cm^2	Radiation-Induced Embrittlement, Creep Capability
	Alloy 617		
	Haynes 230		
Safety Rod Thimble (if used)	Same as Vessel	Up to 1050 K 10^{22} n/cm2	Irradiation Embrittlement, Creep Capability
	Refractory Metal		Irradiation Embrittlement, Creep, Dissimilar Material Joining
Radial Reflector	BeO	Up to 900 K 10^{21} n/cm^2	Irradiation Swelling and He Gas Release, ^{6}Li poisoning, Be/BeO Handling Restrictions
	Be		
Shielding	Water	Up to 500 K	Thermal Management
	Be	Up to 800 K	Be Handling Restrictions during manufacturing
	B$_4$C		
	LiH		Neutron and gamma swelling vs. temp. and irradiation
Shielding and Reflector Canning	Steel or Ni Superalloy	Same Range as shielding	
	Titanium Alloy		
Loop Piping	Alloy 617	300-900 K	Maintenance of internal insulation @ 900 K, Joining
	Haynes 230		Maintenance of internal insulation @ 900 K, Joining
Insulation	Porous Metal or ceramic	Up to 1150 K	Thermal conductivity, Loop Material Compatibility
	Ceramic Fiber		Thermal conductivity, Loop Material Compatibility
Insulation Liner	Mo Alloy	Up to 1150 K	Fabricability, Compatibility with insulation, embrittlement
	Superalloy		Compatibility with insulation
Turbine Casing (scroll)	In-792	Up to 1150 K	Creep Capability, Dissimilar Materials Joining (to piping)
	Mar-M-247		
	Alloy 617 or Haynes 230	Up to 900 K	Requires internal insulation
Turbine Wheel	In-792	Up to 950 K	Creep capability, Carburization/Decarburization/Deoxidation
	Mar-M-247		
Compressor	Ti-Al-V	400-600 K	Compatibility w/ gas loop
	Superalloy		
Shaft	1018 Steel	400-900 K	
	Superalloy		
Alternator Magnets	Sm-Co	400-450 K	Loss of magnet strength, compatibility with gas loop
Electrical Insulators	Ceramic or Glass	400-450 K	Hermeticity, compatibility with gas loop
Recuperator Core	Alloy 625/690	600-900K	Thermal Stability at Hot Side Temp, Braze Material concerns
	Carbon/Carbon		Compatibility with other loop components (C transport), Fabricability
Cooler Core	CP Titanium	400-550 K	Compatibility with gas and water loops
	Alloy 625/690		

Program Status At Termination

The Prometheus Project was more ambitious than past space reactor programs because of the power level (200 kW_e), the longer lifetime (up to 20 years), the greater need for system autonomy to support propulsion and science operations in deep space, and flexibility to meet many types of missions. Greater autonomy is needed because of the long delay time in communications between Earth and Jupiter and the long periods of communications blackout.

The Prometheus program concluded that the direct gas Brayton system is well suitable for deep space and surface missions and that the technologies are within reach for a first space reactor mission. Material performance and current uncertainty in characterizing material performance limited the selection and design decisions at the time of program termination. Therefore, additional work was needed in quantifying fuel materials, including cladding materials, in an irradiation environment. Since the reactor has a fast spectrum, some irradiation testing on candidate materials must be performed in fast spectrum test reactors.

A major development problem identified is the current lack of nuclear facilities within the U.S. for fast spectrum testing. Experimental Breeder Reactor-II and the Fast Flux Test Facility (FFTF) test reactors that formerly could be used for this purpose have been decommissioned   Also critical experiment facilities have been largely decommissioned. Currently, no critical facilities exist to perform benchmark critical experiments. Also, facility problems exist for reactor mock-up tests and full-up reactor testing of prototype and flight systems.

Chapter 8

Fission Surface Power (FSP)

Introduction[1]

Future human missions to the moon and Mars are projected to require high power in difficult environments where sunlight is limited and reliability is paramount. Power levels to sustain a crew of four and provide for operations are estimated to be in the 10's of kilowatts. The moon's 29.5 day rotational period results in a long, cold, lunar night of 354 hours. On Mars, though the night period is only 12 hours, the sunlight at the surface is reduced to about 20% of that of the moon. Power must be provided to sustain surface operations through the night periods as well as support daytime operations. Martian dust storms and missions to the higher latitudes further decrease the available sunlight for solar power. Both lunar and Mars missions will depend on highly reliable and sufficient power!

Nuclear fission is being considered the prime power source candidate for human lunar and Mars surface applications because it offers significant mass and volume savings relative to comparable solar power systems with energy storage. Also, recent power system studies by NASA indicate that nuclear is cost competitive with solar-based power systems.

Under the current NASA Exploration Technology Development Program and in partnership with the Department of Energy, a technology program is underway to develop Fission Surface Power (FSP). The current goal is to develop viable FSP options by 2013 to support an expected flight power system development for lunar outposts and later missions to Mars. The project objectives are as follows:

1) Develop Fission Surface Power concepts that meet expected surface power requirements at reasonable cost with added benefits over other options.
2) Establish a hardware-based technical foundation for FSP design concepts and reduce overall development risk.
3) Reduce the cost uncertainties for FSP and establish greater credibility for flight system cost estimates.
4) Generate the key products to allow decision-makers to consider FSP as a preferred option for flight development.

The system must be designed to be safe during all mission phases. Plans are to launch the reactor cold--no fission products from nuclear operations--to keep radiation below significant levels. The reactor is to be maintained in a subcritical state prior to mission events requiring its use. During operation, radiation shielding would be provided to protect humans and equipment. This shielding could use either in situ resources or earth-delivered materials. After shutdown, the radiation from the reactor would be reduced to relatively low residual levels.

Operational robustness is a necessity. Multi-fault tolerant designs and long-life components requiring no maintenance are considered essential in the design.

Most of all, the FSP affordability design philosophy is founded in principles of conservatism, simplicity, and robustness. These are presented in Table 1 and can be summarized as follows:[2]

- Modest requirements and operating conditions.
- Selection of a well-established reactor concept.
- Significant terrestrial and some space experience.
- Large fabrication experience (low cost).
- Large operational database.
- Reactor design for self-regulation of perturbations with large margins.
- Robust control system.
- Extensive reliance on existing terrestrial and prior space nuclear power systems databases.

Table 1. Design philosophy for FSP. [3]

Conservatism:
- Reactor operating temperature is constrained to less than 900 K, a relatively low temperature that allows use of stainless steel and other non-refractory, well-characterized materials for power system structure
- Maximal use of known materials and fluids with demonstrated compatibility with nuclear power applications
- Generous performance and structural margins
- Large safety factors
- Design based on extensive terrestrial reactor power system experience

Simplicity:
- Modest power and lifetime requirements
- Simple controls:
    - Negative Temperature Reactivity Feedback: assures safe response to reactor temperature excursions
    - Parasitic Load Control: maintains constant power draw regardless of electrical loads and allows thermal system to remain near steady-state
- Slow thermal response
- Conventional design practices
- Established manufacturing methods
- Modular configurations that allow testing of components, subassemblies, and assembled system

Robustness:
- Safe during all mission phases
- High redundancy
- Fault tolerance, including ability to recover from conditions such as:
    - Loss of reactor cooling
    - Stuck reflector drums
    - Power conversion unit failure
    - Radiator pump failure
    - Loss of radiator coolant
    - Loss of electrical load
- High technology readiness level components
- Hardware-rich test program
- Multiple design cycles and hardware iterations

Requirements

To demonstrate technology readiness by the early 2020's, the development of FSP systems must be initiated prior to complete definition of lunar surface system requirements. A set of requirements has been formulated to accommodate a broad range of planetary missions in a single design. Projected mission and environmental variations have been characterized by prior exploration architecture studies, based on and supplemented by data from robotic spacecraft to both the moon and Mars and the Apollo missions to the moon.

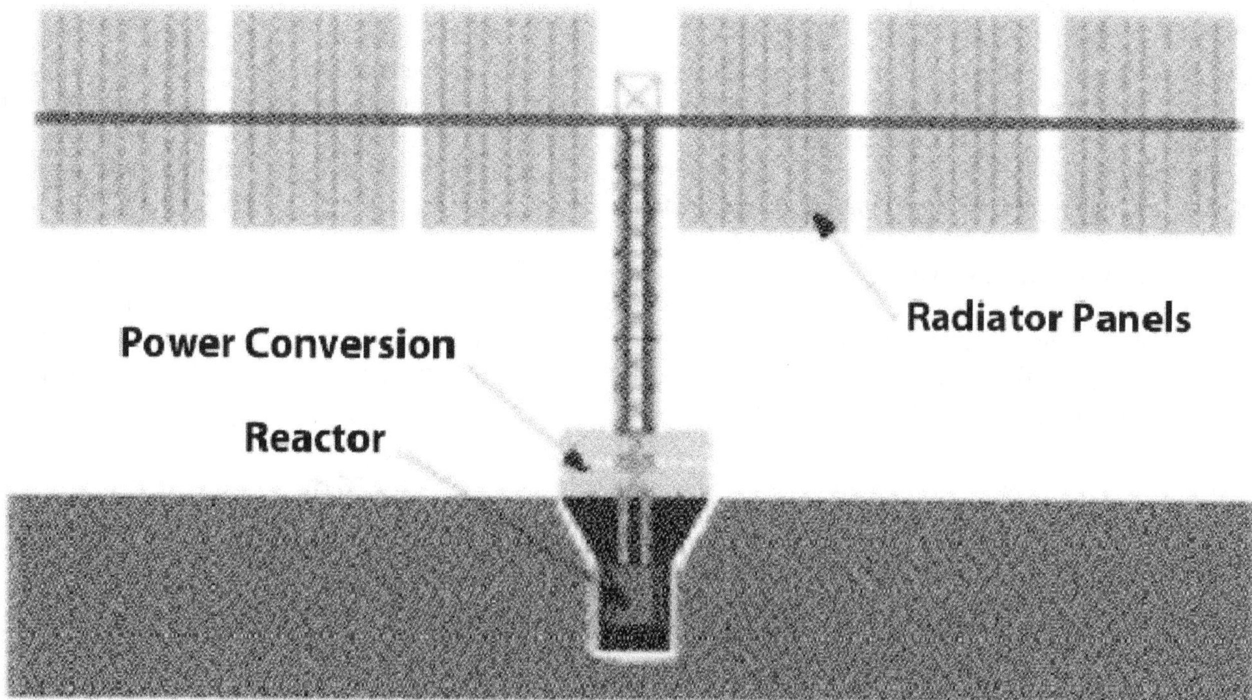

Fig. 3. Emplacement configuration. *Source:* **Lee Mason, David Poston, and Louis Quall**

The shielding geometry for the two configurations are shown in Fig. 4.

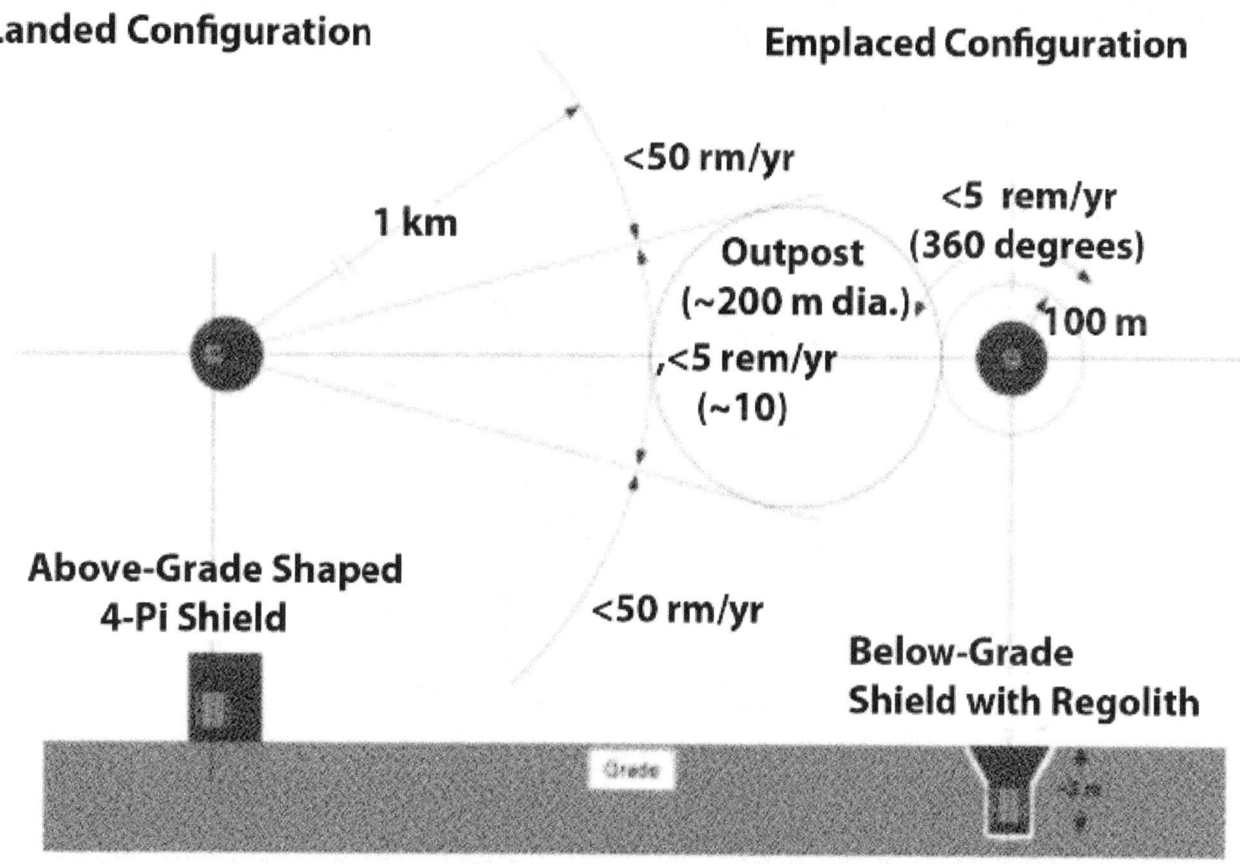

Fig. 4. Shield geometry assumptions. *Source:* **Lee Mason, David Poston, and Louis Quall**

The landed configuration might be the choice if suitable construction equipment is not available for the emplacement configuration. However, there is a large mass penalty--about a factor of two. Currently, emplacement has been selected as the baseline configuration with the landed configuration retained as an option. With this, a notational concept layout has been developed (see Fig. 5). The deployed span is approximately 34 m tip-to-tip above grade. The radiators bottom edge is approximately 1 m above the surface to minimize the potential for dust on the radiator surfaces. The reactor is located at the bottom of a 2 m excavation with an upper plug shield protecting the equipment above from direct radiation.

The stowed configuration, shown in Fig. 6, is approximately 3 by 3 by 7 m. The reactor is located at the bottom with the Stirling power converters and radiator pumps mounted on a 5 m truss structure that attaches to the top face of the plug shield. The two symmetric radiator wings are deployed via a scissor mechanism from the truss, similar in concept to the International Space Station radiators. Cavity radiators are provided to remove waste heat from components within the excavated hole.

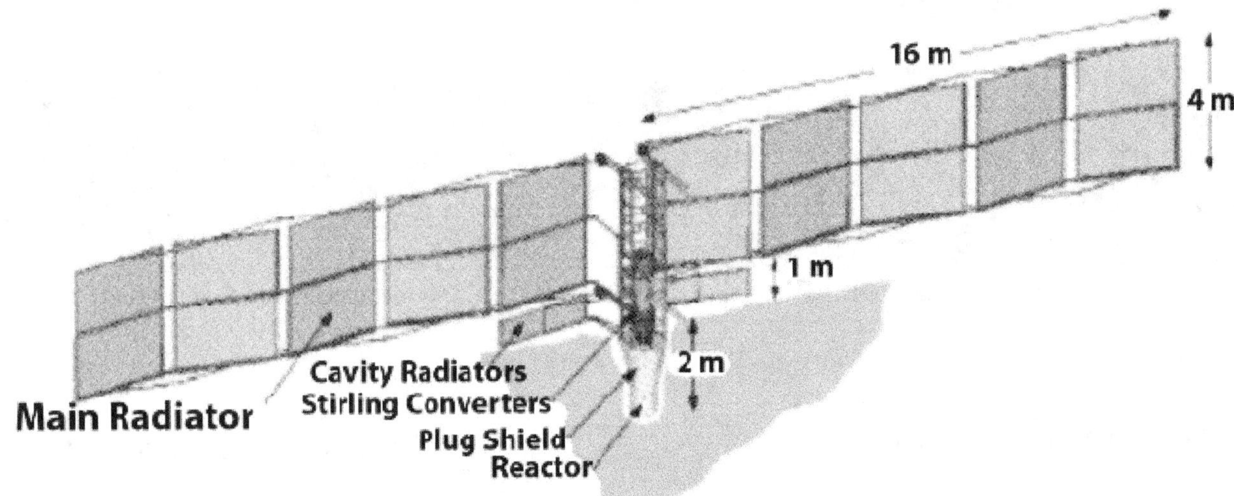

Fig. 5. Reference concept layout. *Source:* Lee Mason, David Poston, and Louis Quall

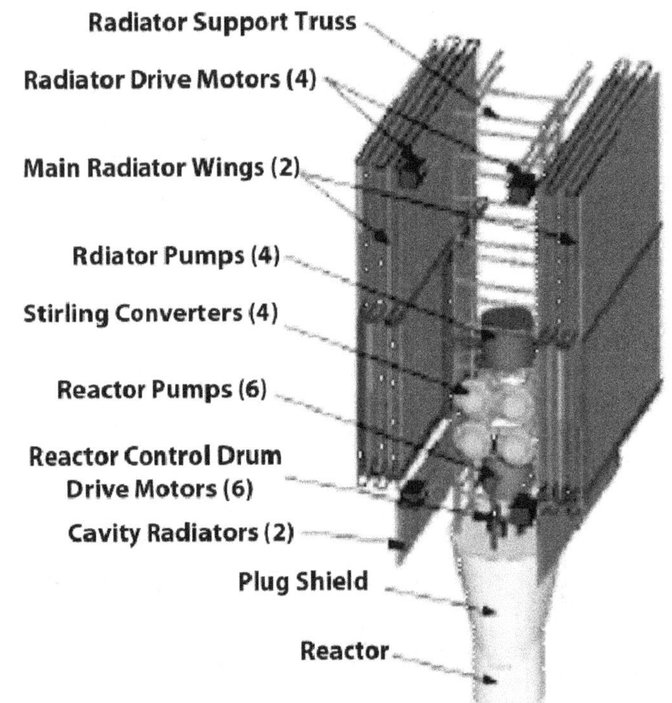

Fig. 6. Reference concept in stowed configuration. *Source:* Lee Mason, David Poston, and Louis Quall

A preliminary schematic for FSP is shown in Fig. 7. The schematic indicates the reference fluid selections, operating temperature, component redundancy levels, and overall PCAD architecture. The reactor (Rx) produces approximately 175 kW_t with a peak clad temperature of 900 K. The IHX are NaK-to-NaK heat exchangers to isolate the primary coolant NaK from the Stirling converters. The heat exchangers allow the temperatures in the reactor and Stirling converters to be optimized independently. Each intermediate NaK loop includes redundant pumps and services two Stirling converters at a supply temperature of 880 K. The Stirling hot end temperatures are designed to be 830 K. Each Stirling engine has a dedicated cooling loop with a 400 K water exit temperature. The effective radiator temperature is 380 K. The use of redundant components and parallel fluid loops permit the system to produce partial power in the event of failures.

Each Stirling converter includes two axially opposed Stirling heat engines and two linear alternators. The alternators deliver 6 kW_e each at 400 Vac and 100 hz to a Local Power Controller. The controller is located 100 m or more from the reactor. It converts the 400 Vac to 120 Vdc for distribution to the users. Forty-eight kilowatts provides margin for electrical losses ($\sim$ 3 kW_e) and system parasitic loads ($\sim$ 5 kW_e) in meeting the 40 kW_e user loads. A 5 kW_e solar array and 10 kWh battery are included for startup and backup power.

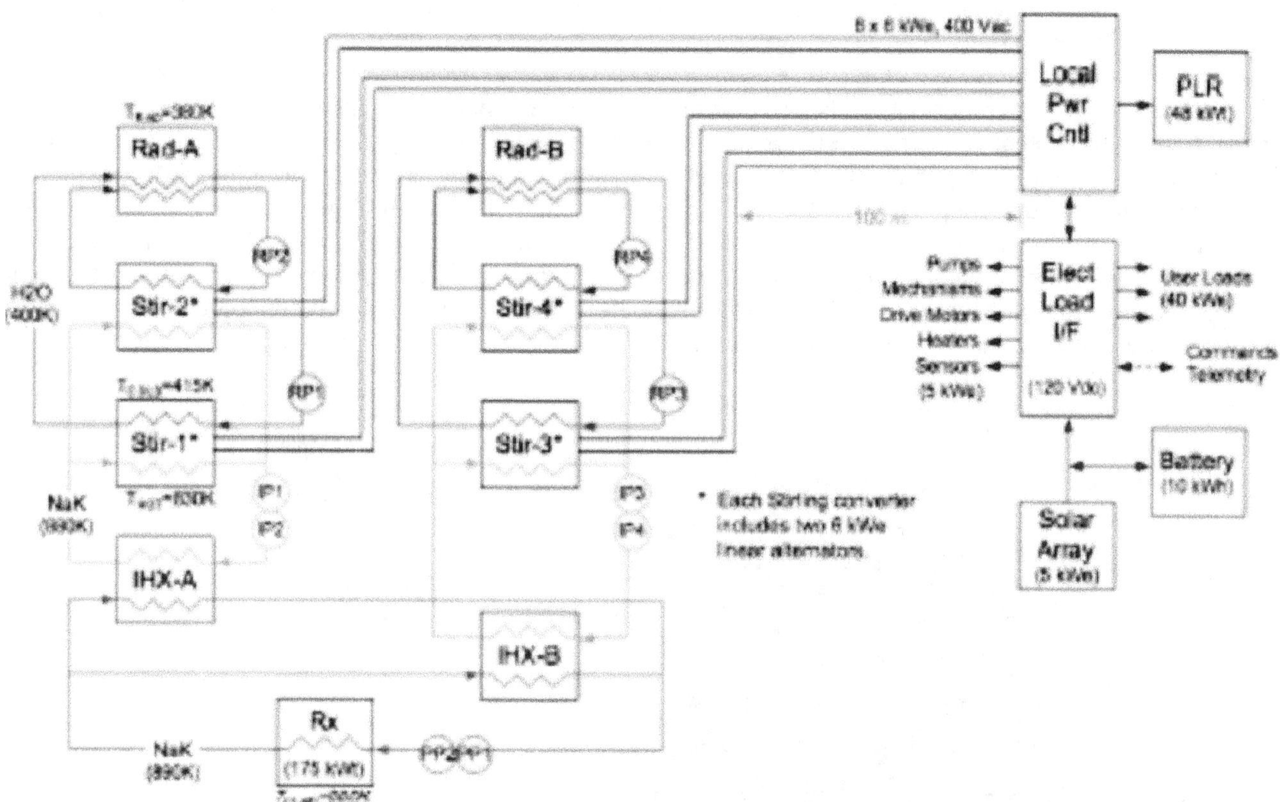

Fig. 7. Reference concept system schematic. *Source:* **Lee Mason, David Poston, and Louis Quall**

Reactor Module[6]

The reactor design philosophy places the emphasis on minimum technical risk. This led to the selection of the stainless steel/UO_2 reactor with pumped NaK coolant concept because the materials and components are well characterized. In particular, the SS-316 stainless steel was selected because it is readily available, low cost, easily workable, and an extensive reactor industry data base already exist. Planned fuel burn-up of less than 1 percent over 8 years alleviates concerns of fuel swelling due to buildup of fission product gases.

The core contains 85 fuel pins with a 1.75 cm pin outside diameter and a stainless steel (SS-316) clad thickness of 0.051 cm. The UO_2 fuel is 93% enriched and contains 1 molar percent of natural gadolinia (Gd_2O_3). The gadolinia provides fuel stability and provides a spectral poison to help maintain subcriticality during water immersion scenarios. The fuel is assumed to be 95% theoretical density, with a nominal 0.0065 cm assembly gap between the fuel and clad (cold/BOL). Fuel swelling is small (~ 0.8% in the peak pellet) and the gap between the pellet and clad grows with temperature--the SS-316 clad expands at a greater rate than the UO_2 fuel.

The cold/BOL fuel pin height is 38 cm. Within the fuel pin there is 8 cm of BeO pellets above the fuel and 5 cm of BeO pellets below the fuel. The extra BeO at the top is to help augment the shielding. A small expansion region is located at the top of each pin to serve as a fission gas plenum; however, the fission gas production/release at the expected burnup and temperature is not expected to cause stress/creep concerns in the cladding.

The average cladding temperature is 870 K with peak temperatures of 898 K. The peak cladding fast fluence is $5.0 \text{ X } e^{21}$ n / cm^2, which is below the threshold of significant ductility loss. The peak center-line fuel temperature during nominal operation is 1,095 K, average fuel center-line is 1,010 K, and overall average fuel temperature is 955 K.

The reference reactor core geometry, shown in Fig. 8, uses a triangular pitch pin-lattice arrangement. A tie-structure holds the pins axially and radially on one end, but allows the pins to float axially on the other end. At low power the pin spacing is very tight (P/D = 1.02). This is beneficial because: (1) it allows the void fraction to be low enough so that internal safety rods, or other measures, are not needed to maintain flooded subcriticiality, and (2) it keeps the potential reactivity effects of pin movement small, even if the spacing mechanisms should fail. Wire wrap is used to help maintain spacing and promote interchannel mixing. Clearance in the assembly between the wire and adjacent pin is 0.0076 cm. This provides ample flow even if pins are clumped together and keeps reactivity effects small (on the order of cents). Flow is highly turbulent (Re = 24,000), and the film temperature drop in the coolant is only a few degrees K. This makes the design very tolerant to any thermal-hydraulic changes caused by pin movements. The relatively large pins reduce the likelihood of pin deflections and the potential tolerance stack-up that could significantly change flow characteristics.

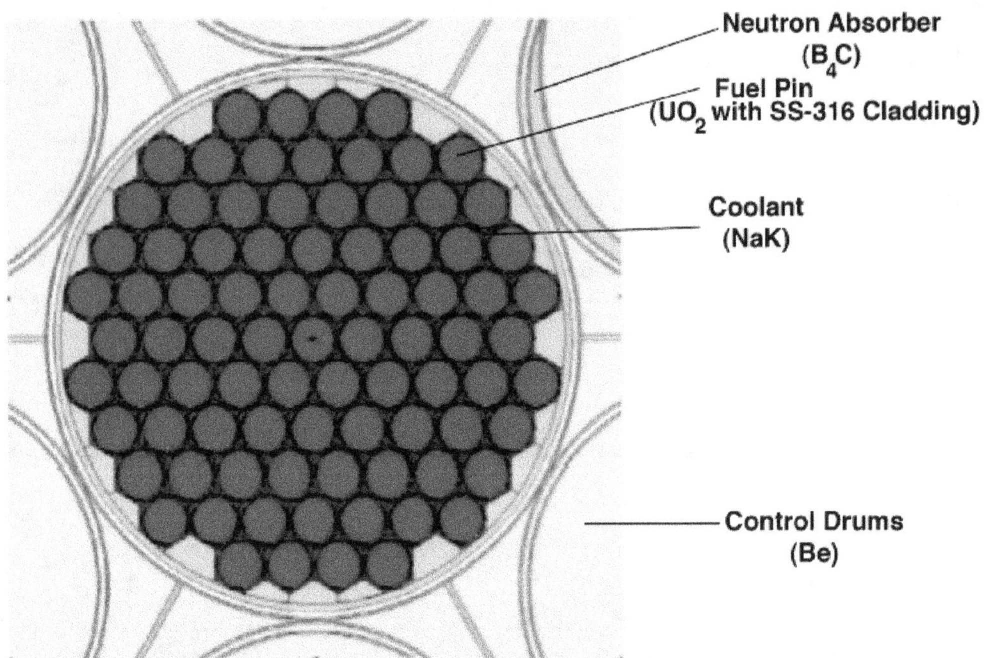

Fig. 8. Radial core schematic. *Source: David I. Poston, et al*

The coolant for the reactor is NaK-78. NaK liquid metal reactor coolant has been used in fast spectrum reactors since the 1960s. Flow is fed to the bottom of the reactor via piping that travels between the radial reflector and shield. This allows the external lattice flow area and hydraulic diameter to be large (minimizing pressure drop), and more importantly brings the radial reflector and control drums closer to the core and removes a potential flooded region in the reactor.

Coolant enters the core at 840 K and exits with a mean outlet temperature of 890 K. NaK is baselined because of its low freezing point of 262 K and significant experience in other space power systems (for instance flight reactors SNAP-10A, RORSATs, and Topaz I). It is liquid at room temperature, and radiative heat losses at 262 K are sufficiently small to require minimal heating (if needed) in cold space or shaded regions. One potential problem with NaK is radiation activation. To isolate the NaK reactor coolant from the power conversion subsystem, an intermediate pumped-loop heat exchanger has been added. The reactor coolant NaK flow rate is ~ 4 kg / s and the primary loop pressure is ~ 5 psi. A pump efficiency of ~ 10% would result in a power requirement of ~ 2 kW_e. It is anticipated that two pumps capable of full flow will be in the primary loop and two primary-to-intermediate heat exchanger loops (currently it is assumed that all 6 pumps are identical, i.e., same unit design for primary and intermediate loops), along with an accumulator, insulation, and potentially impurity traps and/or heater elements to prevent freeze if needed.

The reactor core is enclosed in a cylindrical vessel made of SS-316. The vessel is 0.25 cm thick. A cylindrical vessel is used to simplify structural and mechanical design and fabrication. The vessel thickness was sized to meet 1/3 ultimate and 2/3 yield stress criteria during the postulated worst-case transient (unmitigated loss-of-flow) in terms of coolant/vessel temperature and pressure. Also, SS-316 was selected because the peak vessel fluence is well below the significant damage thresholds.

A radial reflector surrounds the reactor core. Within the reflector are six control drums. The reflector is composed of Be metal in a SS-316 can. Though Be is a heavier option than BeO, Be is less susceptible to radiation and temperature induced swelling and cracking. Data suggest that the swelling will be less than 1%. Control drums are baselined also because the reactor can more easily be integrated into a 4π radiation shield. The control drums are composed of Be in a stainless steel can, with a small arc of B_4C absorber to provide control. Each drum has a dedicated motor and drive mechanism. A clearance gap of 2 mm is provided between the drums and reflector to prevent contact that might hinder drum movement. The drums are designed to meet the criteria in Table 2. This includes 2% excess reactivity margin at end-of-life (EOL), with 0.5% margin in the case of a stuck drum at EOL.

Table 2. Reactor criticality requirements.

Description	Requirement
Drums out (BOL-EOL, Cold-Warm)	>1.02
One drum stuck in (BOL-EOL, Cold-Warm)	>1.005
Drums in (BOL-EOL, Cold-Warm)	<0.95
Dry sand surround, radial reflector off	<0.985
Dry sand surround, drums in	<0.985
Freshwater flood, on concrete, radial reflector off	<0.985
Freshwater flood, on concrete, drums in	<0.985
Seawater flood, wet sand surround, radial reflector off	<0.985
Seawater flood, wet sand surround, drums in	<0.985

In regard to radiation shielding, the reference design approach is to emplace/bury the system in regolith. Fig. 9 shows the reactor enclosed in a shield that may or may not be needed when buried in regolith. The shielding requirements include:
- · Dose to unshielded astronauts at closest outpost location 5 rem / y,
- · Fast neutron fluence to Stirling alternators 1 MRad and 1 x e^{14} n / cm^2,
- · Gamma dose to Stirling alternators 2 mrad.

The neutron shielding material is B_4C contained in stainless steel. Boron carbide shielding has been used extensively in terrestrial applications. B_4C was selected over LiH or H_2O despite being less mass effective because it may be simpler to develop and lifetime qualify. The upper shield contains six vertical penetrations for the control drum shafts and penetrations for inlet and outlet NaK piping. These penetrations should not cause significant streaming paths because they are filled with NaK and the drum shafts.

B_4C has the benefit of providing not only neutron shielding but also acts partially as a gamma shield. As a result there is only a small mass penalty for using stainless steel (as opposed to W or deleted U) to meet the gamma shield requirements.

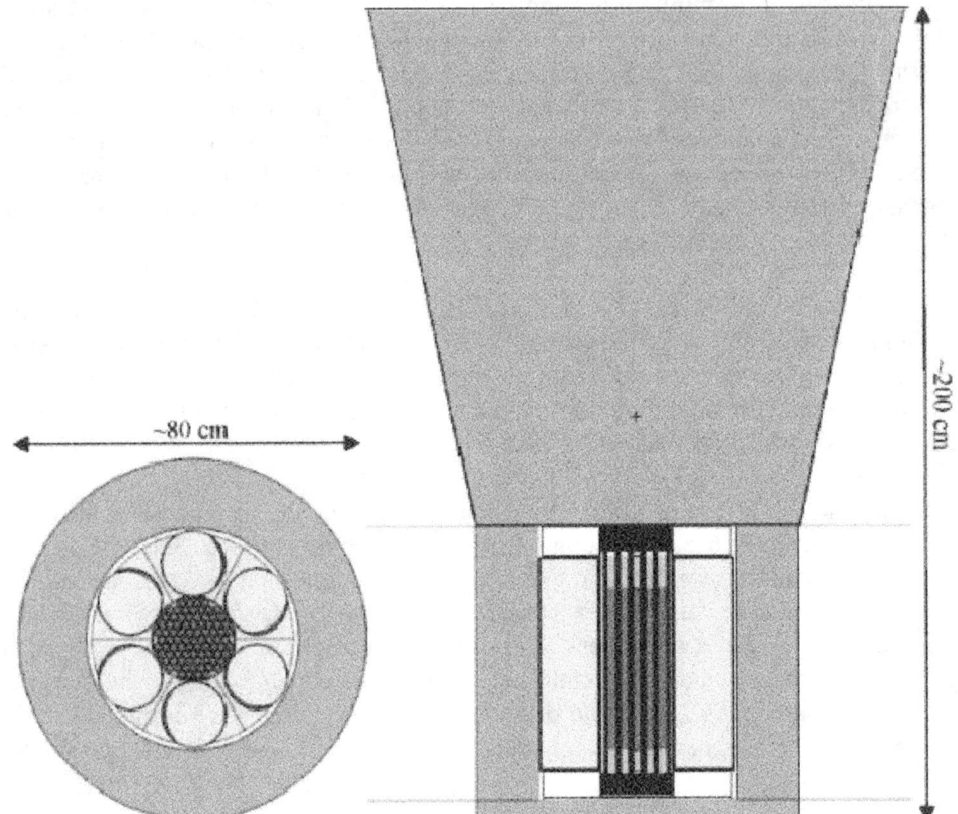

Fig. 9. Layout of reactor and shield. *Source: Lee Mason, David Poston, and Louis Qualls*

The reactor control strategy is to keep the instrumentation and controls as simple as practical. The reference reactor control philosophy is to perform only one function after startup and initial operations--to control reactivity via drum movement based on reactor temperature. Sufficient reactivity is provided to complete the mission even if 1 of the 6 control drums cannot be moved from its launched/stowed position. Once the reactor has reached criticality, only 1 drum has to remain working through end-of-life to provide the reactivity needed. Even if all drums failed, there would be only a modest drop in temperature throughout the lifetime, and full power could possibly be provided for the full lifetime. The drum rates will be limited to prevent significant overpowering and temperature during startup or spurious drum movement while still allowing reactor startup in less than a day. A significant advantage of the low thermal power of the system is the low decay-power and low adiabatic heat up rate. The reactor will require little, if any, control response to possible transients (loss of PCS load, loss of flow, spurious reactivity insertion).

Power Conversion Module[7]
The FSP system uses four Stirling converters that produce approximately 12 kW$_e$ each, for a total of 48 kW$_e$. The Advanced Radioisotope Power System program is the process of qualifying a power system using Stirling engine electric power converters for space flight. The same technology will be scaled for this application. The Stirling converters are comprised of two, axially opposed free piston engines coupled to linear alternators (pictured in Fig. 10). The unit is approximately 1.2 m in length. The Stirling engine is heated by pumped, single-phase NaK from the intermediate heat exchangers. Each intermediate heat exchanger feeds two Stirling engine units and transfers heat from the NaK into the Stirling heater heads (refer to Fig. 7). The intermediated heat exchanger assembly includes a single centered inlet manifold, flow annulus, and two exit manifolds that form a jacket over the monolithic heater heads. This configuration provides uniform heating with reasonable fluid velocity and

negligible pressure drop. The heater head on the Stirling converter is made of IN718 and operates at 830 K hot-end and 415 k cold-end. This is well within the Stirling engine data base. The converter efficiency is ~ 28% and the linear alternators deliver 6 kW_e each at 400 Vac, 100 hz, single-phase to the Local Power Controller.

A Power Management and Distribution (PMAD) is used to adjust the voltage to mission needs of 120 Vdc.[8] This includes the transmission cabling, power controls, and electrical bus interface. A shunt is used to compensate for user load variations allowing the Stirling converters to operate in a constant power mode. A parasitic load radiator is provided in the PMAD Module to dissipate excess electrical power not consumed by the outpost loads. The gross Stirling power output of 48 kW_e is sufficient to account for FSPS auxiliary loads such as pumps, motors, etc, and PMAD losses and still provide a net power to the outpost of 40 kW_e. Within the habitat module, a remote operations console is located to monitor the health of the power plant.

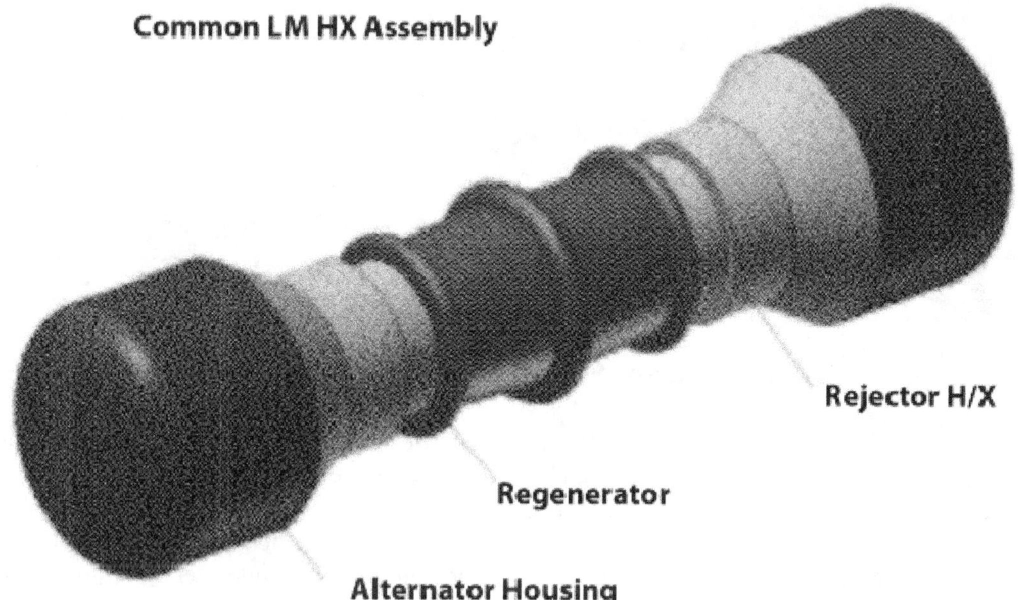

Fig. 10.   Twelve kilowatt-electric dual-opposed Stirling convertor concept. *Source: Lee Mason, David Poston, and Louis Qualls*

Free Piston Stirling Engine flight development is underway as part of the Advanced Stirling Radioisotope Generator (ASRG) program. This development is directly applicable to the FSP design. The ASRG (Fig. 11) will produce 151 W_e power at the beginning-of-mission and to have a design life of 14 years plus three years in storage. It includes two Stirling engines that are coaxially aligned and in opposed operation to minimize piston-induced vibration. The ASRG include:
- an integrated, single fault-tolerant controller based on active power factor correction,
- autonomous operation and fault isolation from the space vehicle,
- one-half generator operation capability in case of an engine failure, and
- operation in vacuum of deep space and in planetary atmospheres.

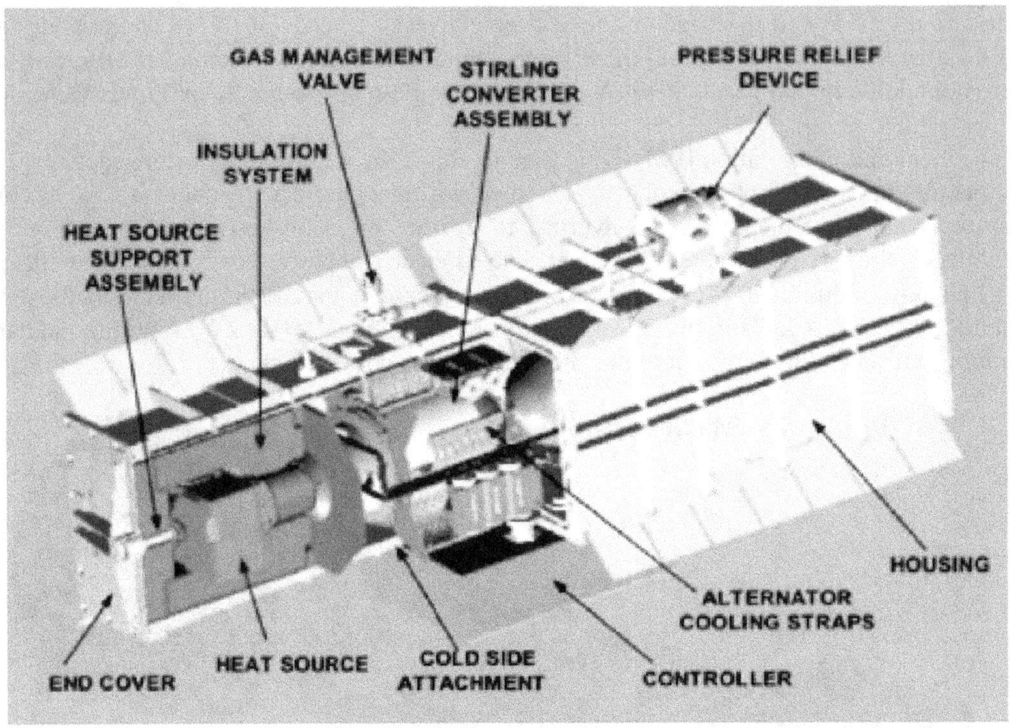

Fig. 11. Advanced Stirling Radioisotope Generator.

A schematic of the Stirling convertor is shown in Fig. 12[9] with an external view in Fig. 13. The Advanced Stirling Converter currently part of the Engineering Unit uses a heater head made of Inconel 718. This limits the hot-end temperature to about 925 K. With the use of MarM-247 superalloy (% weight Mo 0.65, W 10.0, Ta 3.3, Ti 1.05, Cr 8.4, Co 10.0, Hf 1.4, Zr 0.055, B 0.15, balance Ni) for the heater head material, the allowed operating temperature can be increased to 1,125 K. With this hot-end temperature and a temperature ratio of 3:1, the efficiency of the convertor can reach nearly 40% from heat input to alternating current output from the alternator.[10]

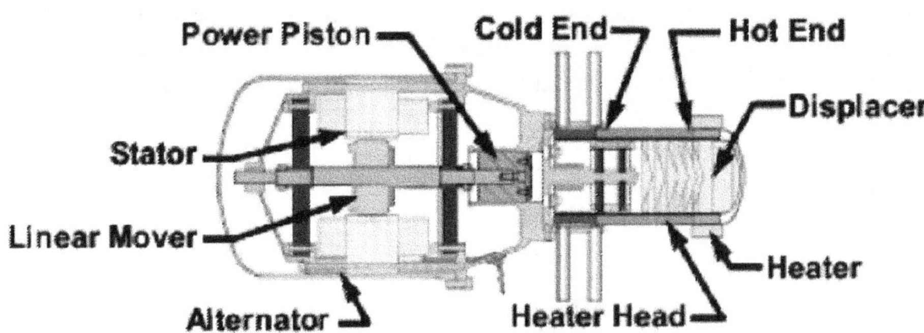

Fig. 12. Stirling Converter Assembly.

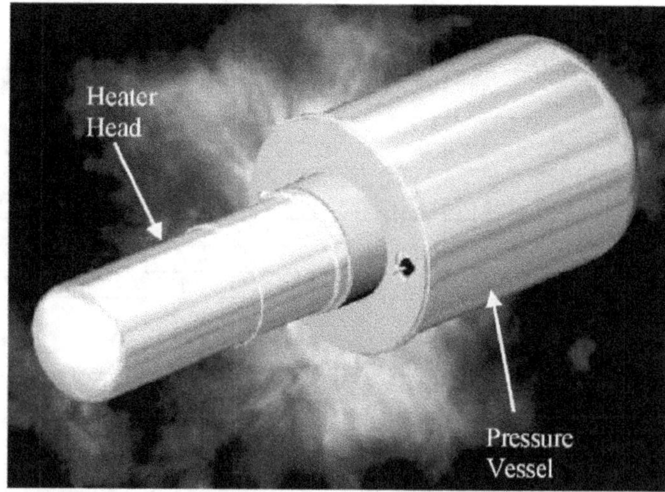

Fig. 13. External view of the Advanced Stirling Converter (ASC-2)[11]

The Advanced Stirling converters have evolved over time. Table 3 list different versions of ASC machines. The FTB (Frequency Test Bed) was used to investigate performance at power levels about 80 W_e operating at high frequency (> 100 HZ). These achieved a record setting 36% efficiency at a temperature ratio of 3.0. The ASC-1 units included the use of higher temperature heater head materials capable of > 17 y lifetimes and developing hermetic sealed units. The ASC-2 further evolved the convertors with lighter weight units being the objective.[12]

Table 3. Different versions of ASC machines.[13]

	Units	Head Material	Head Temperature (C)	Hermetic	Additional Design Evolution
FTB	2	Stainless Steel	650	No	
ASC-1a	2	MarM-247	850	No	
ASC-1b	2	MarM-247/IN-718	850	No	Inertia welded head
ASC-0	2	IN-718	650	Yes	Brazed displacer dome/body joint
ASC-1HS	2	MarM-247/IN-718	850	Yes	Improved piston/gas bearing/center port configuration
ASC-E	3	IN-718	650	Yes	Improved piston sensor, Inconel rejector pressure wall for reliability
ASC-2	4	MarM-247/IN-718	850	Yeas	Diffusion bonded external acceptor

The ASC convertors are currently being run to accumulate hours and operating experience as shown in Table 4. The Engineering Unit will incorporate a convertor with an Inconel displacer and operate at 925 K. Therefore, priority has been given to operating at these conditions.

Simulated launch load vibration testing in both axial and lateral directions indicate that the generators meet Qualification level and Qualification plus 3 dB levels. The result are given in Table 5.

Table 4. Accumulated runtime on ASC-1 convertors.

| | ASC Accumulated Hours** as of April 16, 2007 | | | | | Thermal cycles to |
	Total	≥ 650 C	≥ 750 C	≥ 800 C	850 C	≥ 650 C
ASC-1 #1	126.0	97.5	46/5	46.5	44.5	35
ASC-1 #2*	86.0	64.5				17
ASC-1 #3	279.0	177.1	90.0	82.5	75.0	57
ASC-1 #4	135.0	83.9	37.2	36.2	32.1	42
ASC-0 #1	1151.0	876.0				55
ASC-0 #2	1155.0	880.0				52
Total	2932.0	2179.0	173.7	165.2	152.6	258
* Engine 2 has Inconel displacer **Operating hours do not include time during heat leak tests						Thermal cycles include heat leak tests

Table 5. Launch vibration levels during testing at PWR (in both axial and lateral directions).

	Level	Time (minutes)
Workmanship	6.8 g_{rms} random	1
Flight	8.7 g_{rms} random	1
Qualification	12.3 g_{rms} random	3
Qualification + 3dB	17.5 g_{rms} random	1

An electrically heated engineering model of the Advanced Stirling Radioisotope Generator (ASRG) has accumulated over 3,700 hours of duration testing as of June 2009. The hot-end operated at a temperature of 913 K (640 C). The test unit delivered a stable power of 135 W_e.

Heat Rejection Module

The heat rejection module includes pumped water heat transport and two-sided composite radiator panels with titanium-water heat pipes (see Fig. 7). Included in the power plant schematic are four water heat transport loops and two radiator wings (two loops per wing). A schematic of one wing is shown in Fig. 14. Each radiator wing includes 10 sub-panels, each measuring approximately 3.5 m wide by 1.75 m tall. Total heat rejected is approximately 120 kW_t and the total two-sided radiator area is 175 m^2. This includes a 10 percent area margin. The sizing of the radiator is based on the worst-case, equatorial sun angles and the use of a low-absorptivity, low-emissivity mylar surface apron to reduce the thermal contribution from the surface. The effective radiator sink temperature is 250 K.

The radiator assembly receives heated water at 400 K from the Stirling converters and returns the water to the Stirling engines at 370 K. A dedicated pump and volume accumulator is included on each heat rejection loop. Each loop rejects about 30 kW_t. The wings can be folded for transport with individual radiator panels connected by flexible interconnects in an accordion type arrangement. The panels themselves use a sandwich construction with aluminum honeycomb filler and regularly spaced titanium-water heat pipes between two ploymer-matrix composite facesheets. The facesheets interface with the circular heat pipes supported in graphite saddles to provide a good thermal interface. The manifold used to transfer heat from the pumped water heat transport loops to the water heat pipes is protected from micrometeoroids by a bumper shield. The evaporator section of the heat pipes is bent to maximize heat transfer area and provide efficient packaging with the two coolant channels.

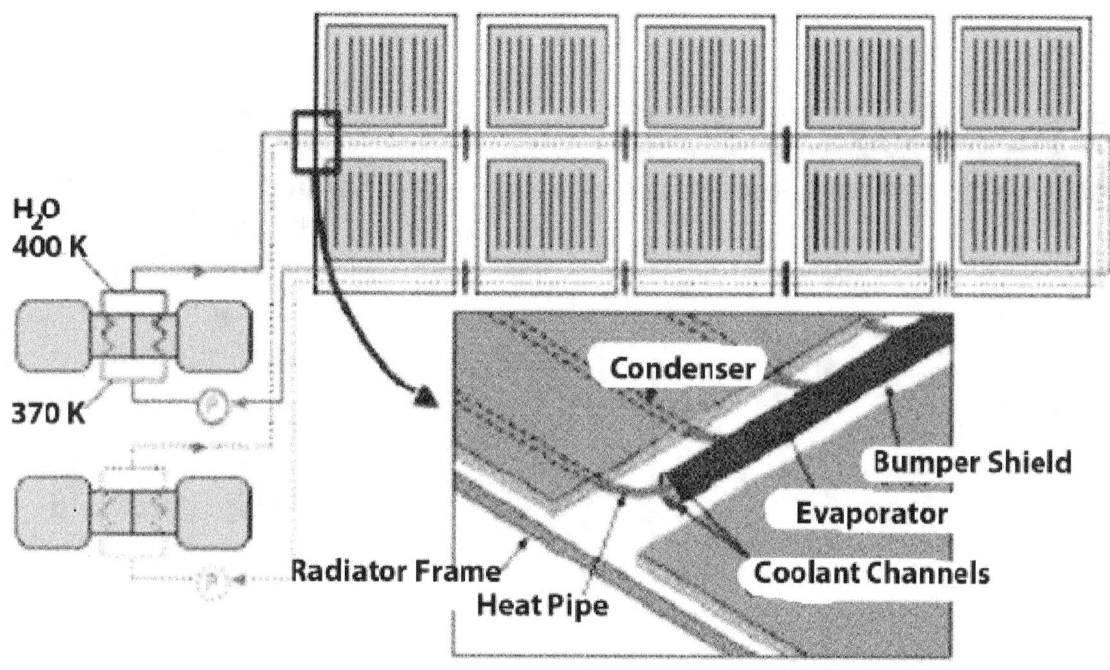

Fig. 14. Heat rejection and radiator panel concepts. *Source: Lee Mason, David Poston, and Louis Qualls*

System Mass[14]

The total system mass is on the order of 4,100 kg; allowing a design margin of 20 percent for uncertainties at this stage in the design process the mass is 4,928 kg. A breakdown of the mass is presented in Table 6. The mass breakdown shows the reactor is 22 percent, the shield 41 percent, power conversion 8 percent, heat rejection 15 percent, and PCAD 14 percent (including the solar array and battery). The heaviest element, the shield, has the greatest potential for mass reduction either by substituting other materials or relaxing the radiation criteria for the Stirling engines.

Variations in system mass in the power range from 10 - 50 kWe is shown in Fig. 15. Both emplaced and landed configurations are included. An 80 percent mass penalty would occur at 40 kWe for the landed configuration compared to the emplaced configuration.

Table 6. Reference concept mass summary. *Source: Lee Mason, David Poston, and Louis Qualls*

	Mass, kg	Comment
Reactor	913	93 percent enriched UO2, NaK coolant, SS–316 cladding/structure, Be drum reflectors, 1 primary and 2 intermediate loops, 6 EM pumps, 175 kWt, 900K peak clad temp, cavity radiators
Shield	1676	B4C and SS–316, 1.2-m-thick axial plug, 1.2 by 1.5 m elliptical face, <2 Mrad and 1×10^{14} n/cm^2 at Stirling converters, lunar regolith augmentation, <5 rem/yr at 100 m radial distance
Power conversion	344	Free-piston Stirling, 4 dual-opposed converters, 8 linear alternators × 6 kWe, 100 Hz, T_H = 830 K, T_C = 415 K
Heat rejection	615	Pumped H$_2$O coolant, 4 independent loops, 400 K inlet temp, composite radiator panels with Ti/H$_2$O heat pipes, scissor deployment, mylar surface apron, 175 m^2 total area
Power conditioning and distribution	559	400 Vac distribution, 100 m cabling, 120 Vdc user bus, parasitic load control, comm/telemetry link, 5 kWe solar array, 10 kWh battery
Subtotal	4107	
Margin	821	20 percent
Total	4928	

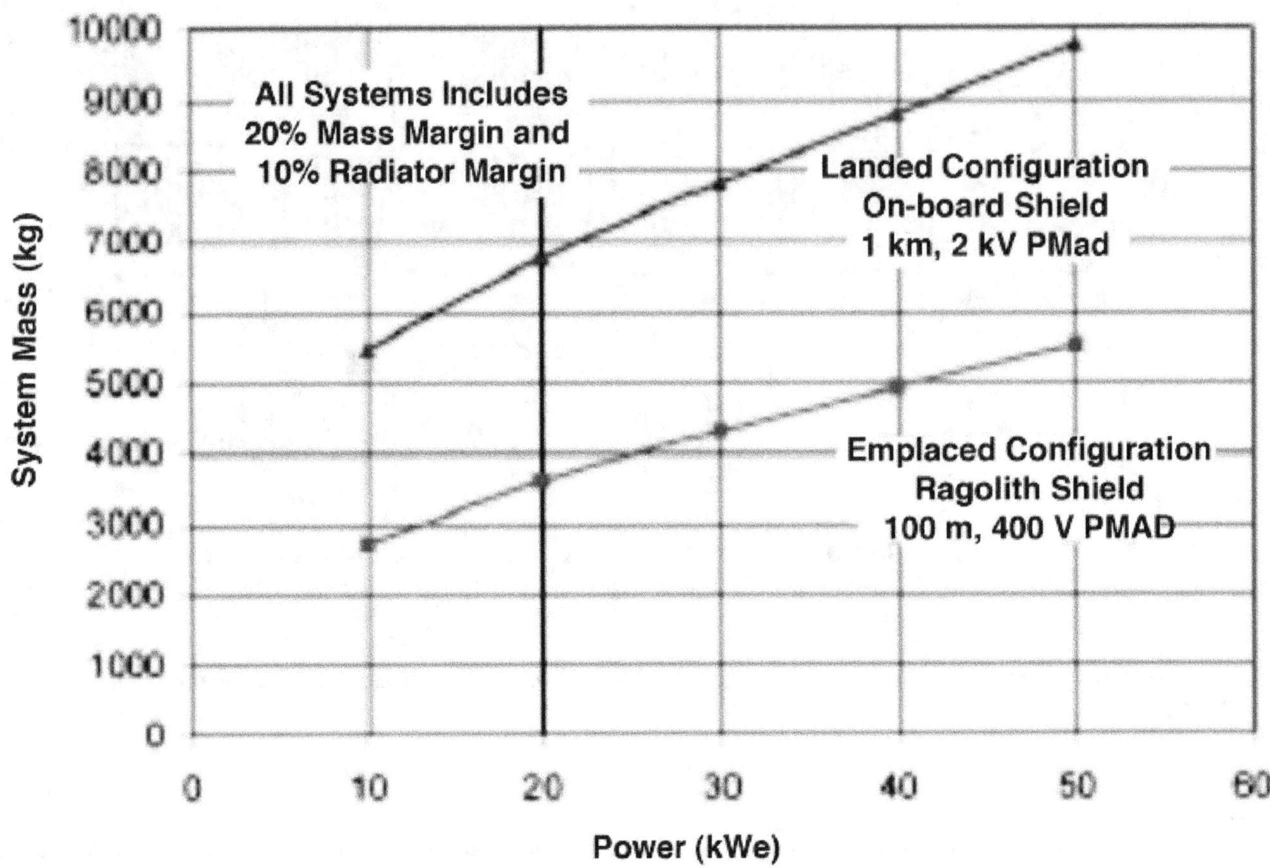

Fig. 15. System mass scaling for emplaced and landed configurations. *Source: Lee Mason, David Poston, and Louis Qualls*

Development Status[15]

In 2007, the Affordable Fission Surface Power Study was completed. This study was a rigorous ground-up estimation of the cost of development of a first-unit lunar fission surface power system by a NASA/DOE team with inputs from industry space power system developers. The estimate cost is about $1.4 billion to develop, flight qualify and deliver the first FSP flight unit. Subsequent flight units would cost on the order of $215 million each. Also establish was that the FSP system could be designed to be extensible to Mars surface missions.

In mid-2008, a preliminary reference concept was completed. This design, described above, is the basis for the project's ongoing technology development and demonstration efforts. In addition, the Brayton converter is being developed as an alternative to the Stirling converters.

Currently, underway are a series technology demonstration projects as part of NASA's Pathfinder Program. The ones relevant to FSP are illustrated in Fig. 16. If the design decision is made to continued FSP development, the full scale component and system demonstrations will follow.

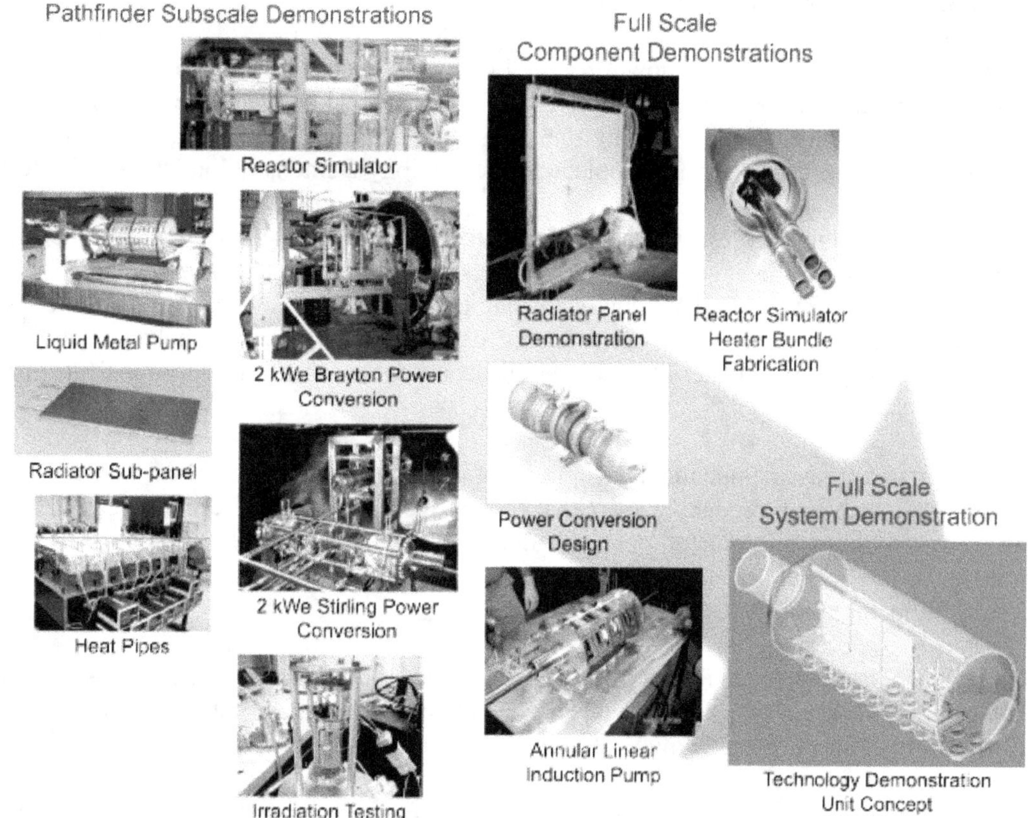

Fig. 16. Element demonstration projects.

Reactor Module Technology Development

The reactor technology selected represents relatively low technology risk and can be easily demonstrated with minimal technology development effort.   The proposed reference reactor is based on experience from the Experimental Breeder Reactor and the Fast Flux Test Facility. The selected concept was based on using existing experience and avoid an extensive fuel development and qualification effort.

Primary areas of concern in need of early focus are: (1) the primary heat transfer loop which connects the reactor to the power conversion system; (2) shielding; and (3) instrumentation and controls. To meet the reactor module development objectives, the following activities are planned:

- Conduct design studies and experimental testing on reactor component technologies that are applicable to the reference FSP system;
- Define the reactor fuel type and qualification process;
- Develop prototypic test articles and evaluate performance (e.g., reflector drives, instrumentation, shielding materials, fuel pin simulators);
- Characterize reactor interfaces and evaluate interactions across reactor module interfaces;
- Evaluate radiation environment effects on material properties and component life;
- Provide realistic test environment for evaluation of primary loop components and behavioral phenomenon;
- Accurately simulate the heat transfer and fluid flow conditions expected in the primary loop;
- Develop prototypic test articles and evaluate performance (e.g., pumps, heat exchangers, sensors);
- Characterize primary loop interfaces and evaluate interactions with mating components; and
- Establish safe and reliable liquid metal handling procedures.

Power Conversion Module Technology Development

Risk reduction consists of the gradual development, buildup, and testing of a non-nuclear system level Technology Demonstration Unit (TDU), shown in Fig. 17. Key objectives of the TDU include: validation and refinement of FPS requirements, characterization of system-level performance, physical and electrical characteristics, cost and risk. Leading up to the TDU are a progression of component and subsystem development and tests. Current plans are to complete the Technology Demonstration Unit (TDU) testing in 2014.

Near term plans are to assemble and test sub-scale systems for future full scale development units. The units will be integrated with representative heat source simulators. Existing equipment will be leveraged where currently available such as the 2 kW$_e$ Brayton, 30 kW$_e$ dual-Capstone Brayton, 2 kW$_e$ Stirling, 50 kW$_e$ Alternator Test Unit and lunar PMAD testbed. Focus will be on reducing development risk in such key areas as heat source integration, electrical integration, materials integration, component performance, transient operations, and autonomous controls.

Technology activities underway include the development of a reactor simulator and heat transfer loop. Recently, a Thermal Simulator capable of producing over 2 kW$_t$ for over 100 hr has been fabricated. The simulator represents one element of a 37-element core simulator for the (TDU). For the heat transfer loop, an electromagnetic pump has been fabricated. Electromagnetic pumps have been used on terrestrial power plant, but none have been manufactured for over 15 years.

A NaK heated pair of Stirling engines that delivered 2.4 kW$_e$ were successful tested and thus the detailed design and fabrication of a full scale 12 kW$_e$ Stirling power conversion unit has been initiated. Also, a 2 kW$_e$ Brayton unit, originally developed for the Solar Dynamic Flight Demonstration Project, has been successful tested.

After selection of the power conversion subsystem, 12 kW$_e$ power conversion demonstrator unit(s) will be developed. Whether only a single option or a primary option with a backup will be developed depends on available funding.

The Stirling engine is located close to the reactor and thus will be exposed to high radiation doses. Radiation tolerance testing has been performed on the alternator without degradation for gamma levels up to 20 times the expected gamma radiation dose for the life of a Free Piston Stirling System.

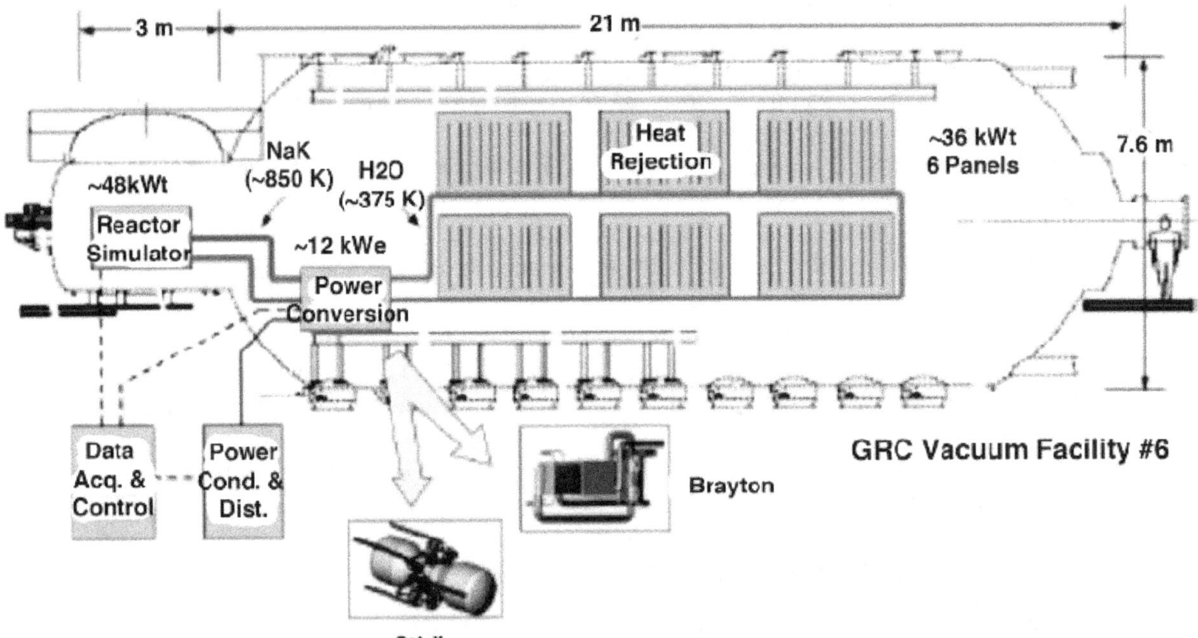

Fig. 17. Technology Demonstration Unit notational layout. *Source: Donal T. Palac, Lee S. Mason and Scott Harlow.*

Heat Rejection Module Development

The Heat Rejection module technology development is to develop and test radiator components and related materials for use in heat rejection systems for Brayton and Stirling based power conversion subsystems. Included are plans for a Radiator Demonstration Unit (RDU) and titanium-water heat pipes development. Full scale radiator panels will be used to validate long term thermal/structural performance.

An RDU was tested in 2009 and successfully demonstrated the rejection of 6 kW_t of heat as designed under simulated lunar conditions. This RDU included a water manifold to deliver heat to the evaporator ends of the heat pipes, the heat pipes themselves, and the 1.7 m tall by 2.7 m wide composite radiator panels that radiate the heat distributed through the panel by the heat pipes. Next planned is a demonstration of a 36 kW_t heat rejection system.

Summary

Lunar outpost, and later missions to Mars, will require 10's of kilowatts of electric power. Natural environmental restrictions imposed by the two locations make it necessary to use self-contained power sources--nuclear power being the only one available in the time frame of interest. The availability of a heavy launch vehicle, such as Ares V, or its equivalent, which will be necessary for the manned exploration program to the moon and Mars. This reduces the restrictions on mass and volume that was true on many previous fission power system developments. Thus, the Fission Surface Power system designers could select the power plant configuration with emphasize on relatively low technical risk and minimum development cost and still meet the surface power requirements.

The reference Fission Surface Power selected is based on: a uranium oxide reactor fuel with decades of operational experience, well-characterized stainless steel construction, and low temperature liquid metal (NaK) coolant for heat transfer from the reactor to the power conversion system based on a coolant that has been used on all reactors flown in space up until this time; power conversion systems that have decades of development experience; and a relatively low temperature, low risk radiator using pumped water with titanium-water heat pipes. The reference design incorporates redundant components and parallel fluid loops in the system to produce partial power in the event of failures. The temperatures in the reference design are relatively low compared to other space power systems developments eliminating the need for refractory metals. The reactor peak clad temperature is only 900 K with a NaK outlet temperature of 890 K. Stirling hot end temperature is 830 K. The radiator operates at 400 K water exit temperature and has an effective temperature of 380 K.

It is believed that the FSP strategy has led to an affordable design concept. This is summarize in Table 7.

Table 7. SPS preliminary basis for affordability. *Reference Donald T. Palac, et al 2010.*

Power level and design life	• 40 kWe, 5 to 8 years
Design approach	• 900 K liquid-metal cooled reactor with UO_2 fuel (terrestrial design basis), approximately 1 MWt thermal power level • Stirling power conversion with 850 K input, ~10 kWe/Stirling engine • 400 K water radiators (ISS-derived), <200 m^2 • 400 V transmission, 120 V bus (ISS-derived) for loads
Technology needs	• Liquid metal primary loop and Stirling hot-end interface • End-to-end system performance test (TDU) • Reactor criticality benchmarking tests
Launch and Startup	• Up to two units delivered on a single lunar lander • Reactor startup after installation and crew inspection
Mission and Environment	• One of several power sources for crew and equipment; backup power and crew availability provide contingency options • Technology and concept design extensible to Mars surface missions • Lunar day/night cycle, 50 to 350 K sink, accommodation of dust

Chapter 9

Alternate Space Reactor Concepts

In addition to the power systems already discussed, a number of other reactor power plant designs were studied to meet potential space power requirements. Some of the most interesting of these concepts are discussed in this chapter. They include: the Rankine Cycle Liquid Metal Fast Reactors, the Medium Power Reactor Experiment (MPRE), the High Temperature Gas-Cooled Electric Power Reactor (710 Reactor Program), the Advanced Space Nuclear Power Program (SPR), fluidized bed reactors, heat pipe reactors (SPAR/SP-100 Program) and the Multi-Megawatt Power Plant Program. In addition, there is included a discussion at the end of this chapter of bimodal power systems with the ability to provide limited thermal nuclear rocket propulsion plus long-term electric power.

Rankine Cycle Liquid Metal Fast Reactors

Between the early 1960s and 1973, programs to develop liquid metal cooled, fast neutron spectrum reactors coupled with potassium Rankine cycle power converters existed. The SNAP-50/SPUR component technology program (discontinued in 1965) was followed by the Advanced Liquid Metal Cooled Reactor Program (which continued until 1973). The SNAP-50 program was to develop 300 - 1,200 kW_e power plants with a 10,000 hour lifetime.   The 300 kW_e design included a 2.2 megawatt-thermal reactor integrated with a potassium working fluid Rankine cycle power converter. This same technology was planned to grow to 1,200 kW_e by enlarging the reactor to 8 - 10 MW_t and using four of the 300 kW_e potassium Rankine cycle converter modules.[1, 2, 3, 4]

A potassium Rankine cycle nuclear reactor schematic for a 300 kW_e is presented in Fig 1. The reactor has a fast neutron spectrum in the range of 0.35 MeV to 0.55 MeV. The fuel options were highly enriched (93%) uranium-235, either as a carbide or nitride . The preference seem to be the nitride fuel form. The pressure vessel, the core support structure, and the fuel cladding used columbium alloy. At temperatures lower than 810 K, parts of the support system substituted titanium. The primary loop working fluid was lithium. The lithium fluid entered the reactor, flowing upward through an annular passage that was formed by the outer boundary of the core and the pressure vessel wall. This cooled the core boundary and the pressure vessel prior to the lithium entering the center of the core from the upper plenum. The nominal operating conditions for the 300 kW_e power plant are summarized on Fig. 2.[5] A schematic illustrating a possible arrangement of the components in a spacecraft is shown in Fig. 3. Additional details about the layout of the SNAP-50 reactor core are presented in Fig. 4.[6]

Lithium presented a unique design problem in that the lithium needed to be melted prior to operating the reactor. This was accomplished by enclosing the pressure vessel in a columbium alloy jacket that provided a channel for high temperature helium to preheat and maintain the core and pressure vessel at 645 K prior to its powered operation. This jacket also provided for an inert gas atmosphere for oxidation protection of the pressure vessel. Thermal insulation outside the preheat jacket and around the upper and lower regions of the pressure vessel was used to minimize the temperature of the support structure, the reflector, and bearings and components below the reactor.

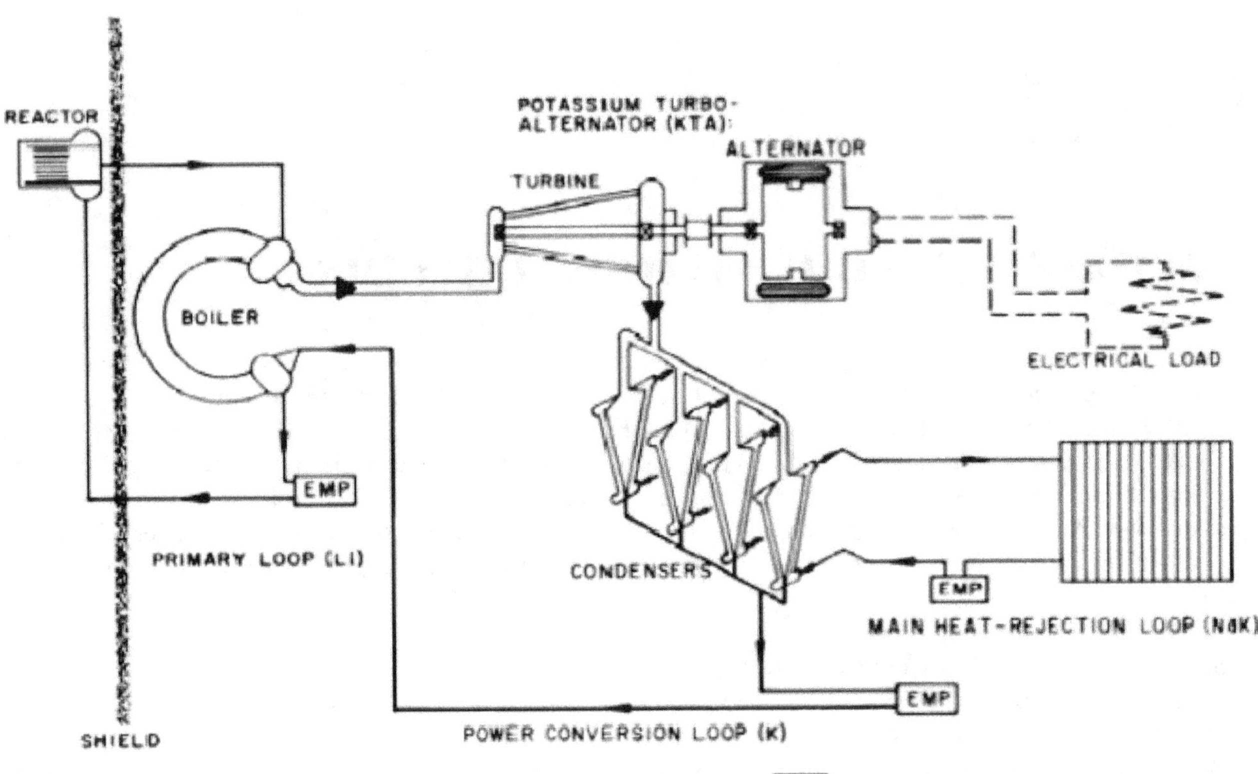

Fig. 1. Schematic diagram for advanced potassium Rankine cycle nuclear power plant. *From J. A. Heller, T. A. Moss, and G. J. Barna,1969.*

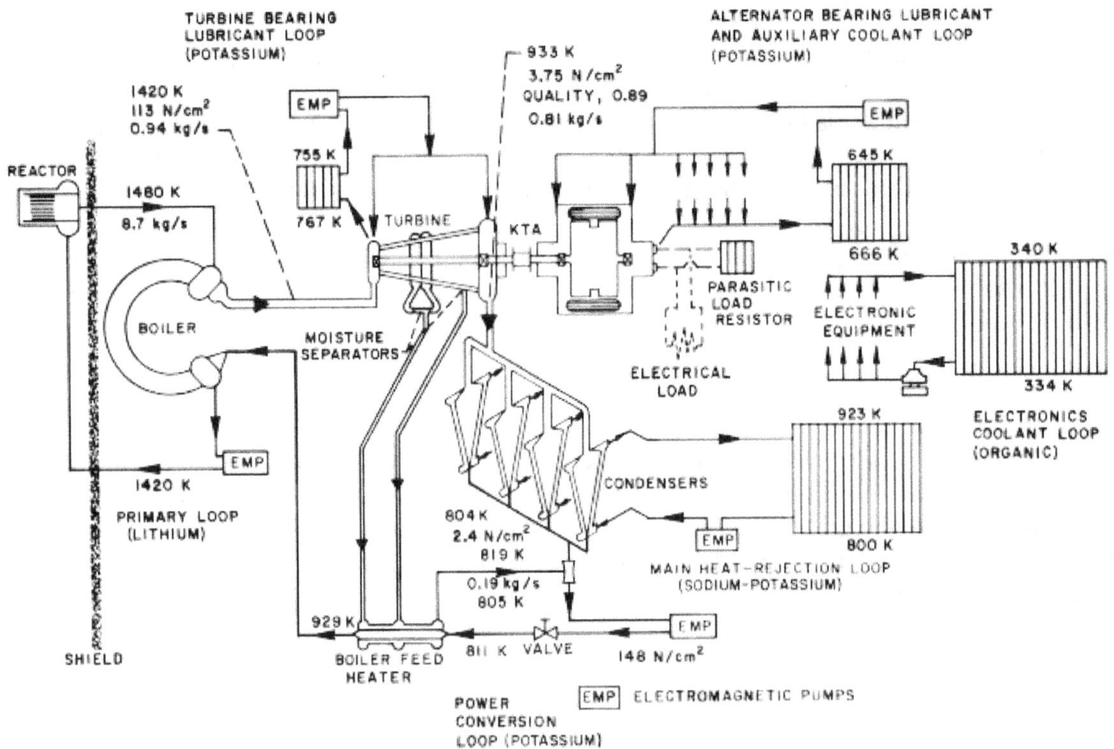

Fig. 2. Nominal operating conditions for an advance 300 kWe potassium Rankine cycle nuclear reactor . *From J. A. Heller, T. A. Moss, and G. J. Barna, 1969.*

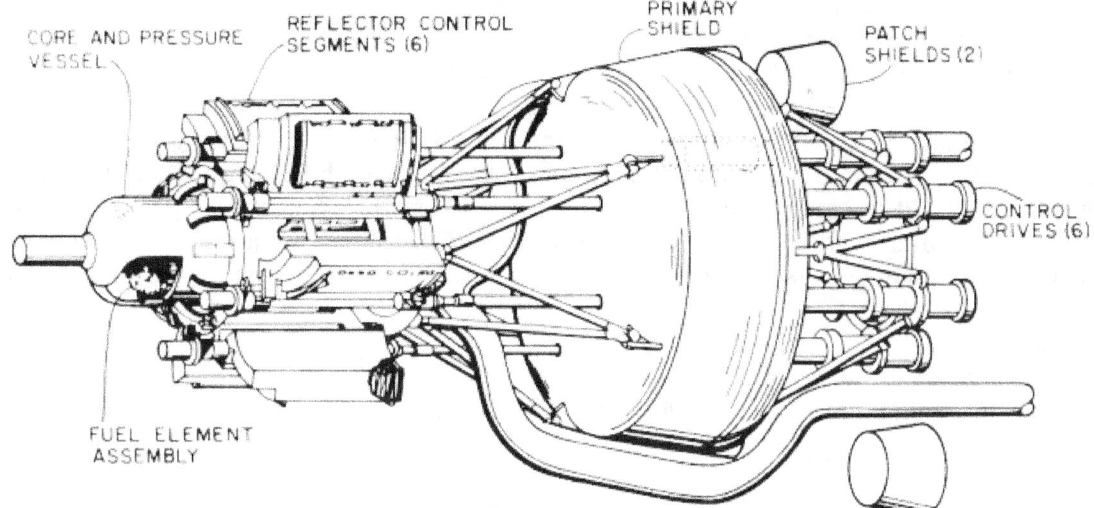

Fig. 3. SNAP-50 (PWAR-20) reactor and shield configuration. *From John E. Allen, 1965.*

Six moveable reflector segments were arranged around the core so that in the fully closed position they completely surrounded the reactor core. Each segments consisted of a columbium alloy container filled with blocks of beryllium oxide. These were connected to the reactor and reflector support structure by means of flexure bearings at each end. Reactor control was achieved by positioning the reflector segments to vary neutron leakage from the core.

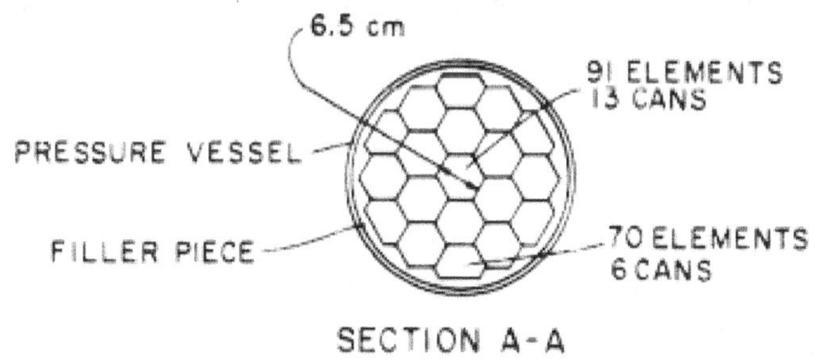

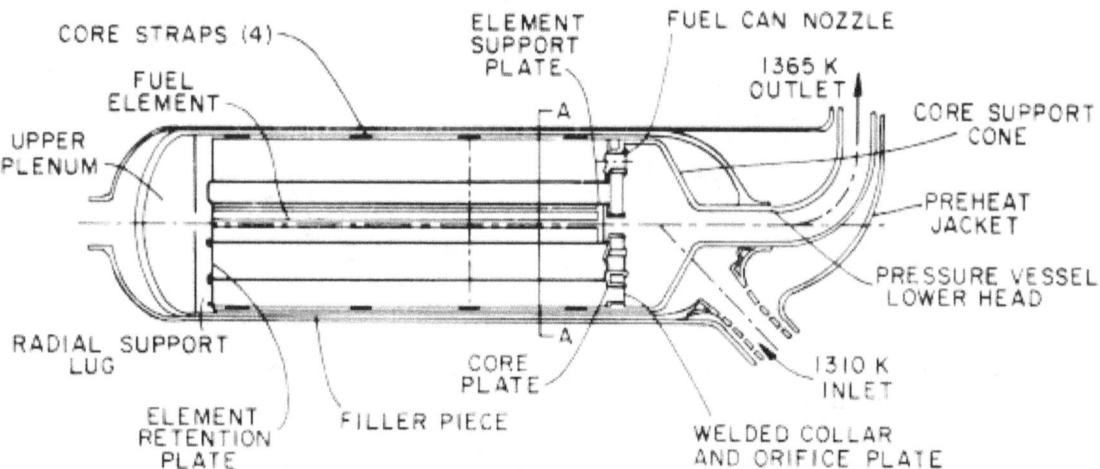

Fig. 4. SNAP-50 (PWAR-20) reactor core and pressure vessel. *From H. J. Branch, 1964.*

The candidate fuels uranium nitride (UN) and uranium carbide (UC) were extensively tested. The UN fuel test exhibited no adverse effects after nearly 6,000 hours at an average cladding temperature of 1,350 K and a burnup of 1 percent of the total number of uranium atoms. In other tests at higher temperatures, UN also performed well. At temperatures between 1,420 K and 1,480 K for 1,690 hours at 1.2 percent burnup, fuel swelling was less than 1 percent and gas release less than 12 percent. The swelling rate remained quite low as burnup rates progressed beyond 1 percent, but a sudden surge of gas release occurred between 1.5 and 1.6 percent burnup. Based on calculations of gas pressure associated with this release, which was about 20 percent of the total gas produced, the strength limit of the cladding material was approached. Consequently, UN fuel burnup beyond 1.6 percent would require stronger cladding materials, such as tungsten-rhenium (W-Re) or lower gas release rates.[7]

Uranium carbide fuel was also tested. However, an equipment failure at 2,500 hours limited the amount of data obtained. The fuel demonstrated good performance to 2,500 hours. However, UC exhibited higher swelling rates than did the UN fuel.

There was an extensive testing program on non-nuclear components with some components tested to 10,000 hours. However, complete systems were never tested. Electromagnetic (EM) pumps, with their relative simplicity of design, were used in both the primary and boiler feed loops. A cutaway sketch of the primary helical induction pump appears in Fig. 5. This pump is analogous to an induction motor except that the rotor has been replaced by a liquid-metal-filled annular passage. Liquid metal enters the pump at the left into an annular passage. The rotating magnetic field imposes a force on the liquid, causing it to rotate circumferentially in the annulus. With flow constrained within the helical channel in the annulus, the fluid is forced to move from one end of the pump to the other. The static pressure of the fluid increases along the helical channel with the high pressure fluid making a 180 degree turn at the end of the annulus and exiting the pump through the outlet pipe. The design is quite similar for the boiler feed EM pump, but the primary loop pump has an additional central coolant passage to meet higher operating temperature requirements. Table 1 summarizes the design specifications for these EM pumps. Efficiencies of 19 percent (potassium) and 16 percent (lithium) were calculated.

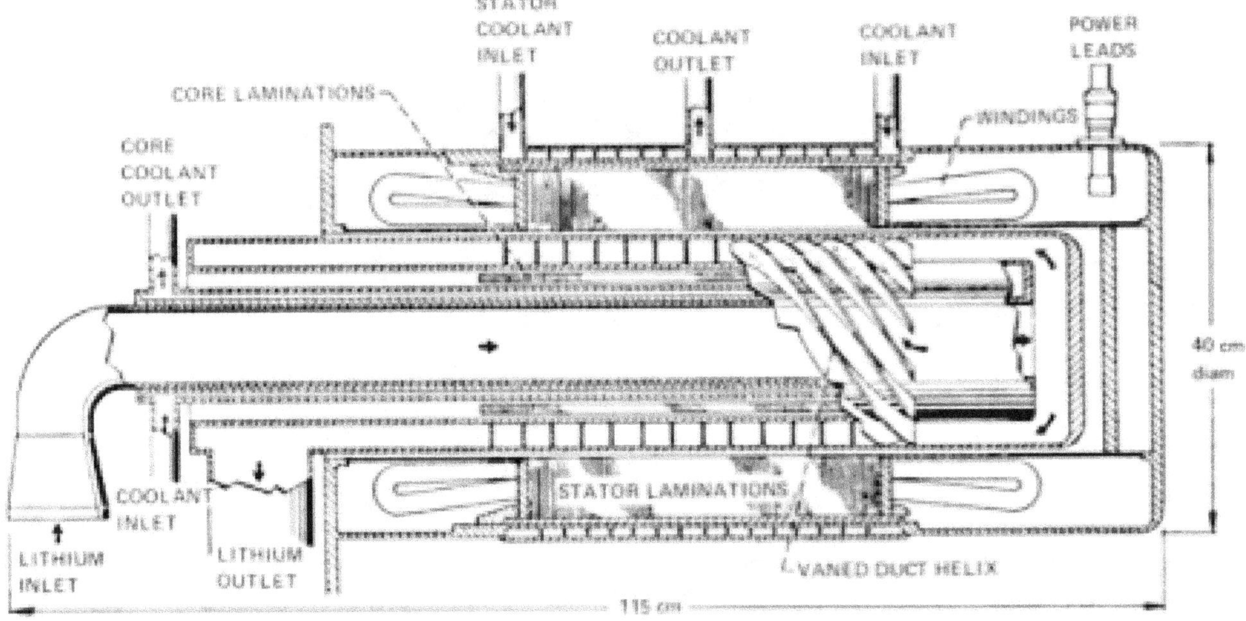

Fig. 5. Cutaway view of a lithium working fluid, helical induction pump for an advanced Rankine cycle reactor system. *From J. A. Heller, T. A. Moss, and G. J. Barna, 1969.*

Table 1. Electromagnetic pump design specifications.

Fluid	Primary-loop	Boiler-loop
	Lithium	Potassium
Fluid inlet temperature (K)	1,425	811
Flow rate (kg/sec)	13.6	1.47
Developed pressure rise (N/cm^2)	14	166
Net positive suction head (N/cm^2)		4.8
NaK coolant inlet temperature (K)	700	730
Coolant temperature rise (K)	310	280

The boiler was used to transfer thermal energy from the primary reactor loop to the secondary power conversion loop. For this Rankine cycle design, superheated potassium vapor temperature was 1,420 K. The primary reactor loop lithium temperature was 1,480 K. The boiler had to be compact, provide stable performance during startup and at operating conditions, and be structurally resilient to withstand system and environmental stresses over many thousands of operating hours.

The designed boiler contained 31 tubes (approximately 230 cm long) housed in a 15.2 cm diameter shell (reference Fig. 6). Flow through the boiler was such that the potassium working fluid flowed in a single pass through the inside tubes, while the primary loop lithium flowed in the space between the tubes and shells in a countercurrent direction. The tantalum alloy T-111 (Ta-8W-2Hf) refractory metal was used in the design. The boiler's arc shape was used to facilitated packaging and minimized stresses. Each boiler tube contained a swirl-generating insert, such as a helical vane or wire coil. These inserts augmented boiling heat transfer processes and helped make the boiler's operation insensitive to gravitational effects. The potassium working fluid started as a liquid, experienced nucleate boiling, entered a transition boiling region, and finally emerged as a superheated vapor from the boiler.

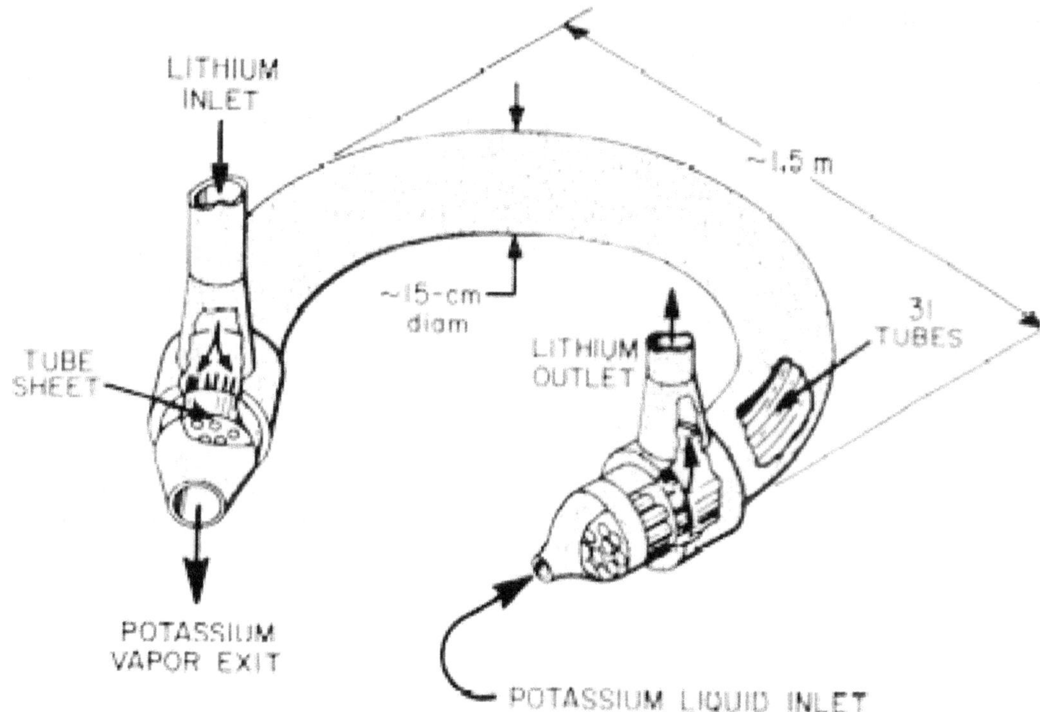

Fig. 6. Once-through 2 megawatt-thermal potassium boiler, advanced Rankine cycle reactor system. *From J. A. Heller, T. A. Moss, and G. J. Barna, 1969.*

The superheated potassium leaves the boiler and was expanded in the turbo-alternator. The turbo-alternator incorporated a I0-stage axial-flow turbine flexibly coupled to a I0-pole homopolar alternator in the design. Support for each turbine and alternator rotor were a pair of potassium-lubricated, pivoted-pad journal bearings. Also, a double-acting potassium pivoted-pad thrust bearing in each component supported forward and reverse loads. The unit operated at 19,200 rpm and produced a net electrical power output of 428 kW_e. Dimensionally the unit was 1.83 m long and 0.61 m diameter at the alternator, and had a mass of 590 kg. One major design consideration centered around keeping the "moisture" of the potassium in the turbine as low as possible--by the moisture of a thermodynamic mixture is meant the fraction of mass present in the liquid phase. If the moisture of the potassium working fluid reached 20 percent, catastrophic turbine blade damage would result from liquid droplet impact erosion. To resolve this problem an interspool liquid separator was added to the design. It swirled the entire flow exiting the first stage by the upper set of vanes to centrifuge the liquid to the wall. A slot in the wall scavenged the liquid and some vapor and returned the condensed mixture to the boiler feed heater. A lower set of vanes straightened the vortex flow prior to its entry into the second turbine stage. All the blading, the rotor discs, and the tie-bolts were fabricated from the molybdenum alloy TZM (Mo-0.6 Ti-0.1 Zr-0.035 C).

Representative mass estimates for the 300 kW_e advanced Rankine cycle space power system are presented in Table 2. This system had an efficiency of 19 percent and a specific mass of 32 kg / kW_e (minus the electrical equipment).

Significant material and component testing had been performed on non-nuclear components. Materials must resist the corrosive effects of lithium and potassium, exhibit long term strength above 1,365 K, and possess good weldability and ductility. It was demonstrated in tests lasting beyond 10,000 hours that refractory alloys are compatible with alkali metals at temperatures above 1,475 K if careful procedures are followed. Though corrosive attack by alkali metals is greatly accelerated in the presence of oxygen, refractory alloys containing hafnium or zirconium as getters can be used to tie up the oxygen. This led to the use of tantalum alloy T-111 (Ta-8W-2Hf) as the principal high temperature containment material for the primary (lithium) loop, the boiler, the secondary (potassium) loop and high temperature portions of the turbo-alternator. In March 1970, an endurance test of 10,000 hours was successfully completed with T-111 alloy in a corrosion study loop.[8]

Heat transfer components in a prototype, single boiler tube and a three-tube NaK cooled potassium condenser demonstrated very stable operation.

EM induction pump technology using a prototype boiler feed pump was successfully demonstrated in a 10,000 hour endurance test. This tested pump operated at a potassium temperature of 810 K, a 1.5 kg / s flow rate, and a 1.6 MPa pressure, with the NaK stator coolant supplied at 700 K. The pump voltage was 135 volts at 60 cycles and the pump had a design efficiency of 16.3 percent.

In addition, a 5,000 hour erosion test and liquid extraction performance tests were performed on a three-stage turbine using potassium vapor as the working fluid. These tests were considered to have verified design assumptions and to have indicated that a three to five year lifetime for the turbo alternator was achievable. The measured effectiveness values of the rotor extraction device generally exceeded 30 percent and the effectiveness of the vortex separator approached 80 percent for separator vapor velocities of 9.3 m / s. The alternator bore seal was considered to be perhaps the component of least engineering certainty.

At the time of program termination, high temperature Rankine cycle technology components had been demonstrated and a point was reached where a breadboard system could be tested with a high probability of success.

Table 2. Mass estimates for nominal 300 kW$_e$ advanced Rankine cycle space power system.

	kg		kg
Reactor (2 MW thermal, 3.8 at.% fuel burnup, 50,000-hr life) and reactor controls, drive, and structure	1,810	Alternator coolant loop (K):	
		Radiator panel	145
		Pump	59
		Expansion tank	23
Primary loop (Li):		Piping	36
Boiler	136	Potassium inventory	(a)
Pump	317	Total weight	263
Pipe and insulation	145		
Expansion tank	82	Turbine coolant loop (K):	
Lithium inventory (0.18 m^3) (72.5 kg)	(a)	Radiator	34
Total weight	680	Pump	23
		Expansion tank	23
Power conversion loop (K):		Piping	16
Turboalternator	590	Potassium inventory	(a)
Condensers (4)	110	Total weight	96
Boiler feed pump	136		
Condensate heat exchangers (2)	46	Electronics cooling loop (DC-200):	
Separator	23	Radiator	124
Potassium inventory control	46	Pump	11
Potassium injection tank	114	Expansion tank	11
Potassium inventory	68	Piping	48
Vapor piping	114	Inventory	46
Liquid piping	23	Total weight	240
Total weight	1,270		
		Electrical equipment:	
Main heat-rejection loop (NaK):		Speed control	46
Radiator panels (4)	982	Voltage regulator-exciter	92
Pumps (4)	217	Internal power conditioning	92
Expansion tanks (4)	181	Parasitic load resistor	270
Piping	195	Interconnecting cable	23
NaK inventory	(a)	Controls and instrumentation	23
Total weight	1,575	Total weight	546
Shadow shield (for unmanned spacecraft, 10° half-cone angle, ~46 m to payload)	2,700	Structure (assumed to be 10 percent of engine weight)	890
		Total estimated weight	10,070

---

[a] Included in above weights.

Medium-Power Reactor Experiment (MPRE)[9]

The Medium-Power Reactor Experiment (MPRE) program started in 1958 had also selected a potassium Rankine cycle electric power conversion system. This effort was aimed at developing space nuclear power plants in the twenty to several hundred kilowatt-electric range with a minimum service life of 10,000 hours. The design had a single-loop potassium Rankine cycle system. Activities were pursued on this concept until 1966. Initial technical questions centered around the compatibility of stainless steel with boiling potassium and the heat transfer capability of boiling potassium under forced convection for a reasonably small size core. By 1964, progress was sufficiently favorable to support the initiation of work on a reactor experiment of 1 megawatt-thermal, a power level corresponding to about 140 kilowatts-electric. When the program was terminated in 1966, both the one MW$_t$ reactor and its test facility had been designed. Also, a series of component and system test rigs had been designed, constructed, and operated to demonstrate and validate the design concept. Technical questions still to be resolved included the control of free-liquid surfaces under microgravity conditions and a reliable potassium vapor turbogenerator unit.

One of the design objectives of the MPRE (shown schematically in Fig. 7) was to develop a design that would be a major simplification over the commonly considered multi-loop systems. Potassium was used as the working fluid, partially because it has sufficiently low neutron activation rates. This would permit the coolant to be pumped beyond the radiation shield. Thermal energy was added to this working fluid in the reactor core, with boiling occurring in the upper segment of the core. This was considered acceptable because potassium has little effect on reactivity and irregular potassium boiling in the core would not adversely affect nuclear power control. Above the core was a vapor separator designed to give the exiting potassium vapor a quality of approximately 99 percent. An expansion tank was located below the vapor separator.

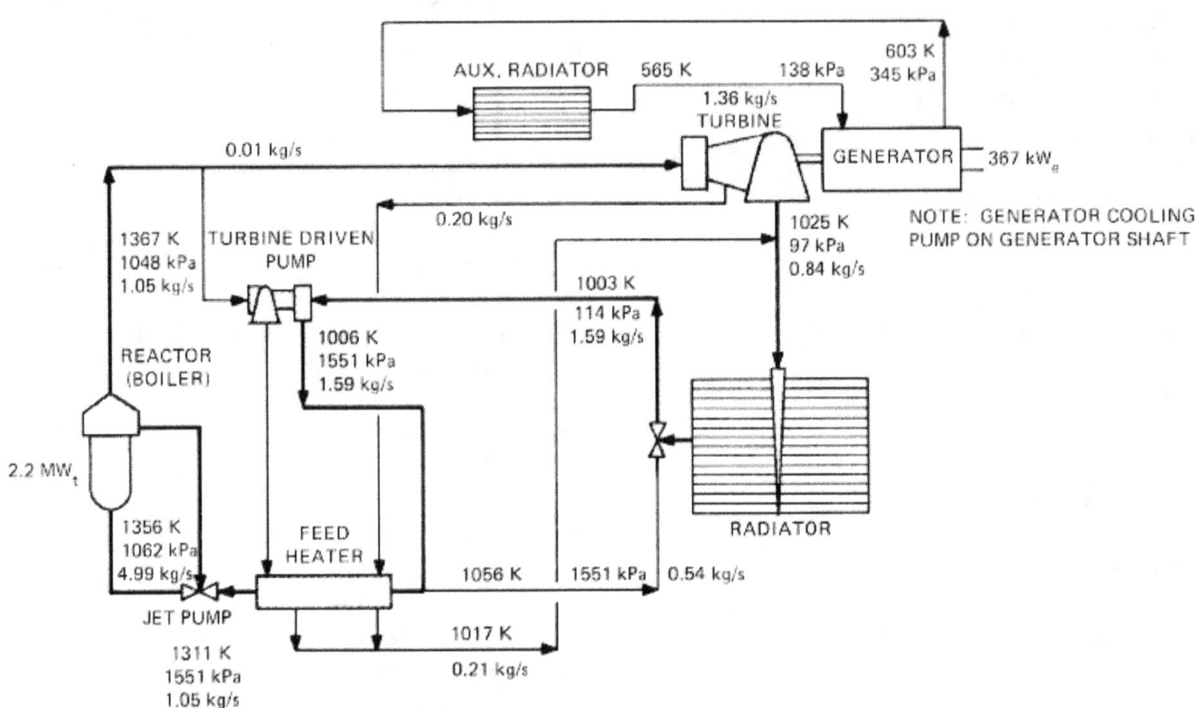

Fig. 7. Medium power reactor experiment (MPRE) showing single loop potassium Rankine cycle. *From A. P. Fraas, 1967.*

After exiting the reactor, the potassium vapor was expanded through the main turbine to generate electricity. A regenerative feed heater was included to reduce the size of the radiator, increased cycle efficiency, and minimized thermal stresses in the boiler recirculation system. The potassium was condensed in the radiator.

Jet pumps, driven by a portion of the feed pump discharge flow, were to facilitate removal of condensate from the condenser under microgravity conditions. These were installed near the condenser outlet manifold. To achieve a high heat transfer coefficient in the reactor, use was made of the recirculating boiler effect. The potassium liquid removed from the vapor separator was recirculated through the core by means of jet pumps driven by the boiler feed supply. To simplify the bearing and shaft system designs, an independent centrifugal feed pump (used to pressurize the fluid) was driven by a free turbine.

The reactor core of the MPRE (see Fig. 8) consisted of bundles of fuel rods with fully enriched UO_2 pellets in a stainless steel capsule (see Fig. 9). The fuel rods were 1.27 cm diameter and arranged in an equilateral triangle pattern with a center-line spacing of 1.43 cm. This configuration created a coolant passage volume fraction of 0.283. Surrounded the core was a reflector using beryllium oxide (BeO). Its dimensions were 8.26 cm thick on the side and 12.70 cm thick on the bottom. As shown in Fig. 8, the side reflector was constructed of a stacked series of 3.05 cm thick BeO discs. As depicted in Fig..9, a 7.6 cm long stack of BeO pellets were positioned above the UO_2 in each fuel rod and served as a portion of the top reflector. A 10.9 cm void space above the BeO pellets provided a reservoir for fission gas accumulation. Also in this void space was a mechanism to restrain axial movement of the BeO and UO_2 pellets consisting of a piston and molybdenum spring. For reactor control the lower portion of the BeO reflector was divided into four independent quadrants that could be moved axially. These adjusted reactivity and together were worth 5 percent in reactivity change (i.e., 5% in $\delta k/k$).

Surrounded the reflector was a thin, neutron absorbing layer of 0.1 percent boron steel. This thin layer of steel reduced the amount of neutron activation products and the need for gamma ray shielding. The layer of boron steel also tended to isolate the core from the environment in the event of a launch abort and reactor core immersion in water.

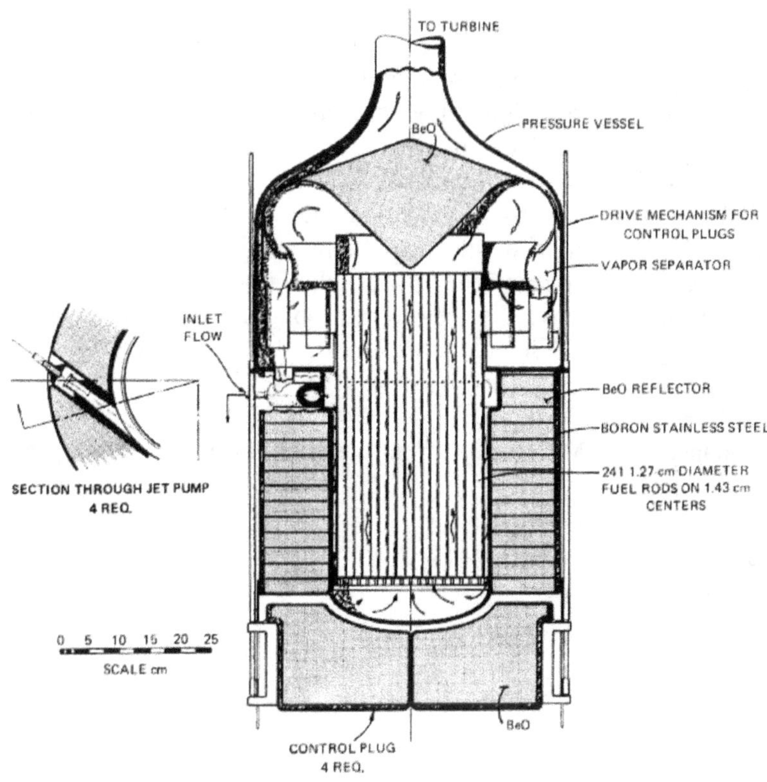

Fig. 8. Reactor schematic for MPRE. *From A. P. Fraas, 1967.*

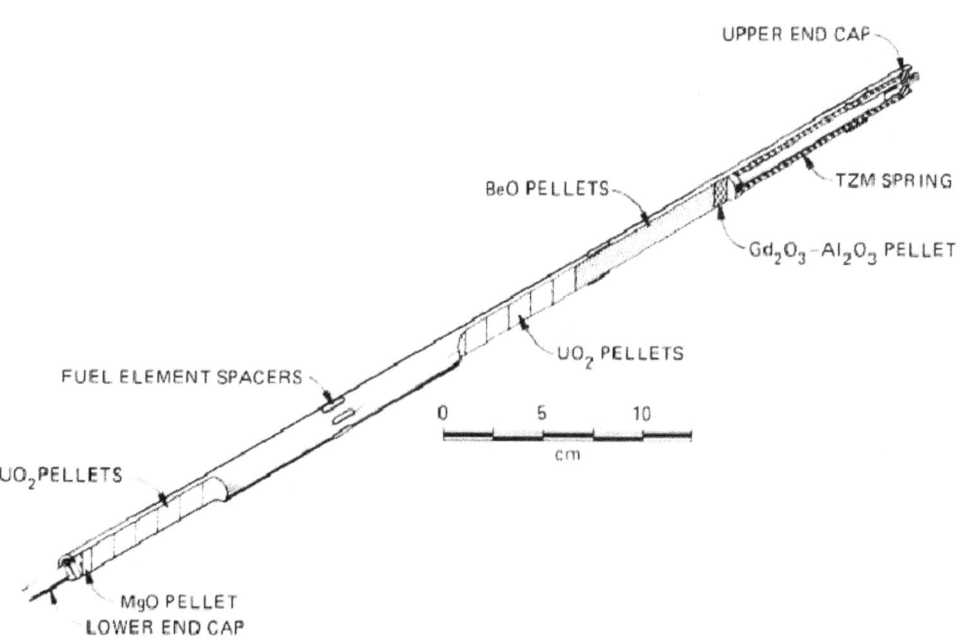

Fig. 9. Fuel element for MPRE. *From A. P. Fraas, 1967.*

Examining the coolant flow path in the MPRE design, the potassium working fluid entered the reactor core through the bottom, rose between the fuel elements, and emerged from the core at a vapor quality of about 20 percent. The fluid proceeded through the vapor separator where the liquid was skimmed off and discharged to the expansion tank directly beneath the vapor separator. Next, the potassium vapor passed radially outward at the bottom of the vapor separator, then upward around the outer perimeter, and finally out the vapor exit pipe at the top. Here, the liquid potassium entered a set of four jet pumps, operating in parallel beneath the expansion tank. These jet pumps extracted liquid potassium from the expansion tank and discharged it to return to the core inlet.

Table 3 summarizes key MPRE parameters. The program resolved many important technical issues. Significantly, the compatibility of stainless steel with boiling potassium was demonstrated in over 160,000 hours of component and systems tests, including 9,000 hours testing of reduced-scale mockups of the MPRE system. One mockup operated for 2,500 hours. Calculations and critical experiments demonstrated that boiling potassium reactors have good nuclear stability and control characteristics. Boiling flow stability and flow distribution were demonstrated in electrically heated mockups. Nucleation sites were used to maintain smooth nucleate boiling conditions. Otherwise, explosive boiling of the potassium could be a major problem. In the area of heat rejection, a lightweight, tapered-tube, direct condensing radiator was developed and shown to give a uniform flow distribution with good flow stability. For bearings, a potassium lubricated, tungsten carbide bearing was operated in a bearing testing rig for 4,000 hours with 350 starts and stops under load with no loss in performance or detectable surface damage.

Table 4 shows the performance of the MPRE for various electric power levels and operating temperatures.

Table 3. MPRE design parameters for stainless steel fuel element.

Reactor Operating Conditions

Thermal Power (MW)	1
Coolant	Potassium (boiling)
Inlet Temperature (K)	1,105
Outlet Temperature (K)	1,111
Inlet Pressure (kPa)	207
Outlet Pressure (kPa)	200

Reactor Geometry

Fuel Element Arrangement	Equilateral triangular pitch
Moderator	None in core
Diameter (equivalent) (cm)	22.3
Distance Across Corners (cm)	24.1
Length (cm)	29.5
Side Reflector Thickness (cm)	7.6

Fluid Flow Data

Core Flow Passage Equivalent Diameter (cm)	0.5
Vapor Flow at Outlet (kg/s)	0.51
Liquid Flow at Core Outlet (kg/s)	2.02
Total Liquid Flow Entering Core (kg/s)	2.53
Vapor Quality Leaving Core (%)	20

Heat Transfer Data

Power Density (core average) (W/cm^3)	79.5
Heat Transfer Surface Area (m^2)	2.8
Fuel Element Cladding Temperature (K)	
Average	1,119
Maximum (Hot Spot)	1,133
Fuel Central Temperature (K)	
Average	1,588
Maximum	1,755

Fuel Element

Configuration	Rods
Fuel Form	UO$_2$ pellets in metal capsules
Number of Fuel Elements	241
Overall Length (cm)	50.5
Fueled Length (cm)	29.5
Fuel Enrichment (%)	93
Cladding Material	Type 316 stainless steel
Cladding Thickness (cm)	0.05
Fuel Element Outside Diameter (cm)	1.3
Burnup at End of Life (% of uranium atoms)	0.76

Core Composition, Volume Fraction

Stainless Steel	0.11
UO$_2$	0.60
Void Inside Fuel Elements	0.01
Coolant Flow Passages	0.28

Control Elements

Reactivity Control Method	4 movable BeO plugs (quadrants of the bottom reflector)
Control Plug Motion	Vertical displacement parallel to axis of reactor; total travel 20.3 cm
Vertical Height of Control Plugs (cm)	21.6
Sheath Material	Type 304 stainless steel
Combined Worth of Control Plugs (%δk/k)	5.0

Pressure Vessel

Material	Type 304 stainless steel
Shell Outside Diameter (cm)	26.7
Wall Thickness (cm)	0.64
Overall Length (with expansion tank and vapor separator) (cm)	86.4

Physics

Mass of UO$_2$ (kg)	78.0
Mass of U-235 (kg)	63.8
Effective Delayed Neutron Fraction	0.0075
Median Energy For Fission (keV)	840
Prompt Neutron Lifetime (sec)	3.2×10^{-7}
Reactivity Worth of Control Plugs	0.05
Isothermal Temperature Coefficient of Reactivity $((1/k) \cdot (dk/dT)) (K^{-1})$	-1.6×10^{-5}
Fuel Reactivity Coefficient $(\Delta k/k)/(\Delta m/m)$	0.545
Loss in k, Fuel Burnup For 1 MW-year	0.0040
Reactivity Effect Of Potassium Filling Coolant Passages (relative to normal operating condition)	0.0014
Maximum Excess k From Control Plug Insertion	
Full Power	0.006
Cold, Clean	0.020
Neutron Flux at Outer Surfaces of Core (neutrons/cm^2-sec)	
Fast ($E = 0.9 - 10$ MeV)	0.41×10^{14}
Thermal ($E = 0 - 1$ eV)	3.4×10^{11}

Radiator

Power To Be Dissipated (kW)	860
Emissivity of Treated Surface	0.92
Fin Efficiency	0.86
Area Required (m^2)	40.4
Radiator Diameter (m)	3.1
Radiator Height (m)	4.6
Vapor Quality (%)	91.7
Potassium Flow Rate (kg/s)	0.45
Number of Tubes	96

Table 4. Characteristics of a series of single-loop boiling potassium reactor systems (based on the MPRE concept).

Reactor Outlet Temperature (K)	1,110	1,110	1,365	1,365	1,365
Net Electrical Output (kW)	140	367	367	1,000	5,000
Reactor Thermal Output (MW)	1	2.2	2.2	6	30
Reactor Core Diameter (cm)	24.4	30.7	30.7	42.9	73.4
Reactor Mass (kg)	341	532	532	1,073	3,468
Radiator Height (m)	4.8	10.9	5.4	9.2	22.3
Radiator Diameter (m)	3.1	3.1	3.1	4.9	10.1
Radiator Mass Without Armor (kg)	423	843	529	1,441	7,180
Meteoroid Armor Mass (kg)[a]	205	314	241	918	6,455
Total System Mass (kg)[b]	1,305	2,602	1,910	5,080	25,385

[a] Armor designed to give 0.1% probability of a meteoroid penetration in 10,000 hours.
[b] Not including shield mass.

The MPRE program successfully demonstrated the world's first potassium vapor turbine, employing potassium lubricated bearings. Operation included 4,000 hours of accumulated testing and 2,500 hours on one unit. In addition, performance of typical jet pumps in the cavitating regime and pump design characteristics necessary for microgravity operation of a Rankine cycle system were demonstrated. Fuel element development included testing four full-scale fuel elements. These were operated in-pile at design power and temperature for a total of over 30,000 hours with no sign of difficulty. In fact, one fuel element operated for over 14,000 hours-the equivalent of three years of normal service.

Shielding studies showed that the effects of neutron and gamma scattering could be kept small from a cylindrical radiator concentric with the reactor. Vapor separator, expansion tank, boiler, and condenser bench tests, including startup and operation under microgravity conditions, were tested satisfactorily. Also, a vapor separator-expansion tank was shown to operate fairly well in microgravity simulation tests conducted onboard a KC-135 aircraft. Finally, an electrically heated mockup (37 percent full scale) of the MPRE was operated for over 3,600 hours with a potassium vapor, turbine-driven feed pump.

Continued development of the MPRE system would have involved: (1) testing fuel elements for up to 20,000 hours; (2) the intentional testing of a defective fuel element with a low fission product release rate; (3) the startup and microgravity operation of an electrically heated, potassium vapor power plant in space; and (4) the endurance testing of nuclear-fueled power plants on the ground.

High-Temperature, Gas-Cooled Electric Power Reactor (710 Program)[10]

From 1962 until 1968, taking advantage of technology developed during the Aircraft Nuclear Propulsion Program, the 710 Advanced High Temperature Gas Reactor Program was conducted. The objective was to develop and demonstrate a high performance, fast spectrum, refractory metal reactor as a heat source for space power and propulsion applications. This program started out as a demonstration of both a closed-loop neon-cooled reactor for electric power generation and an open-loop hydrogen working fluid system for rocket propulsion. In 1963, the open-loop, rocket propulsion objective was deleted, while in 1965 the remaining program objective was reduced to a fuel element technology program.

The 710 Reactor design used a fast spectrum, refractory metal reactor with a hexagonal, cermet fuel element that had multiple tubular flow channels. Fig. 10 is a schematic of the 710 Reactor. The core design could be power flattening by varying either the fuel loading or the hydraulic diameter between fuel elements or within fuel elements. The latter approach resulted in a smaller core size. The fuel elements were mechanically attached to a metallic front tube sheet, using a single point support. This tube sheet, in turn, was attached to the pressure vessel. Because of the fast neutron spectrum, the metallic front tube sheet also served as a reasonably efficient front reflector. If a rear reflector was needed, a thicker header plate could have been used at the back of each fuel element. Surrounded the core was a radial beryllium reflector, consisting of cylindrical sections,. This reflector could be located inside or outside the pressure vessel. Reactivity control was achieved using control drums containing neutron-absorbing boron strips attached to the reflector cylinders and rotating them.

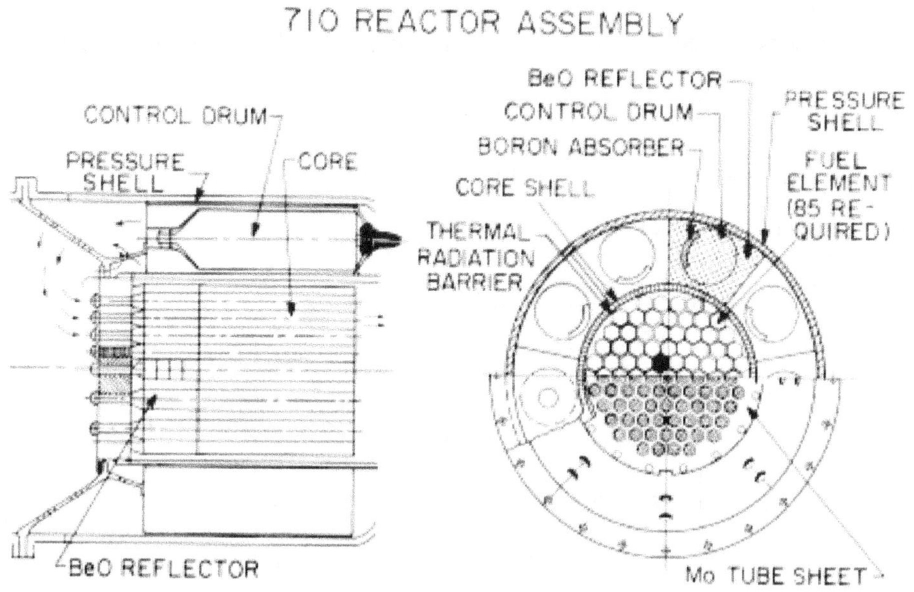

Fig. 10. Longitudinal and cross-sectional views of 710 Reactor. *From General Electric Co., 1968.*

The 710 Reactor concept evolved from a desire for a high performance reactor that could be used for both power and propulsion. The high performance demands established the need for a high working fluid temperature and a high core power density. Electric conversion units considered both Brayton and Rankine cycles. For Brayton cycle systems, a working fluid temperature of 1,475 K was selected for compatibility with uncooled refractory metal turbines. Rankine cycle systems had similar limits. Power densities were 80 to 500 W / cm^3 to meet one to five year system lifetimes.

A cermet fuel was selected with a continuous refractory metal structure surrounding the individual fuel particles. The cermet's body high strength and good thermal conductivity provided partial fuel and fission product retention. This cermet body was than enclosed with a bonded cladding. A tubular fuel element concept was considered advantageous, since tubes provide a flow channel independent of other structural elements, resist distortion, and can be fabricated to close dimensional tolerances. A refractory metal was selected for a structural material to support and contain the fuel based on the refractory metal's high melting point, low vapor pressure, sufficient ductility, ability to be fabricated to close tolerances, strength, and compatibility with nuclear fuels, such as uranium dioxide and uranium nitride. For example, tungsten melts at 3,655 K, a melting point temperature that is higher than U0$_2$ (3,090 K) and UN (2,900 K).

Both uranium dioxide and uranium nitride fuels were under consideration. Uranium dioxide was initially selected as the 710 Reactor fuel because it had the highest melting point of the nuclear fuels under consideration and the lowest vapor pressure. It was also nonreactive in the stoichiometric state with refractory metals, was readily available, and considerable technology existed for the material. Although uranium nitride has a lower melting temperature and higher vapor pressure at elevated temperatures than UO$_2$, UN has greater high temperature creep strength, higher thermal conductivity (by an order of magnitude), higher nuclear density, and is more compatible with lithium (a candidate working fluid). These beneficial UN fuel characteristics result in a potentially smaller core size.

Tungsten cermet fuel elements and various cladding candidates were extensively tested. The results of some of these efforts are shown in Table 5. Consistent quality fuel was obtained with fuel particles encased in 40 volume percent tungsten (W).   Small quantities of rhenium (Re) were found to significantly increase the ductility of the matrix.   The claddings had similar thermal expansion coefficients.   Analysis of the 710 Reactor fuel element indicated that the most likely cause of failure in power generation applications would be strains produced by fission product buildup.

Table 6 lists the characteristics of the 710 Reactor and Table 7 provides reference design conditions for the Brayton cycle system.

Table 5. Status of 710 reactor program fuel element testing (as of 14 October 1966).

Type of Test	Specimen No.	Specimen Geometry[a]	Cladding Material (at.%)	Temp. (K)	No. of Cycles	Time at Temp. (hr)
Static	ASH-7	P.L.-1 channel	W-30Re-30Mo	1,920	65	7,084
	ASH-11	P.L.-1 channel	W-30Re-30Mo	1,920	65	7,084
	E83W	P.L.-1 channel	W-30Re-30Mo	1,920	2	564
	P1W	P.L.-19 channel	W-30Re-30Mo	1,920	48	4,564
	P1T	P.L.-19 channel	T-111	1,920	5	2,000
	P2T	P.L.-19 channel	T-111	1,920	5	2,000
	P5T	P.L.-19 channel	T-111	1,920	5	2,000
	P3T	P.L.-19 channel	T-111	1,920	3	1,600
	P6T	P.L.-19 channel	T-111	1,920	3	1,600
	P7T	P.L.-19 channel	T-111	1,920	3	1,600
	P-20	P.L.-37 channel	Ta	1,920	64	4,500
Dynamic	FS1T	F.L.-1 channel	T-111	1,920	8	2,714
	F2T[b]	F.L.-19 channel	T-111	1,920	7	668
In-Pile	LT-710-1	P.L.-1 channel	W-30Re-30Mo	1,810	31	2,410
	LT-710-2	9 in.-1 channel	T-111	1,810	9	1,095

[a] P.L.—partial-length fuel element: F.L.-full-length fuel element
[b] This was a shakedown test for the "A" section of the first primary dynamic loop.

Table 6. Summary of basic 710 Reactor characteristics.

HIGH PERFORMANCE FUEL ELEMENT DESIGN

Refractory Metal Cermet
- High-temperature capability
- Good strength
- High thermal conductivity
- Adequate ductility
- Fabricability with high fuel loadings
- Bonding of fuel to cladding
- Low dimensional tolerances achievable in fabrication

Wrought Refractory Metal Cladding
- High-temperature capability
- High strength
- Good ductility
- Positive fuel and fission product retention
- Quality verification possible prior to fuel element assembly
- Compatible with highest melting point fuels
- Compatible with liquid metals and hydrogen

Tubular Fuel Element
- Dimensional stability
- Perturbation alleviation (by use of multiple channels per element)
- Flexibility of temperature flattening

SIMPLICITY OF REACTOR CONCEPT

- Single-Point Fuel Element Support
- Series Flow Without Bypass
- Integrated Front Support and Reflector
- Integrated Fuel Element and Rear Reflector (if desired)
- No In-Core Controls
- Only Fuel, Cladding, and Coolant in Core

RESULTANT FUNDAMENTAL FEATURES

- Ease of Temperature Flattening
- Small Temperature Perturbations
- Negative Doppler Coefficient
- Neutronics Insensitive to Structure
- Safe on Water Immersion
- Self-Shielding for Core Gamma Radiation
- Negligible Fission Product Poisoning

VERSATILITY OF APPLICATION

- Power Extensibility
- Coolant Variation
- Temperature, Power, and Life Flexibility

Table 7. Operating conditions for various 710 reactor Brayton cycle systems.

General	COOLANT		
	He-Xe(40)	Neon	Argon
Total electrical power (kWe)	200	200	200
Total power thermal (MWth)	0.872	0.872	0.872
Reactor coolant flow rate (kg/s)	4.85	2.46	4.85
Reactor inlet gas temperature (K)	1,099	1,099	1,099
Reactor outlet gas temperature (K)	1,444	1,444	1,444
Reactor inlet pressure (MPa)	2.74	2.03	2.70
Reactor core pressure drop (%)	0.83	0.83	0.83
Core Thermodynamic and Nuclear Characteristics			
Channel inlet Reynolds number	14,900	7,170	18,260
Channel outlet Reynolds number	12,410	6,010	15,190
Average heat flux (watts/cm^2)	12.9	12.9	12.9
Max. average surface temperature (K)	1,505	1,581	1,631
Max. fuel burnup[a] (fissions/cm^3 of fuel)	8.8×10^{19}	8.8×10^{19}	8.8×10^{19}
Avg. core fuel burnup[a] (fissions/ cm^3 of fuel)	4.6×10^{19}	4.6×10^{19}	4.6×10^{19}
Max. average element burnup[a] (fissions/ cm^3 of fuel)	5.9×10^{19}	5.9×10^{19}	5.9×10^{19}
Max. fuel element[a] (nvt ($L_u > 1$ MeV))	7.90×10^{20}	7.90×10^{20}	7.90×10^{20}
Max. matrix power density (watts/cm^3) matrix	46.1	46.1	46.1

[a] For 10,000 hours operation

Advanced Space Nuclear Power Program (SPR)

Between 1965 and 1968, the Advanced Space Nuclear Power Program (SPR) was conducted to explore fundamental technologies for the optimum means of generating electric power in space at the hundreds of kilowatts to ten megawatt levels. Operating lifetimes for these advanced power systems were defined as tens of thousands of hours. In an effort to obtain the highest possible power-to-mass ratio, program emphasis was placed on materials capable of extremely high temperature operation. Some of these designs are being described for their unique characteristics.

One of the designs, the SPR-4, was a 2 MW$_t$ (375 kW$_e$) Rankine cycle nuclear system introduced sodium heat pipes to transport core thermal energy to a boiler (see Fig. 11).[11] This system consisted of: an integral reactor-boiler operating at 1,500 K (see Fig. 12); a nuclear shadow shield composed of two layers of lithium hydride separated by a layer of tungsten; a turbine-alternator unit; two heat pipe radiators (one operating at 1,035 K and the other at 770 K); and a single electromagnetic pump for circulating the potassium working fluid in the power conversion loop. The system had an overall length of approximately 10 m, permitting launch as an integral unit. The SPR-4 had an overall system efficiency of 18.8 percent and a 20,000 hour design lifetime.

The SPR-4 reactor utilized uranium nitride (UN) as the fuel. Each of the 378 hexagonal fuel elements had a central hole through which a sodium heat pipe was inserted. The heat pipe working fluid was sodium. Structural material was tungsten-rhenium alloy (75% W-25% Re). Uranium-235 enrichment was varied in two radial zones. In this design, two potassium boilers, one on each end of the cylindrical core, were integrated with the reactor (see Table 8 and Fig. 12). The condenser regions of the heat pipes protruded into the boilers. The boiler potassium working fluid was boiled on the outside surfaces of the heat pipes. Heat pipes operate in a manner that thermal energy was transferred from the reactor core to the potassium boilers at essentially isothermal conditions. Potassium flow rate was varied to ensure that the working fluid departed the boilers as a saturated vapor. After the potassium vapor passed through the turbine, it was condensed in a large, low mass, flat radiator that contained 2,200 heat pipes (see Fig. 11).

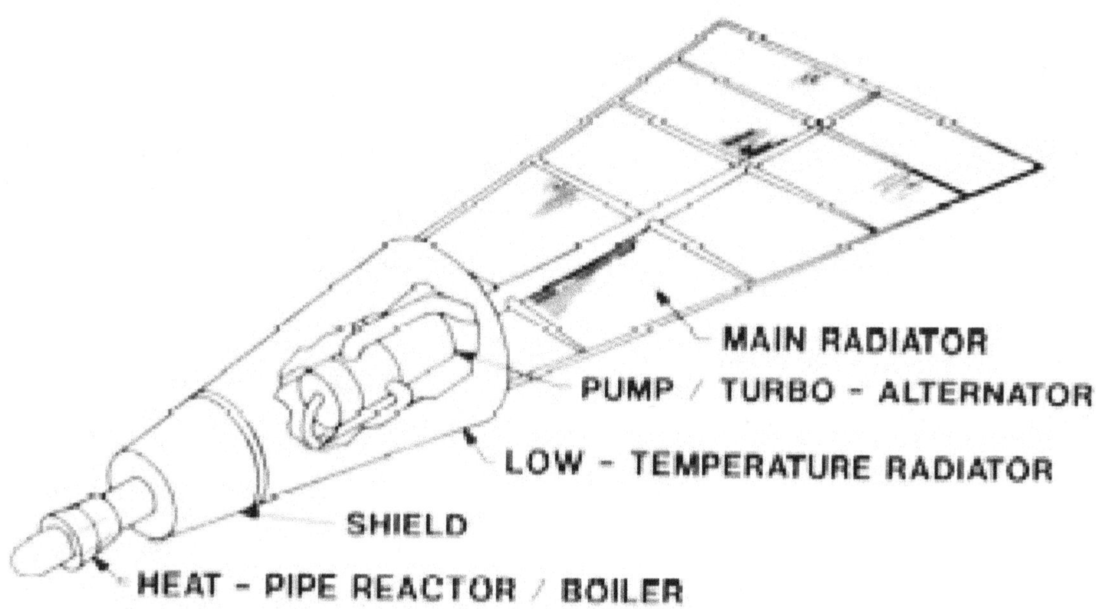

Fig. 11. SPR-4 space nuclear electric power plant. *From John H. Pitts and C. E. Walters, 1970.*

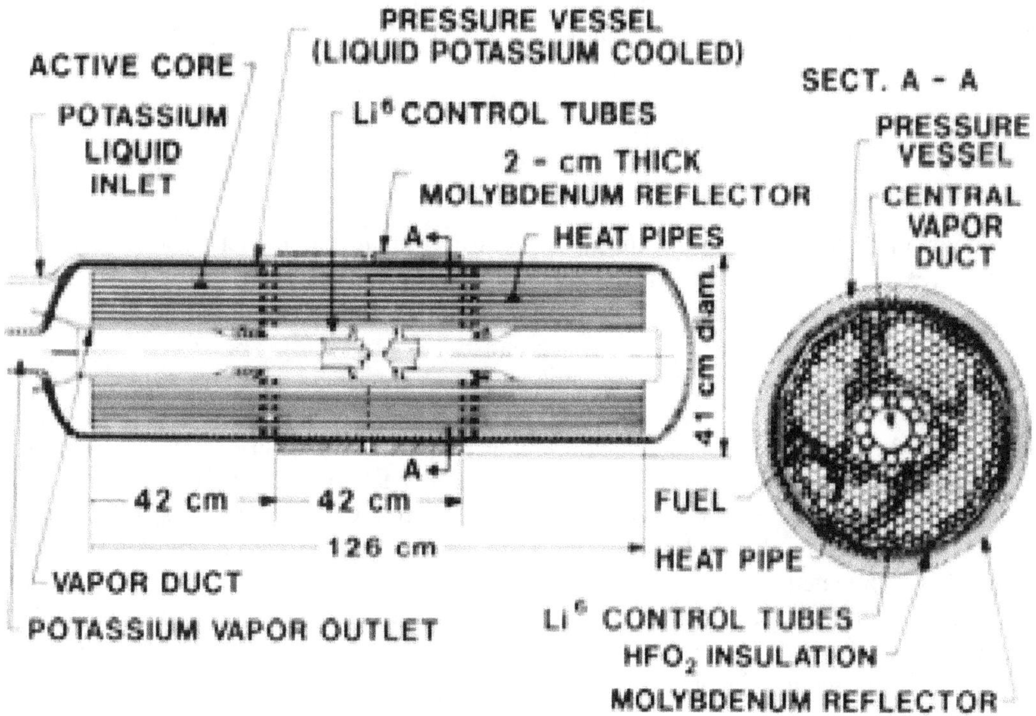

Fig. 12. Schematic diagram for the SPR-4 integral reactor-boiler. *From John H. Pitts and C. E. Walters, 1970*

Table 8. Operating characteristics of SPR-4 integral reactor-boiler.

Fuel Atom Burnup (%)			Volume Fractions	
—average	0.68		—fuel	0.616
—maximum	1.08		—heat pipe metal	0.062
Fuel Enrichment (fraction of ^{235}U)			Intermediate Tube Spacer Volume Fraction	0.016
—average	0.93			
—maximum	0.99		Heat Pipe Fluid Volume Fractions	
			—heat pipe liquid	0.117
Average Fuel Power Density (W/cm^3)	100		—heat pipe vapor	0.150
Radial Peak-to-Average Power Density Ratio (at most severe transverse cross section)	1.18		Swelling Allowance Volume Fraction	0.039
			Total Fueled Core Volume (liters)	32.4
Overall Peak-to-Average Power Density Ratio	1.59		Total Heat Pipe Fluid Flow Rate (kg/s)	0.566
Average Neutron Energy (Mev)	0.39		Heat Pipe Fluid Saturation Temperature (K)	1,500
Prompt Neutron Lifetime (sec)	2.4×10^{-8}		Heat Pipe Fluid Saturation Pressure (bar)	11.0
Fueled Core			Heat Pipe Inward Radial Heat Flux	
—length (cm)	41.9		—average (W/cm^2)	46.4
—mean outside diameter (cm)	33.9		—maximum (W/cm^2)	85.8
Mean ^{6}Li Control Zone Outside Diameter (cm)	12.8		Maximum Heat Pipe Axial Heat Flux Based On Vapor Volume (kW/cm^2)	7.0
Potassium Vapor Return Duct Inside Dia. (cm)	6.3			
			Nominal Peak Fuel Temperature (K)	1,533
Pressure Vessel Inside Diameter (cm)	36.2			
			Component Masses (kg)	
Side Reflector			—shield	1,400
—inside diameter (cm)	37.2		—reactor-boiler	900
—outside diameter (cm)	41.2		—turbine-alternator	400
			—main radiator	400
Number of Fuel Elements	378		—pump	200
			—miscellaneous	450
Number of Heat Pipes	756		Total:	3750

Reactivity was regulated by means of a dual control system in a manner that each system could independently control the SPR-4 reactor throughout its lifetime. These consisted of a 2 cm thick, axially translating molybdenum reflector and centrally located, in-core liquid lithium-6 control circuits.

Another concept, SPR-5, was an out-of-core thermionic system designed for 10 MW$_e$ output.[12] The reactor was cooled from one end by a set of heat pipes with a structure of tungsten enclosing a silver working fluid (Ag-W heat pipes) that extended the full length of the reactor and terminated shortly outside the cooled end. These core heat pipes transferred thermal energy to a second set of Ag-W heat pipes that were set in a matrix of high temperature shielding material, namely BeO with Eu$_2$O$_3$. The thermal energy removal system thus also acted as the main shadow shield. At the other end of the shield, the heat was transferred by radiation across a void to a set of heat pipes that terminated as the emitters of an array of thermionic diodes (see Fig. 13). This void electrically isolated the emitters, while avoiding the use of high temperature insulation as much as possible. Waste thermal energy was removed from the diodes by lower temperature lithium-niobium (Li-Nb) heat pipes that extended to make up the main radiator.

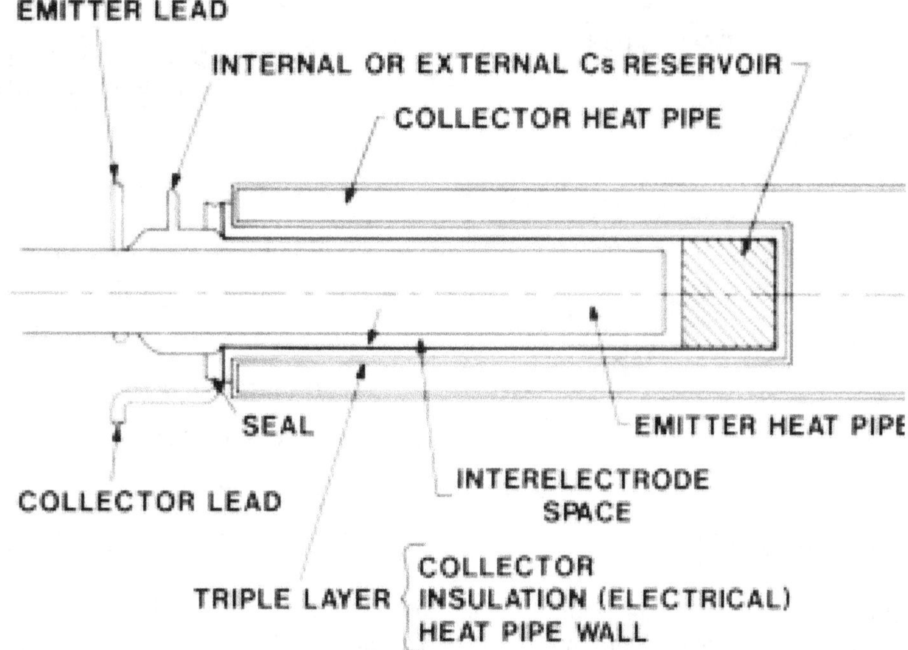

Fig. 13. SPR-5 cylindrical diode for out-of-core thermionic converter system. *From John H. Pitts and C. E. Walters, 1970.*

The reactor coolant temperature was selected to be as high as material properties would permit. The selection of 2,273 K was based mainly on the results of a life test of the silver-tungsten heat pipe. This test showed negligible corrosion after 1,000 hours of operation at 2,223 K. The overall SPR-5 system efficiency was selected to reflect the anticipated performance of a thermionic diode, operating at 2,000 K emitter temperature. The SPR-5 reactor generated 60 MW_t, which yielded a net electrical output of 10 MW_e. The total specific mass of the system was estimated at 4 - 7 kg / kW_e.

Another concept in this study, SPR-6, consisted of a 57 MW_t reactor, a nuclear radiation shield, and a potassium working fluid Rankine cycle power converter. The SPR-6, power plant concept is illustrated in Fig. 14.[13] Figure 15 depicts various state points for the SPR-6 system. In this design concept, high purity lithium was heated in the reactor and pumped through a heat exchanger that then boiled the potassium working fluid in the power generation loop. The power generation loop included four turboalternator units that produced 2.5 MW_e each of useful power. Waste heat was rejected by means of a heat pipe radiator. System components outside the reactor were exposed to reduced radiation doses by the lithium hydride (LiH) shadow shield. Reactor control was provided by either in-core liquid control circuits or by axial translation of the side reflector. Once a nearly neutral coolant mixture had been achieved in the reactor loop, each control method was independently capable of controlling the SPR-6 reactor.

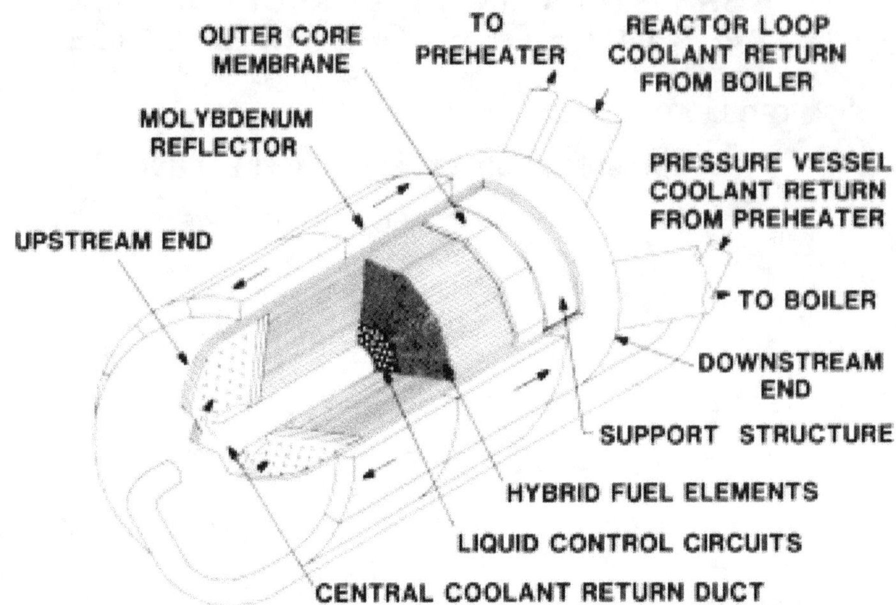

Fig. 14. Cutaway drawing of SPR-6 reactor. *From John H. Pitts and C. E. Walters, 1970.*

The SPR-6 reactor fuel was uranium mononitride (UN), primary loop coolant was lithium, and structural material was 75% tungsten-25% rhenium alloy. Compatibility of these materials at system operating temperatures was experimentally demonstrated. Characteristics of the SPR-6 reactor are given in Table 9. The SPR-6 core consisted of a closely packed bundle of 8,000 fuel elements and some 4,000 coolant channels. These were supported on the downstream end by a 30-cm thick tungsten structure and on the upstream end by a spline attachment that provided for radial and axial growth. The downstream support structure also served as a radiation shield. As shown in Fig. 14, the active portion of the SPR-6 core was an annular region that surrounded a central coolant return passage and a liquid control region. The core outer boundary formed a dodecagon, accommodating the fuel element hexagonal packing pattern. Thin, polygon-shaped tungsten rhenium (75% W-25% Re) membranes enclosed the fueled core region at its inner and outer radii.

The boiler used tantalum tubes arranged in a triangular configuration and separated from each other. This permits the liquid lithium coolant to circulate around the tubes. Pressure in the lithium loop was regulated to minimize the pressure differential across the boiler tube walls. To reduce neutron activation of the natural potassium working fluid to acceptable levels, the boiler was located on the payload (i.e., protected) side of the radiation shield.

A flat, rectangular heat pipe radiator was used to reject waste thermal energy to space. These heat pipes operated in parallel in order to provide radiator system redundancy. The potassium working fluid flow rate through the main heat pipe radiator was varied by adjusting the power to three mechanical pumps. A liquid metal storage expansion tank was needed in each loop to maintain the necessary system pressures.

SPR-6 specific mass was calculated to be 7 kg / kW$_e$. This assumed a high fuel burnup capability, highly efficient heat pipe radiators, a compact radiation shield, and high strength alloys. An optimum reactor coolant temperature of approximately 1,650 K was selected for the system.

As part of the Advanced Space Nuclear Power Program, the vaporization behavior of UN powder at nitrogen pressures of 17,000, 13, and 0.4 pascals was experimentally investigated . Experimental results are given in Fig. 16.

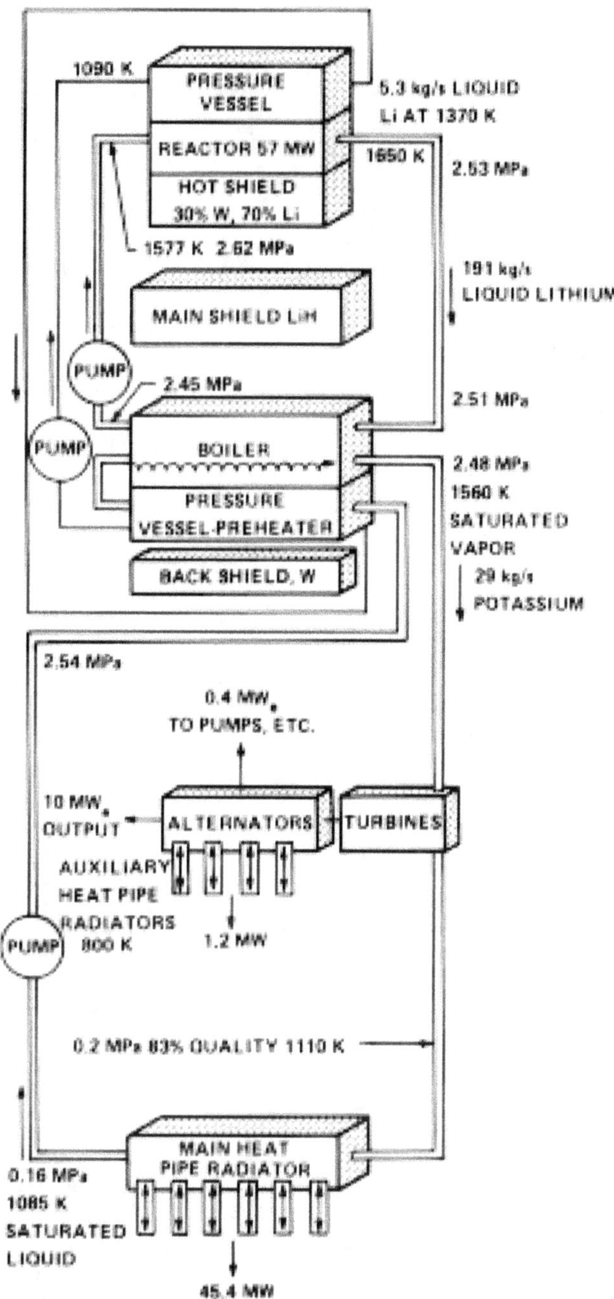

Fig. 15. SPR-6 system schematic. *From John H. Pitts and C. E. Walters, 1970.*

Table 9. Characteristics of the SPR-6 reactor.

Fuel Atom Burnup (%)		Primary Coolant Volume Fraction	0.234
—average	2.1	Interstital Fluid Volume Fraction	0.093
—maximum	3.3		
		Swelling Allowance Volume Fraction	0.057
Fuel Enrichment (atom fraction of ^{235}U)		Total Fueled Core Volume (m^3)	0.228
—average	0.70		
—maximum	1.0	Reactor Coolant Flow Rate (kg/s)	191
		Reactor Coolant Outlet Temperature (K)	1,650
Average Fuel Power Density (W/cm^3)	575		
		Reactor Coolant Temperature Rise (K)	73
Radial Peak-to-Average Power Density	1.12		
		Reactor Coolant Inlet Pressure (bars)	26.2
Overall Peak-to-Average Power Density	1.57		
		Reactor Pressure Drop (bars)	0.9
Average Neutron Energy (MeV)	0.25		
		Core Reynolds Number	5.9×10^4
Prompt Neutron Lifetime (ns)	81		
		Reactor Coolant Velocity Inside Core (cm/s)	740
Fueled Core			
—length (cm)	87.0	Average Heat Flux At Coolant Channel Surface	
—mean outer diameter (cm)	62.1	(W/cm^2)	117
Liquid Lithium Control Zone Outer Diameter (cm)	22.7		
		Average Centerline Fuel Temperature	46
Reactor Coolant Return Duct Diameter (cm)	15.3	(degrees K above local coolant temperature)	
Pressure Vessel Inside Diameter (cm)	65.0		
		Component Masses (kg)	
Side Reflector		—shields	33,000
—inside diameter (cm)	67.4	—main radiator	13,400
—outside diameter (cm)	71.4	—alternators (including auxiliary cooling radiators)	7,200
		—reactor	4,200
Fuel Radius (cm)	0.216	—boiler-preheater	1,800
		—structure	1,500
Fuel Element Cladding		—turbines	1,400
—inside radius (cm)	0.225	—reactor loop piping	1,000
—outside radius (cm)	0.250	—lithium coolant pump	1,000
		—potassium working fluid pump	600
Coolant Tube Inside Radius (cm)	0.220	—lithium fluid	600
		—power conversion loop piping	500
Number of Fuel Tubes	8,064	—potassium fluid	400
		—controls	300
Number of Coolant Tubes	4,032	—pressure vessel cooling loop pump	100
		—miscellaneous	3,000
Fuel Volume Fraction	0.433	Total:	70,000
Cladding (plus coolant) Tube Volume Fraction	0.183		

Irradiation experiments of UN fuel capsules were still in the early stages when the program was terminated. Although none of these reactors were ever flown in space, UN fuel irradiation capsules had been exposed to 1.4 X 10^{22} fissions / cm^3 for 2,229 hours at 1,675 - 1,775 K with little or no apparent swelling (i.e., less than I percent). However, one fine grained UN capsule experienced approximately 3 percent swelling.[14] In addition, at program termination compatibility tests were also underway that exhibited negligible corrosion of the W-Re-Mo capsule with circulating lithium after 1,004 hours at temperatures near 1,675 K.

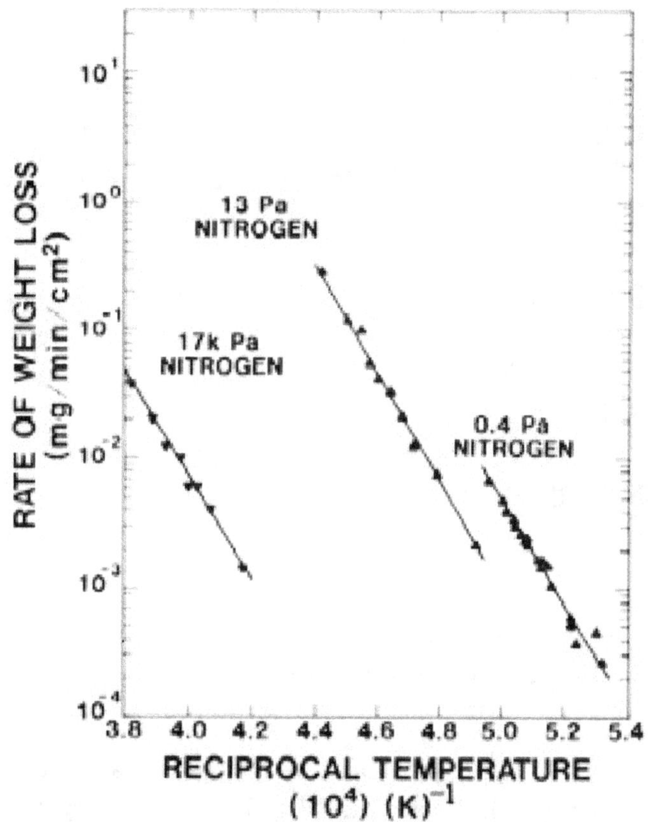

Fig. 16. Rate of mass loss for UN$_{1-x}$ as a function of reciprocal temperature. *From. Lawrence Radiation Laboratory, "Quarterly Report, October Through December 1967*

Heat Pipe Power Plant *(SPAR/SP-100)**

In 1979, a program entitled Space Power Advanced Reactor (SPAR) was initiated to provide 10 - 100 kW$_e$ power for a period of 7 to 10 years. A heat pipe reactor with thermoelectric power conversion was selected for the design. The power plant was subsequently renamed SP-I00. Because the space reactor power program initiated in 1983 also took the name SP-I00, the designation SPAR/SP-I00 is being used to distinguish the heat pipe power plant design.

A diagram of the selected heat pipe power plant configuration is shown in Fig 17. The power plant consists of the nuclear subsystem that included the nuclear reactor heat source and radiation attenuation shield and the power conversion subsystem that included the thermoelectric converters and heat rejection elements. The SPAR/SP-100 heat source is a solid core nuclear reactor that includes heat pipes to extract thermal energy from the core and transport the thermal energy to the thermoelectric converters.[15] The evaporation section of the heat pipe is located in the reactor core, while the condenser section is located within the electric power conversion elements. A major advantage of this heat pipe-in-core design concept is that the need for mechanical pumps and compressors is eliminated. Furthermore, many parallel heat transport elements are incorporated into the reactor design, providing redundant thermal energy transport pathways. This arrangement is designed to eliminate single point failure modes.

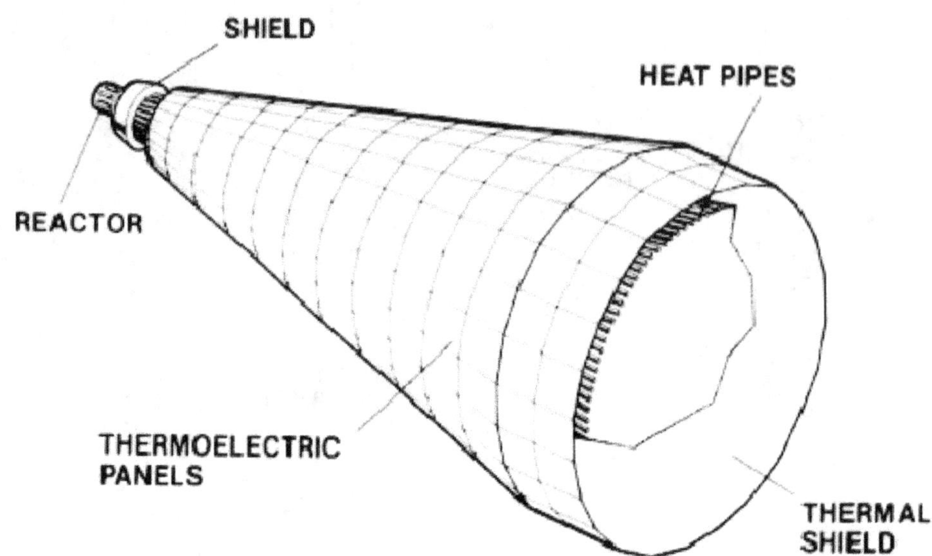

Fig. 17. The heat pipe nuclear power system, featuring a radiatively coupled design between the nuclear reactor subsystem and the thermoelectric power conversion subsystem. *Courtesy of Los Alamos National Laboratory.*

The high power levels, compact size, minimum mass, and long lifetimes in the mission requirements favored the use of a fast neutron spectrum, highly enriched reactor that has a large inventory of fuel in a small volume. A large fuel inventory minimizes reactivity decreases due to fuel burnup during operation. A 1,600 kilowatt-thermal reactor will only burn up about 3.6 percent of the uranium-235 atoms in the fuel in seven years of full power operation. With this typical amount of fuel burnup being a very small percentage of the total fuel inventory, reactor criticality can be maintained throughout a mission.

High temperatures are desired to minimize mass and size of the power plant. However, the operating temperature is constrained by both electrical conversion system material limits and fuel swelling considerations. This resulted in the selection of the reactor heat pipe temperature of 1,500 K. A consequence of this operating temperature led to the selection of the refractory fuel, uranium dioxide (UO_2)--a choice based on materials compatibility with the heat pipes.

Power conversion for the SPAR/SP-I00 concept used an improved version of the silicon germanium highly redundant, space qualified thermoelectric (TE) converters. Thermal energy from the reactor core heat pipes used radiation heat transfer coupling to panels containing TE materials in an arrangement shown in Fig. 18. The thermoelectric hot shoe thermal collectors concentrate this radiant energy. The thermal energy is than conducted through the TE material. Thermal insulation is used to minimize possible thermal energy losses around the thermoelectric material. Thermal energy that cannot be effectively converted to electricity due to second law of thermodynamics limitations is then rejected to space by radiation heat transfer. The heat rejection surface forms the cold shoe of the TE elements. By distributing a sufficient number of these TE elements over a wide area, the cold shoe, in effect, serves as the heat rejection radiator.

As shown in Fig. 18 and 19, the core heat pipes extend from the reactor, bend around to the outside of the radiation shield, and continue to the TE converter region. The radiation shield is located between the reactor and the payload. It serves as a barrier that substantially reduces the amount of reactor-generated nuclear radiation (e.g., neutrons and gamma rays) reaching the payload sensitive portions of the spacecraft or the astronaut crew. This shield contains primarily low atomic weight lithium hydride (LiH) to attenuate neutrons and high atomic weight tungsten (W) to shield against gamma radiation. The shield is configured to minimize mass.

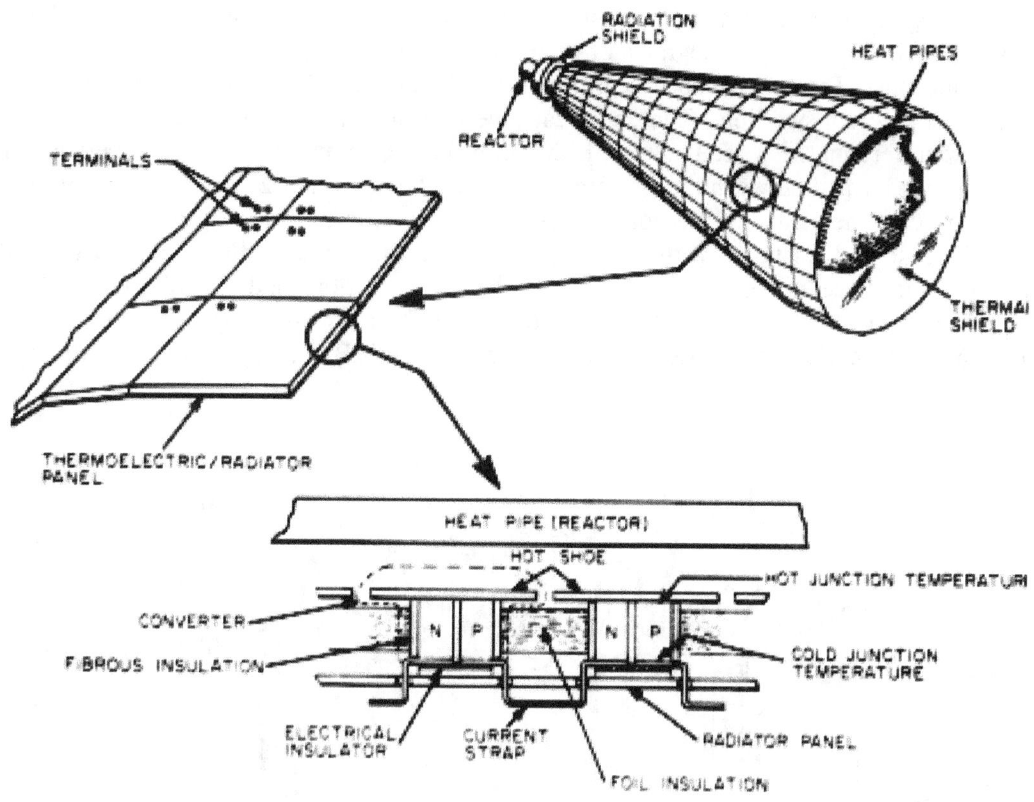

Fig. 18. Schematic of the SP-100 power conversion matrix. *Courtesy of Jet Propulsion Laboratory.*

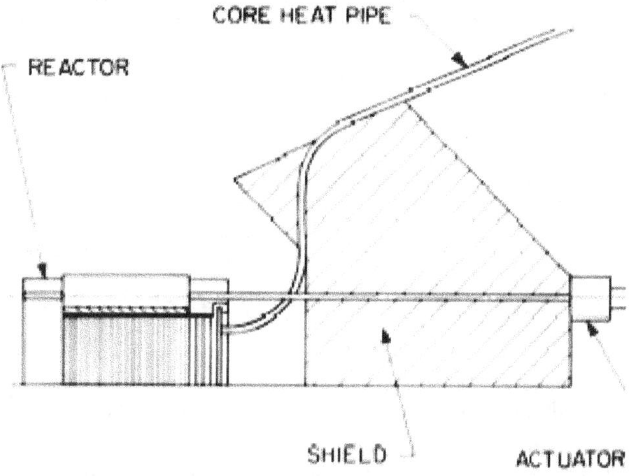

Fig. 19. SPAR/SP-100 nuclear subsystem components. *Courtesy of Los Alamos National Laboratory.*

This design eliminates the need for an independent afterheat removal system for planned or emergency reactor shutdown. This feature results from the heat pipes being self-contained passive heat transport devices. They continue to function even at lower thermal energy input levels and in the absence of prime electric power. The heat pipes have sufficient out-of-core surface area to support radiation heat transfer rejection of the core afterheat.

Fig. 20 is a cutaway view of the reactor. The reactor core region is composed of fuel modules in which the nuclear fuel is arranged in layers between circumferential fins on the heat pipes. A typical fuel module is depicted in Fig. 21. The fins are an integral part of the heat pipe, while the fuel is fabricated as tiles that are inserted between the fins. The fins enhance heat transfer and were added to lower the operating temperature in the UO_2 fuel and reduce fuel swelling. The core heat pipe/fuel modules are arranged in concentric rings. The heat pipes and circumferential fins are made with the alloy molybdenum containing 13 percent rhenium (Mo-13%Re). The fuel tiles located between the fins are highly enriched uranium-235 in the form of uranium dioxide (UO_2). A void space in the center of the core incorporates a neutron poison plug made of B_4C. This plug is inserted before the reactor is installed with its payload onboard the launch vehicle, e.g., at the time the Space Shuttle, and remains in place as a safety feature during launch operations and orbital ascent. The poison plug keeps the reactor subcritical, even if it were immersed intact in water with all its heat pipes broken. Once the reactor and its spacecraft reaches the mission operational location, this B_4C plug was to be removed.

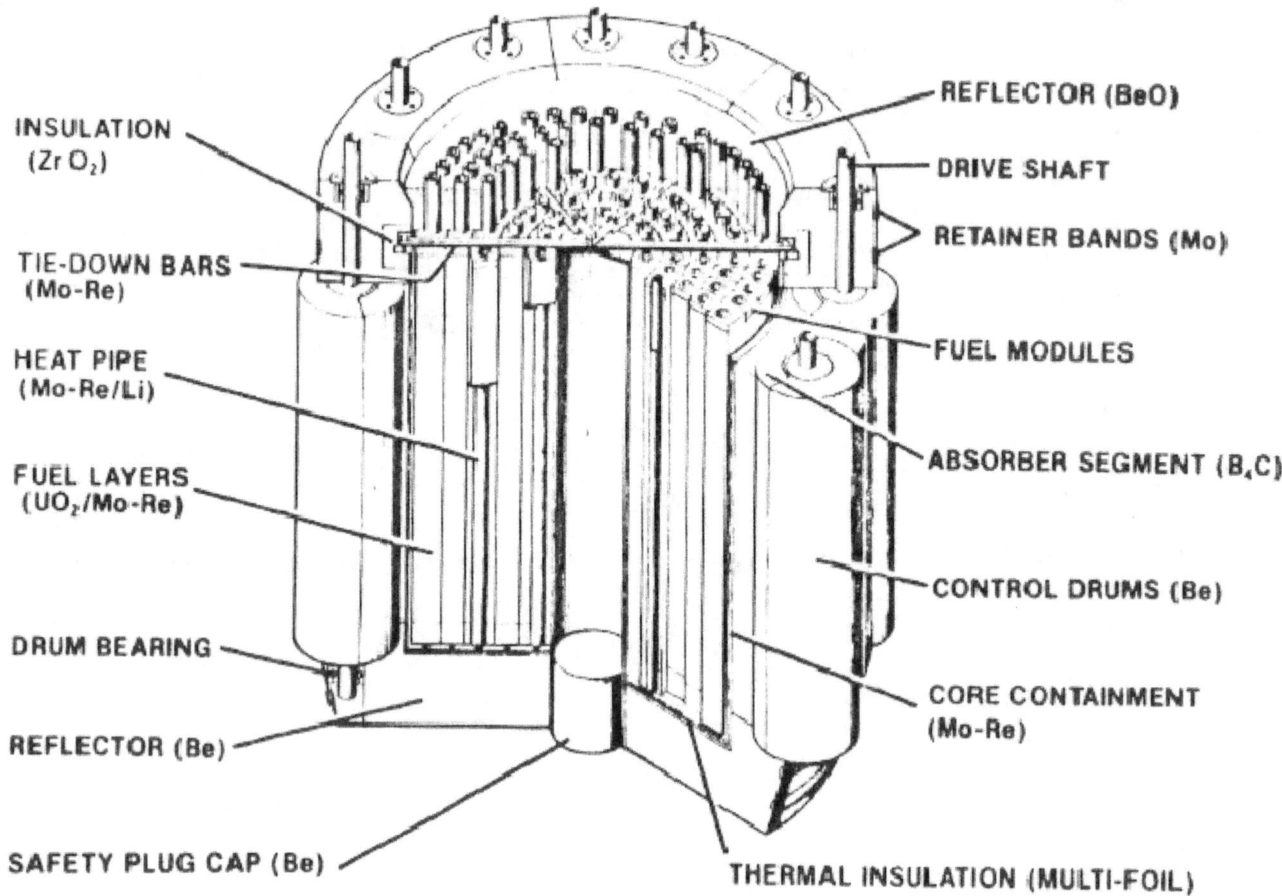

Fig. 20. Heat pipe space reactor. *Courtesy of Los Alamos National Laboratory.*

Surrounding the core is a barrel-shaped containment structure to support the fuel modules--the containment structure is not a pressure vessel. The containment barrel also provides non-compressive support for multifoil reflective insulation layers used to reduce core heat losses. Outside the core containment is located a beryllium reflector used to reduce neutron leakage. Located within the reflector are control drums, rotated by electromechanical actuators. Each drum contains a segment of boron carbide (B_4C), a neutron absorbing material to control reactivity. Movement of this "neutron poison" segment is used to establish the reactor power level.

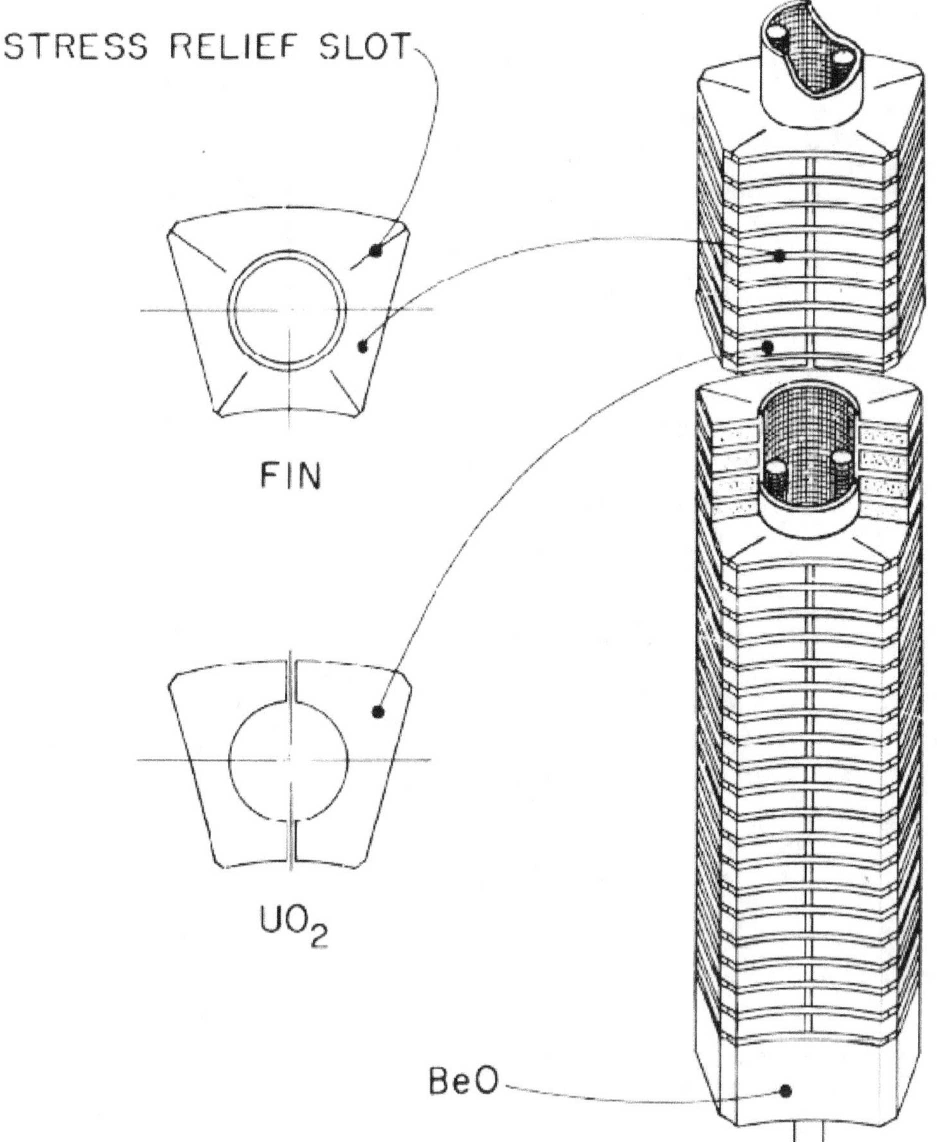

Fig. 21. Typical fuel module. *Courtesy of Los Alamos National Laboratory.*

Table 10 summarizes power plant performance parameters.

Axial and radial power profiles in the core must be account for in the reactor design. In place of varying the isotopic content of the fuel, the quantity of fuel in each row of modules is varied. This is used to achieve a uniform amount of heat generation per fuel module and results in each core heat pipe transporting about the same amount of thermal energy. Within the heat pipe, axial temperature flattening is an inherent characteristic of the heat pipes.

Fuel swelling is function of temperature and operating time. Data on UO_2 fuel swelling are summarized in Fig. 22. Voids are provided in the design to accommodate fuel swelling. The thermal stresses generated by fuel swelling that could result in axial stresses and core distortion are prevented by omitting an occasional fuel tile.

High operating temperature core materials will to some extent evaporate and redeposit themselves. For example, the rate of redeposition of UO_2 and molybdenum is shown in Fig. 23 as a function of temperature. This redeposition was considered in the SPAR/SP-100 design.

Table 10. SP-100 performance parameters.

Output Power (kW$_e$)	
Range	10–100
Nominal	100
Reactor Thermal Power (kW$_t$)	
Range	200–1,60
Reference Design	1,470
Design Life (year)	
Design Power	7
Total	10
Overall Dimensions	
Length (m)	8.5
Diameter (max)(m)	4.3
Radiator Area (m^2)	70
System Mass (at reference design)(kg)	
Reactor and Controls	450
Shield	790
Heat Pipes	460
TE Conversion and Circuitry	250
Thermal Insulation (incl. end panels)	290
Radiator	100
Power Control Subsystem	130
Interface Equipment and Structure	300
Total System Mass	2,770
Specific Power (W/kg)	36

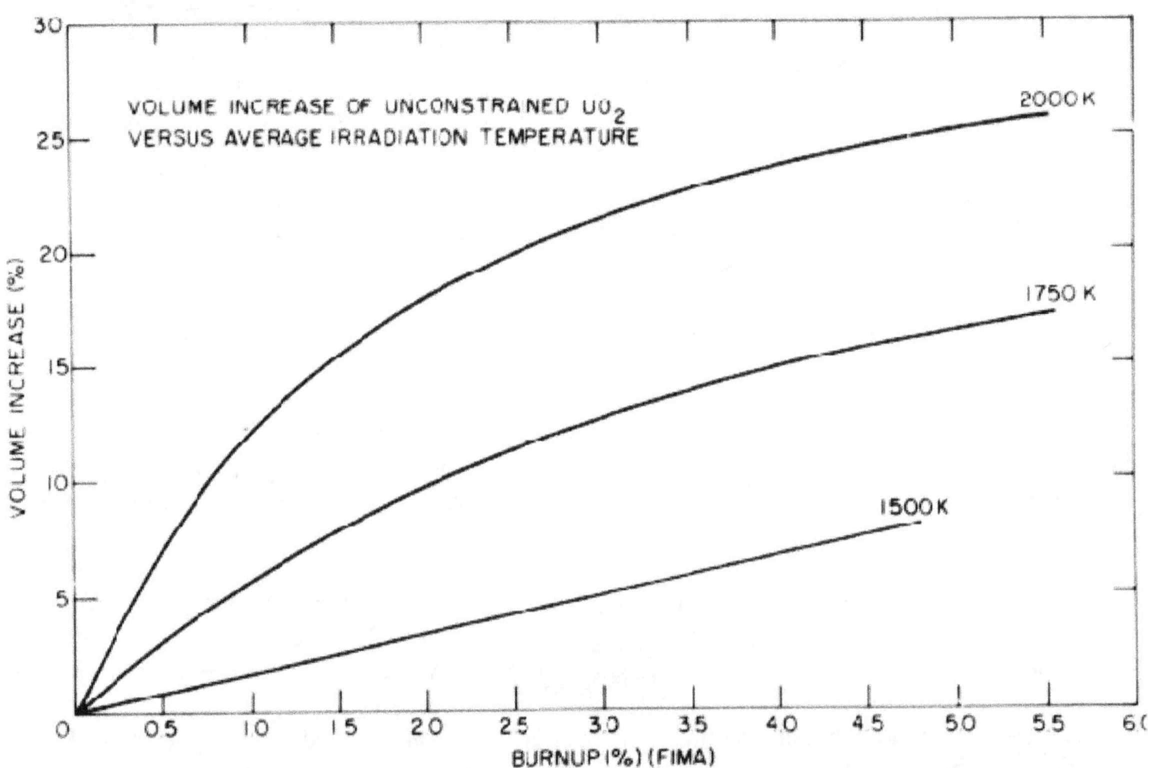

Fig. 22. Fuel swelling behavior of UO$_2$. *Courtesy of W. A. Ranken, Los Alamos National Laboratory.*

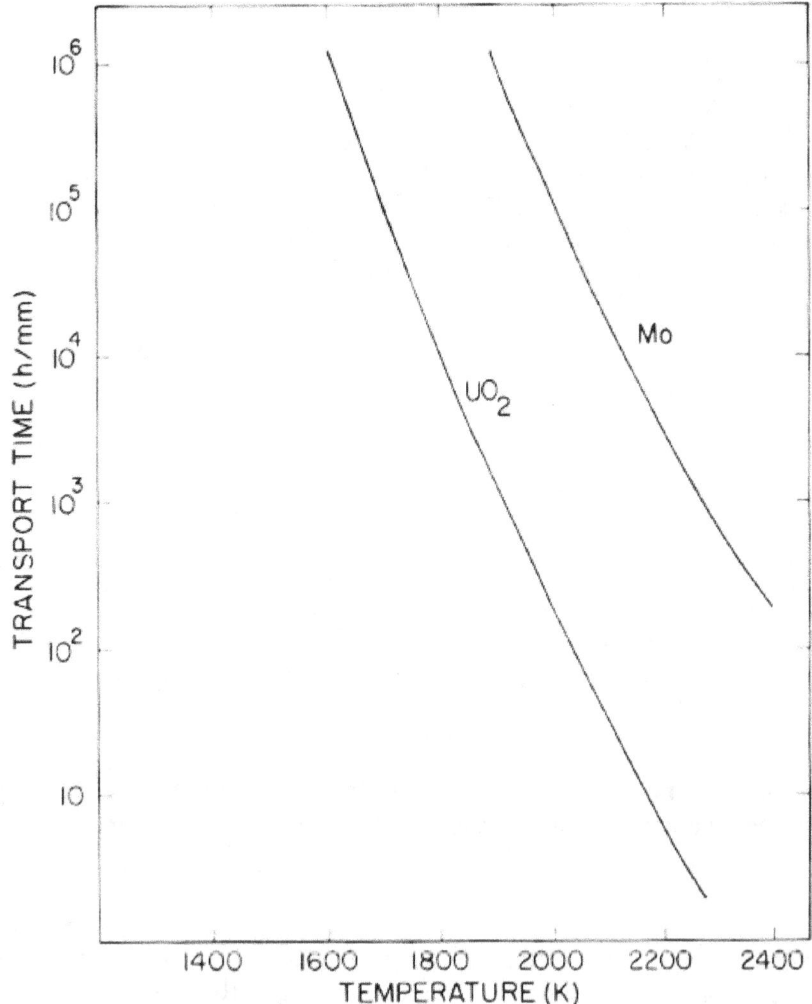

Fig. 23. Self-transport times for UO₂ and molybdenum. *Courtesy of Los Alamos National Laboratory.*

One concern is a possible heat pipe failure. If an individual heat pipe fails, transient heat transfer analysis indicates a peak temperature buildup will occur in a few minutes. This is a result of poor thermal contact between adjacent fuel modules. However, within 24 hours, the UO₂ will redeposit, closing the gap and reduce the fuel temperature back to acceptable levels. Part of the core heat pipe is to provide margins to accommodate a 25 percent overcapacity. Hence, the heat pipes adjacent to an individual failed pipe will transport this additional heat load out of the core. Actually, in the design, an additional 25 percent design margin has been included so that the heat pipe will operate at nearly isothermal conditions and to protect against performance degradation. However, multiple failures of adjacent heat pipes could result in unacceptable consequences; the current design did not guarantee good thermal contact between fuel modules under these unusual circumstances that are considered to have a low probability.

In the reflector design, neutron irradiation-induced swelling and the subsequent loss of thermal conductivity was of particular concern. The circumferential reflector used beryllium, while the reflector section penetrated by heat pipes used beryllium oxide (BeO). Operating temperature of the reflector was about 800 K with cooling by radiation heat transfer to space from its outer surface. The swelling primarily resulted from the formation of helium by (n, α) reactions. Above 575 K, helium gas accumulates into bubbles and swelling proceeds as the bubbles grow. Swelling data are summarized in Fig. 24, normalized to a dose of 10^{20} nvt (10^{24} n / m²). (The term nvt stands for neutron-velocity-time and is a unit of neutron fluence, that is n / cm² or n / m².) As the temperature is increased, no significant swelling occurs up to approximately 825 K. At high neutron fluences (greater than 10^{21} n / m²) and elevated temperatures (above 825 K), helium bubble formation will probably occur and reduce the beryllium's thermal conductivity.

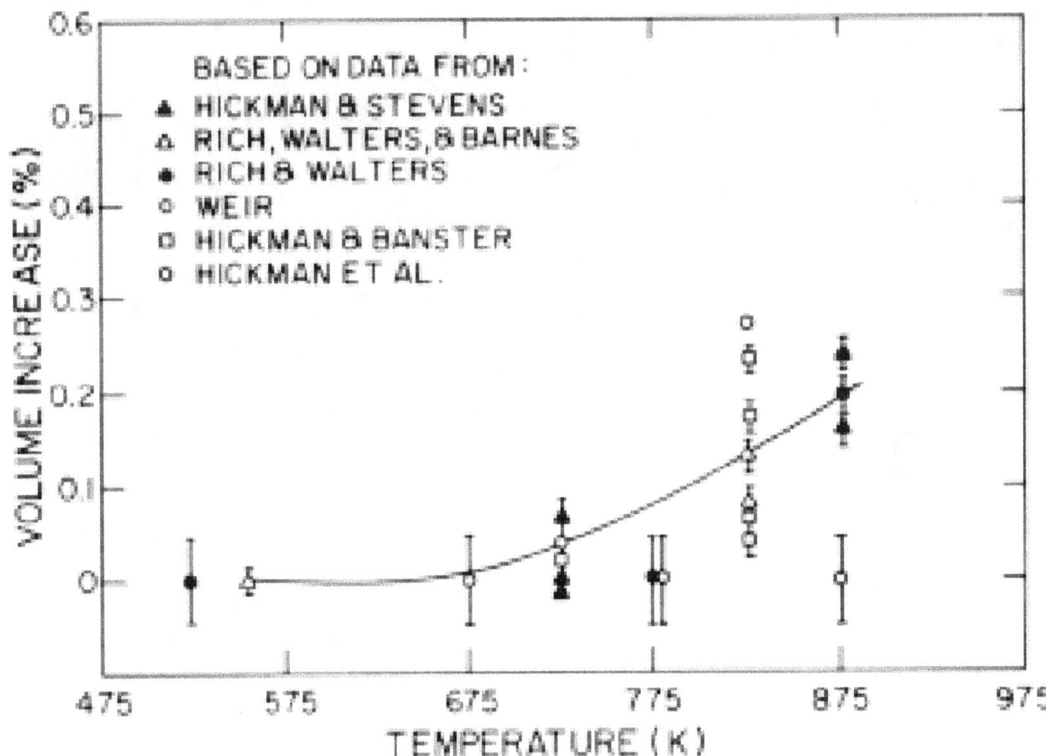

Fig. 24. Volume changes observed in irradiated beryllium as a function of temperature. (The reported values have been normalized to a dose of 10^{24} neutrons/m^2.) *Courtesy of Los Alamos National Laboratory.*

Twelve cylindrical control drums within the reflector would have increased or decreased reactivity by rotating the neutron absorbing material away from or toward the core, respectively. Rotational drum movement used a stepping actuator located away from the core behind the shield. Thus the actuators were in a reduced radiation and thermal environment. Built into the design was approximately 5 percent excess reactivity to compensate for control margin, fuel burnup, temperature defect, and operational contingencies. By enriching the B_4C to 90 percent ^{10}B, a reactivity worth of 12.5 percent was provided. Seven years of operation would result in the reflector increasing in volume by about 2 percent.

A unique feature of the SPAR/SP100 power system is that since there are no pumped coolant loops, there are no pressure vessels to content with in regard to building safety features into the design. So, since the reflector consisted of segments held together by molybdenum retainer bands to provide mechanical integrity to the reactor during launch and orbital operations (see Fig. 25), the reactor was designed to meet aerospace nuclear safety requirements by rapidly oxidizing and disassembling on atmospheric reentry.

Table 11 present some typical SPAR/SP-100 nuclear subsystem performance parameters, including dimensions, fuel swelling, and burn up.

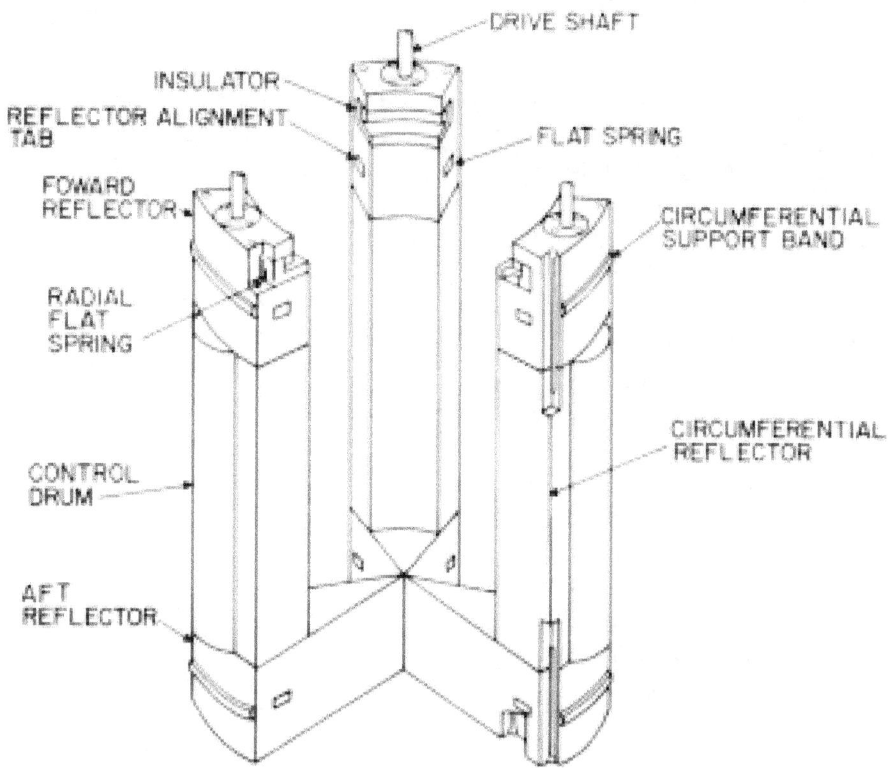

Fig. 25.   Segmented reflector showing drums and banking arrangement that supports upper atmospheric dispersal upon reentry. *Courtesy of Los Alamos National Laboratory.*

Exploring the heat pipe design in more detail, the containment material was a low rhenium (13%) content alloy of molybdenum selected for its excellent high temperature creep strength and its compatibility with the internal working fluid and with the uranium dioxide fuel. This alloy is also ductile at low temperatures and can withstand the launch load environments. Properties of Mo-13% Re are shown in Table 12. The lithium working fluid was selected because of its characteristics at the operating temperature of 1,500 K including its vapor pressure, high heat capacity, low operating pressure, and compatibility with the wall and wick materials. The heat pipe wick design used an axially artery configuration for low pressure drop and to provide a redundant wick structure. This was supported by a circumferential distributive wick to transport the condensing fluid to and from the arteries. Getter materials were added so that residual fabrication impurities and impurities that diffuse through the walls during operation are absorbed and will not react with the container or wick materials. Hafnium, with its affinity for oxygen, was selected for the getter material. Fig. 26 shows the predicted behavior of a typical core heat pipe. The limitations on heat pipe performance in this particular design are the sonic limit and the capillary action that can be achieved with a given wick pore size.

Table 11. Nuclear subsystem characteristics.

Thermal power (MW)	1.5
Heat pipe evaporator temperature (K)	1,500
Number of heat pipes	120
Heat pipe materials	
Container	Mo-13% Re
Working Fluid	Lithium
Getter	Hafnium, Zirconium
Maximum UO_2 temperature (K)	1,730
Fissions per 7 y (fis/cm^3)	8×10^{20}
Beginning-of-life reactivity (K_{eff})	1.05
Burnup 7 y (%)	3.6
Core diameter (cm)	33
Core height (cm)	33
Be reflector average thickness (cm)	9
BeO reflector average thickness (cm)	10
Reactor diameter (cm)	54
Reactor height (cm)	55
Nuclear subsystem mass (kg)	440
B_4C plug mass (kg)	3
Shield fluence at 7 y (nvt)	10^{12}
Gamma dose at 7 y (rad/Si)	10^6
Shield mass (kg)	790
Shield core half angle (Deg)	15
Shield axial thickness (cm)	80

Table 12. Selected properties of molybdenum-13% rhenium alloy.

Physical Properties	Molybdenum-13% Rhenium[a]
Density at 293 K (mg/m^3)	11.89
Thermal conductivity at 1,400 K (W/m K)	~70
Melting point (K)	2,810
Complete recrystallization in 1 h (K)	1,645
Chemical Compatibility	
Lithium at 1,500 K in heat pipe	No reaction experienced
UO_2	No reaction to 2,475 K
Mechanical Properties	
Ductile-brittle transition temperature (K)	140
Projected creep deformation (%)[b]	3×10^{-4}

[a] Electron beam melted alloys

[b] of 15.9 mm o.d. by 14.1 mm i.d. heat pipe after 7 yr at 1,400 K (p ≈ 7 Mpa)

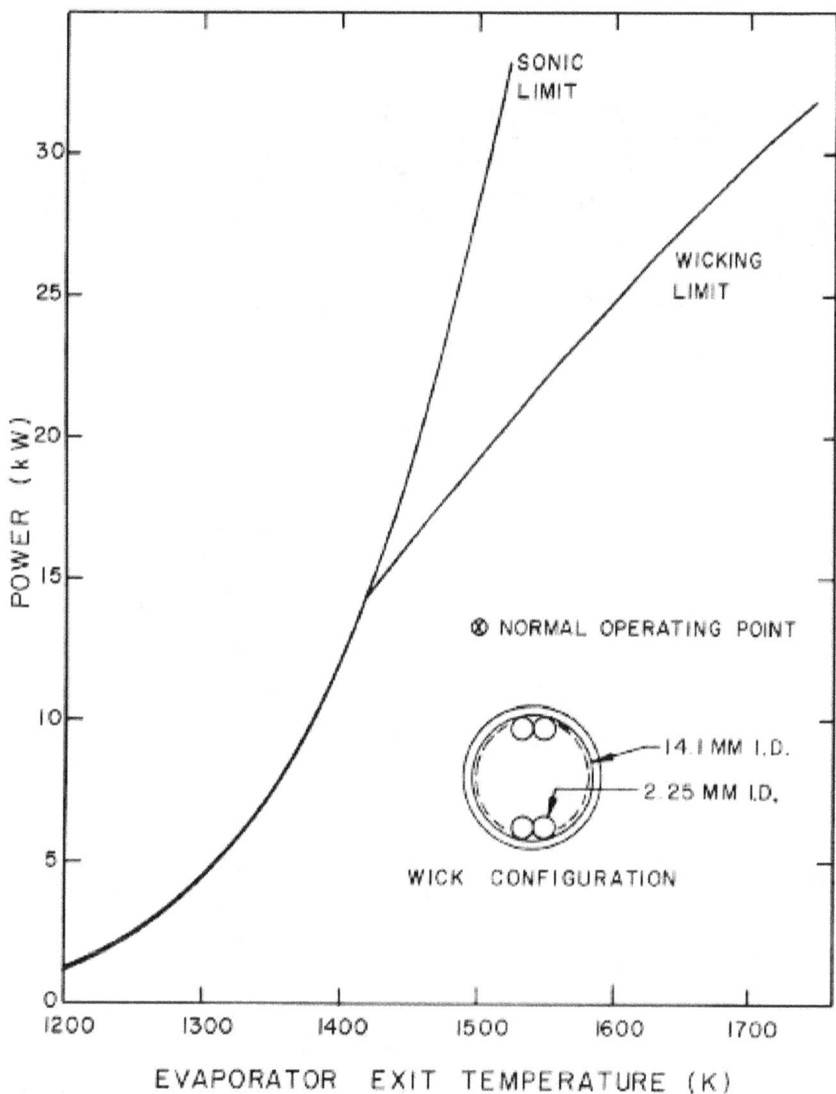

Fig. 26. Typical high temperature heat pipe performance curve. *Courtesy of Los Alamos National Laboratory.*

The SPAR/SP-100 nuclear radiation shield designs were based on the extensive developments in the SNAP program. As such, it used lithium hydride (LiH) for neutron shielding and a layer of tungsten (W) to attenuate gamma radiation. The shield was designed to operate at above 600 K so that reabsorption of radiolytically decomposed hydrogen occurred to avoid swelling. The upper shield temperature is 675 K to avoid hydrogen loss due to excessive thermal dissociation if the casing is punctured by meteoroids. Single failure points were avoided by encapsulating the LiH in a number of compartments. A pressure containment failure in one compartment will deplete the hydrogen in only a small portion of the shield. Tungsten components operate satisfactorily at the LiH temperature level. The shield mass is sensitive to spacecraft radiation level requirements, shield half angle, and separation distance from the center of the core (see Fig. 27).

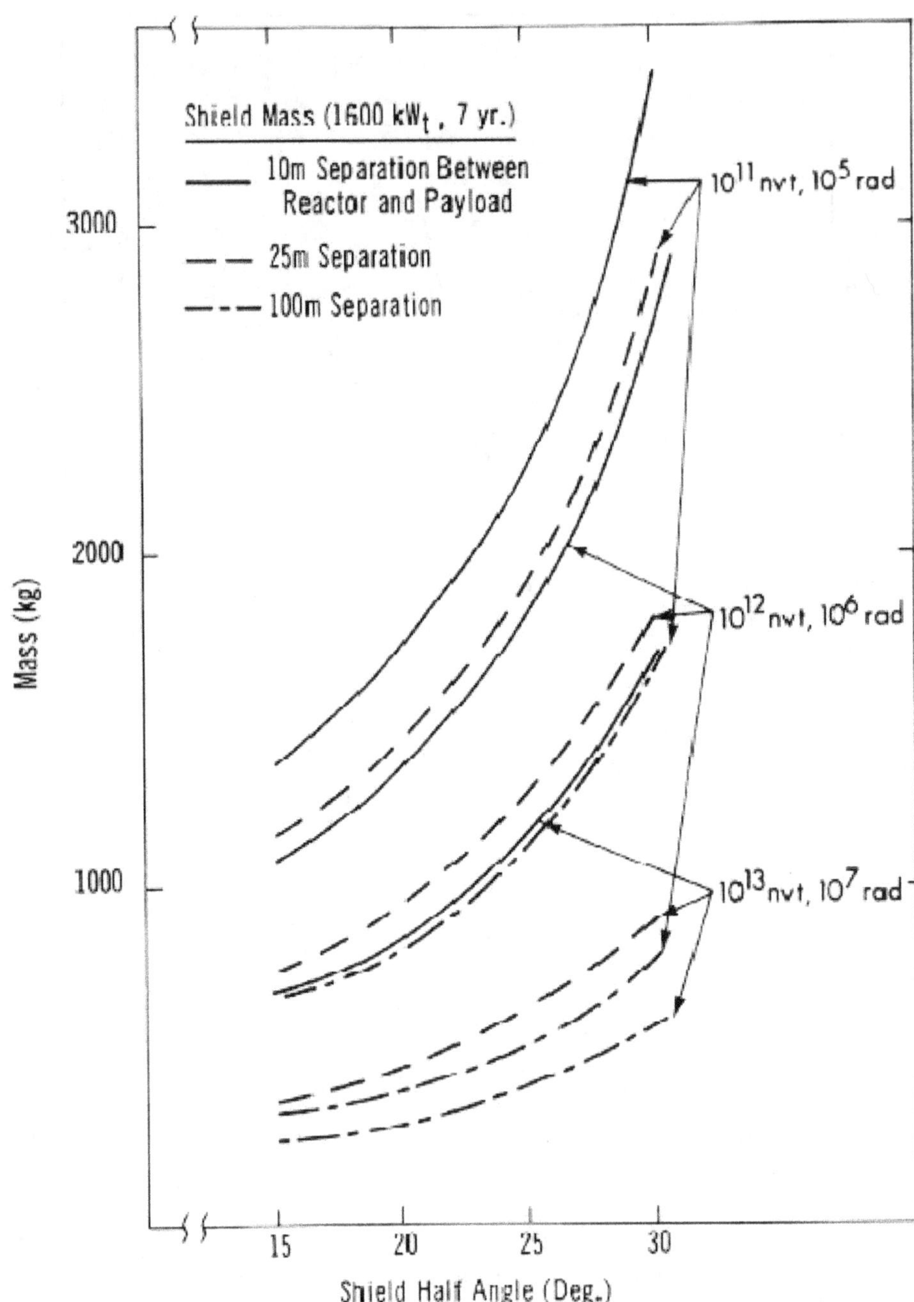

Fig. 27. Shield mass as a function of shield half angle for various radiation attenuation levels and separation distances. *Courtesy of Los Alamos National Laboratory.*

Reactor control was by adjusting the 12 rotating drums that surround the core using functionally redundant reactor closed-loop controllers. For safety, to prevent overheating of high temperature system elements, provision was included in the reactor controller for a power cutback. Also, if an emergency shutdown situation was detected, a shutdown control function scrammed the control drums.

Reactor control during startup sensed temperatures of the core heat pipes. To avoid excessive power and temperature overshoots from the long element time constants, the startup logic sequence was over several hours. Faster startups are possible, if desired.

A representative startup scheme follows:

1. Control drums are ramped relatively quickly from shutdown to a position well below cold critical condition.

2. The drums are then placed on a very slow ramp through cold critical conditions.

3. The system is switched to temperature control when the heat pipe temperatures reach the minimum controllable value,

4. Power level is raised based on the control temperature to achieve normal operation.

Reactor shut down is achieved by driving the drums to their in-position. No additional afterheat removal system is required because the core heat pipes will continue to function as a passive heat removal system.

Meeting aerospace nuclear safety requirements are an important factor in the control system design and redundancy is also valuable in creating systems with good safety and reliability characteristics. Safety features included:

1. Control drums are pinned in their shutdown position with captive-key locks for all non nuclear critical reactor ground handling nuclear operations. For functional testing, one drum at a time can be unlock and operated.

2. A central neutron poison plug of B_4C is installed in the core for launch and orbital ascent. This plug is removed after the nuclear-powered payload has been deployed in space. Removal is by either a teleoperated device or an astronaut doing extravehicular activity (EVA).

3. Control drum actuators include a braking mechanism that lock the drums in position until the brakes are energized. The actuator motors do not have sufficient torque to drive the drums with the brakes engaged.

4. Reactor shutdown can still be achieved even if a number of the twelve drums were to become inoperative.

5. Redundant, parallel-operating, self-test reactor controllers will ramp the control drums to their shutdown position. Each redundant controller has independent energy source that do not share electronic components with the others. The system can be reset to normal operation and restarted after an emergency shutdown.

6. Finally, the control drums are fail-safe with a spring-loaded mechanism to return them to the shutdown condition in the event of the loss of electrical power.

The conversion of the thermal energy to electrical energy was to be accomplished through the use of thermoelectric (TE) converters. Continued development of improved thermoelectric materials in the form of silicon germanium/gallium phosphide (SiGe/GaP) was discussed under SP-100.

Long life heat pipes and fuel module demonstrations are the key technologies needed to validate the heat pipe reactor concept. Short term testing experience provides a high technical confidence that component performance requirements can be met. Heat pipe and fin material, Mo-13% Re, has been shown to be ductile at temperatures well below 200 K and at this low temperature exhibits excellent weld ductility. In fabrication of an in-pile irradiation experiment, fins were successfully machined onto a heat pipe. Compatibility experiments with UO_2 and molybdenum alloy indicate that O_2 diffusion is sufficiently low to avoid serious lifetime limiting effects. Additional work, to fully characterize Mo-13% Re on high-temperature strength (including long-term creep), low-temperature ductility, the effects of impurities, long-term grain growth phenomena, and the effects of neutron irradiation and fission fragments are needed. Heat pipe performance goals have been shown experimentally to exceed the normal and contingency design.

Some high temperature life testing exist, but not in a radiation environment A molybdenum-TZM/sodium heat pipe demonstrated a single heat pipe operation of 53,000 hours at 1,390 K without failing.[16,17] A molybdenum/lithium heat pipe was operated for 25,216 hours at 1,700 K before failure.[18] Since purity of the liquid metal working fluid is of primary importance to high-temperature heat pipe performance and life, means have been developed to assure pure liquid is used in filling the heat pipe. Concern has been expressed about operating heat pipes in a micro-gravity space conditions. No liquid metal heat pipes have operated in a micro-gravity environment; however, low temperature heat pipes have been demonstrated in space since 1967.

Essential fuel swelling data as a function of temperature and burn up exist. The primary experimental fuel data based on sources that most nearly simulate SPAR/SP-100 conditions--that is, fuel in the temperature range of 1,500 to 2,150 K and with a low thermal gradient--are summarized in Fig. 28. These data cover the regions of design interest with expected burnup for seven years operation of 3.6%.

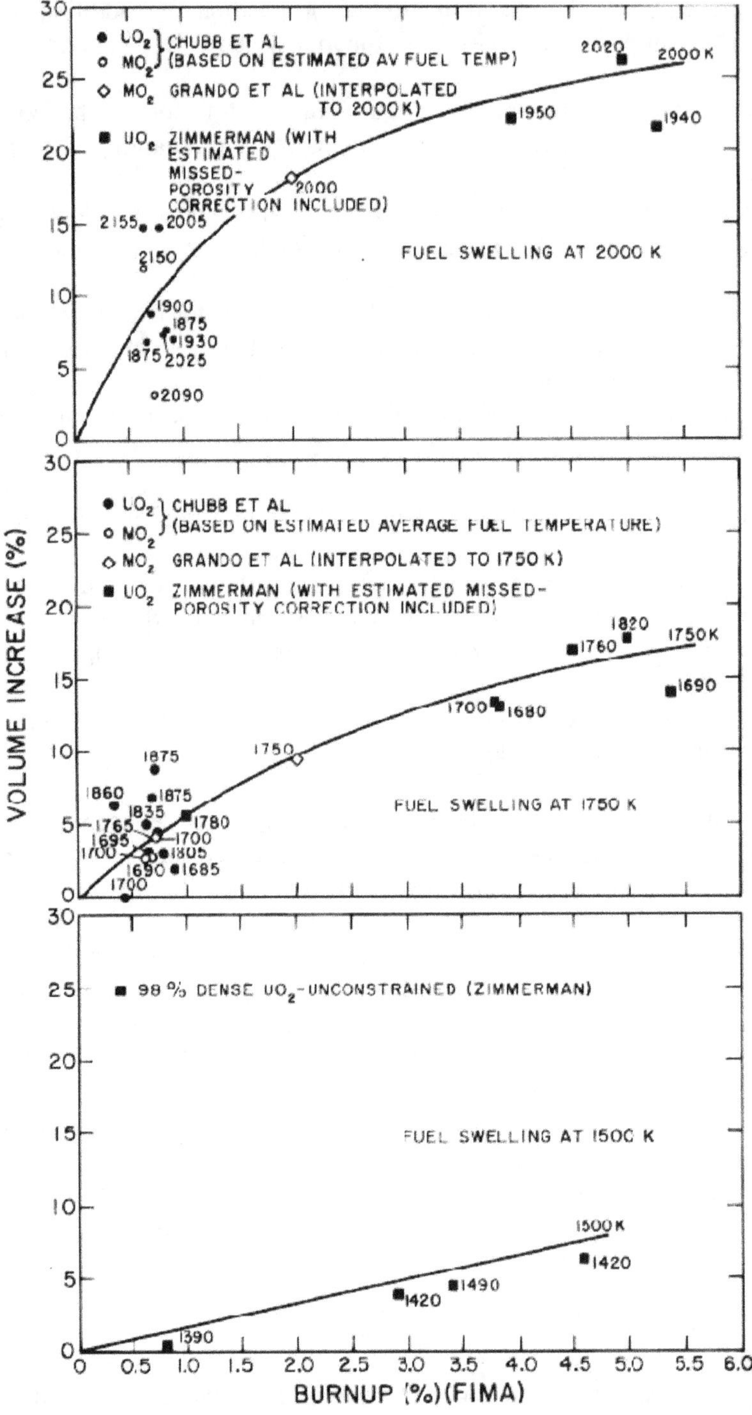

Fig. 28. Recommended fuel swelling data for SPAR/SP-100 systems at 1,500, 1,700, and 2,000 K. *Courtesy of W. A. Ranken, Los Alamos National Laboratory.*

Components, such as the reflector, reactivity control actuators, and shielding incorporate technologies demonstrated from previous space reactor programs. For example, the actuators in the S8DR development had a single unit run over 20,000 h.

Multi-Megawatt Program

Beginning in 1985 and until 1990, a program was underway to develop electrical power in the multimegawatt range for neutral particle beams, free electron lasers, electromagnetic launchers, and orbital transfer vehicles.

The program was discontinued because of a shift in emphasis within the Strategic Defense Initiative. The power requirements were grouped into three categories, as seen in Table 13.

Table 13, Multimegawatt space power system requirements.

	Category I	Category II	Category III
Power requirements (MWe)	10s	10s	100s
Operating time (s)	100s	100s + 1 y of total life	100s
Effluents allowed	Yes	No	Yes
1 Orbit Recharge	No	Yes	No

Six concepts were selected for Phase 1 studies: three for Category I, two for Category II, and one for Category III. The program was terminated prior to Phase II awards. A brief summary of these concepts follows (much of the information is courtesy of Richard Shutters, Multimegawatt Project Office, Idaho National Engineering Laboratory).

For Category 1:
(1) A fast-spectrum, cermet-fuel, gas-cooled reactor derived from the 710 program[19] drives twin counter-rotating open Brayton cycle turbines coupled to super-conducting generators. Table 14 summarizes key parameters.

(2) A fast-spectrum, gas-cooled reactor with a two-pass core heats hydrogen and drives twin gas turbo-generators. Table 15 summarizes key parameters.

(3) A gas-cooled, nuclear rocket derivative based on the Nuclear Engine Rocket Vehicle Application Reactor (NERVA)[20] drives two open cycle, counter-rotating turbine generators. [21, 22] Table 16 summarizes key parameters.

For Category 2:
(1) A space thermionic advanced reactor with energy storage system (STARS) consisting of a liquid-metal-cooled, in-core thermionic reactor coupled to alkaline fuel cells for burst power. Table 17 summarizes key parameters.

(2) A lithium-cooled, cermet-fuel, fast reactor drives a potassium-vapor Rankine cycle with a Na/S battery storage system. Table 18 summarizes key parameters.

For Category 3:
(1) A gas-cooled, particle-bed reactor drives a turbine generator.[23] Table 19 summarizes key parameters.

Table 14. Category I, MMW, Fast-spectrum, cermet-fuel reactor (710 program derivative) with twin counter-rotating open Brayton cycle turbines coupled to super-conducting generators.

Concept Details	
Reactor Inlet Temperature (K)	164
Reactor Outlet Temperature (K)	800
Turbine Inlet/Exit Temperature (K)	1,400/800

Reactor	
Type (derivative of 710 reactor)	Gas-cooled fast reactor
Fuel (86 to 97% enriched U-235)	Cermet (UO_2-Mo),
Materials	Reactor structure, W-Re; pressure vessel, Inconel-750X
Maximum Fuel Temperature (K)	1,083
Coolant	Hydrogen
Burnup	Negligible

Power Conversion	
Turbine/Compressor Type	Conventional axial turbine
Generator Type	Superconducting
Recuperation	No
Materials	Single-crystal Rene N5 for first blades; second stages A286; shaft and rotor astrology; casing HS188; bearings M50 99.999% aluminum stator

Heat Rejection Method	
Main	Effluent
Decay Heat Removal	Hydrogen flow

System Mass (kg)	8,136

Table 15. Category 1, MMW, fast-spectrum reactor with a two-pass core and with gas turbo-generators.

Concept Details	
Reactor Inlet Temperature (K)	150
Reactor Outlet Temperature (K)	1,200
Turbine Inlet/Exit Temperature (K)	1,200/800

Reactor	
Type	Gas-cooled fast reactor with two-pass core
Fuel	UC pins, 316 S.S. clad first pass, Mo-41 Re for second pass
Materials	Vessel is 316 stainless steel
Maximum Fuel Temperature (K)	1,650
Coolant	Hydrogen
Burnup	Insignificant

Power Conversion	
Turbine/Compressor Type	Twin axial turbine counter-rotating; 15,000 rpm
Generator Type	Wound field, non-salient pole, hydrogen cooled
Recuperation	Yes

Heat Rejection Method	
Main	Effluent
Decay Heat Removal	Hydrogen flow

System Mass (kg)	12,887

Table 16. Category I, MMW, Gas-cooled, NERVA-derivative reactor (NDR) with two open-cycle, counter-rotating turbine generators.

Concept Details	
Reactor Inlet Temperature (K)	35
Reactor Outlet Temperature (K)	1,150
Turbine Inlet Temperature (K)	1,150
Reactor	
Type	NERVA derivative, graphite-moderated near thermal
Fuel	Pyrolitic carbon and SiC-coated $UC_{1.7}$ particles in graphite matrix
Materials	Titanium vessel, Zr-C coated graphite internals
Maximum Fuel Temperature (K)	1,200
Coolant	Hydrogen
Burnup	Negligible
Power Conversion	
Turbine/Compressor Type	Two counter-rotating, high pressure axial turbines exhaust to a pair of counter-rotating, low pressure axial turbines. Pressure provided by LH_2 turbopumps.
Generator Type	Hyper-conducting
Recuperation	None except by jacketing the reactor outlet line
Materials	99.999% Aluminum conductors
Heat Rejection Method	
Main	Effluent
Decay Heat Removal	Hydrogen flow through tie tubes to a radiator
System Mass (kg)	10,513

Table 17. Category II, MMW, Space thermionic advanced reactor with energy storage system (STARS) and with a liquid-metal-cooled, in-core thermionic reactor coupled to alkaline fuel cells for burst power.

Concept Details	
Reactor Inlet Temperature (K)	230
Reactor Outlet Temperature (K)	1,130
Main Radiator Inlet/Exit (K)	1,073/1,023
Reactor	
Type	Liquid-metal-cooled, in-core thermionic
Fuel	UO_2 pellets
Materials	W-HfC
Maximum Fuel Temperature (K)	2,520
Coolant	Lithium
Burnup	Negligible
Power Conversion	
Type	Thermionic
Emitter	W-HfC, 2,200 K
Collector	Nb, 1,200 K
Energy Storage	
Type	Fuel cell
Heat Rejection Method	
Main	Heat-pipe panel radiators for steady-state system, expandable radiator for fuel cell system with water.
Decay Heat Removal	Heat pipe radiators
System Mass (kg)	58,152

Table 18. Category II, MMW, Li-cooled-cermet fuel, fast reactor with a potassium-vapor Rankine cycle and Na/S battery storage system.

Concept Details
    Reactor Inlet Temperature (K)   1,500
    Reactor Outlet Temperature (K)   1,660
    Turbine Inlet/Exit Temperature (K)   1,500/1,255
    Radiator Inlet/Exit (K)   1,040 average

Reactor
    Type   Liquid-metal-cooled fast reactor Cermet (UN in W-Re)
    Materials   Pressure vessel and piping, ASTAR 811-C
    Maximum Fuel Temperature (K)   1,870
    Coolant   Lithium
    Burnup   ~7%

Power Conversion
    Working Fluid   Potassium
    Turbine/Compressor Type   Axial turbine
    Generator Type   Wound rotor

Energy Storage
    Type   Sodium-sulfur batteries

Heat Rejection Method
    Main   Carbon-carbon heat pipes
    Decay Heat Removal   Heat pipe radiators

System Mass (kg)   54,544

Table 19. Category III, MMW, Gas-cooled, particle-bed reactor with a turbine generator.

Concept Details
    Reactor Inlet Temperature (K)   220
    Reactor Outlet Temperature (K)   1,050
    Turbine Inlet Temperature (K)   1,050

Reactor
    Type   Gas-cooled, particle-bed reactor
    Fuel UC_2 coated with ZrC and pyrolytic C particles
    Materials   Al vessel, Mo-Re hot frit
    Maximum Fuel Temperature (K)   1,240
    Coolant   Hydrogen
    Burnup   Negligible

Power Conversion
    Turbine/Compressor Type   Axial turbines on counter-rotating shafts, direct coupled generator
    Generator Type   Cryo-cooled, wound rotor alternator, 3-phase output at 20 kV
    Recuperation   None

Heat Rejection Method
    Main   Effluent
    Decay Heat Removal H_2 flow via multiple independent channels

System Mass (kg)   42,000

The major Multimegawatt Program development activities were concerned with fuels. Scoping tests were performed to evaluate the compatibility of UN fuels with W-Re and Mo-Re alloys. The test results showed some problems at high temperatures, but these could be mitigated through control of the UN stoichiometry. Thermodynamic analyses were performed to estimate the chemical compatibility of UC fuels with these alloys. Also, a testing program was performed on two particle bed fuel elements. The elements did not perform as expected. Post-irradiation examinations indicated power/flow matching problems exhibited by nonuniform flow distributions, particle frit chemical and mechanical interactions, and cycling problems.

Some materials development activities were undertaken on Ta-, Mo-, and W-based alloys. Several material lots were produced, but the efforts were terminated before conclusive data were obtained on the alloys.

Bimodal Systems

Combining power and propulsion in a single system offers opportunities to enhance current missions and enable new ones. Enhancements include on-orbit maneuverability and higher payloads on-orbit. Missions enabled by bimodal systems include a number of high power military applications, in addition to several commercial ones. Cost savings can result from: (1) a step down to smaller, less expensive launch vehicles, (2) increased satellite lifetimes, needing fewer satellites and fewer launch vehicles to perform a mission, and (3) increased satellite capabilities enabled by the higher power available.

An example of the advantages of bimodal systems is in transfer of satellites from low earth orbit to geosynchronous or other high earth orbit. Available launch vehicles in 1992, including cost and injected mass to geosynchronous (GEO), are shown in Table 20. Three satellite configurations were evaluated (see Table 21). No launchers are available for the Direct Broadcast High Definition Television (DB HDTV) satellite with its 10,000 - 18,000 kg weight. The proposed mission profile for a bimodal transfer to GEO, shown in Fig. 29, starts with a launch to low earth orbit (LEO) at 185 - 1,000 km. Transfer is made to GEO with a continuous burn using the bimodal power and propulsion system. Orbit circularization and plane change once geosynchronous orbit is attained also uses the bimodal system.[24]

Table 20. US booster and upper stage performance.

Booster / Upper Stage	Launch Cost (FY92$M)	Injected mass at GEO (kg)
Delta II 7925 / PAM-D	50	910
Atlas IIAS	120	1060
Titan III / TOS	188	1360
Titan IV (SRMU) / Centaur	400	5220

Table 21. Geosynchronous satellite platforms and their characteristics.

Satellite	Mass (kg)	BOL Power (kW$_e$)	Lifetime (yrs)	Launcher	Launch date (yr)	Constellation
Advanced Military Communications Satellite	4760	7-9	10	Titan IV / Centaur	2000	3 w/spare
Advanced Early Warning Satellite	2000	3-5	10	Atlas II AS	2000	3 w/spare
DB HDTV	10000	20-100	10	NA	Post-2000	1 (100kW$_e$)

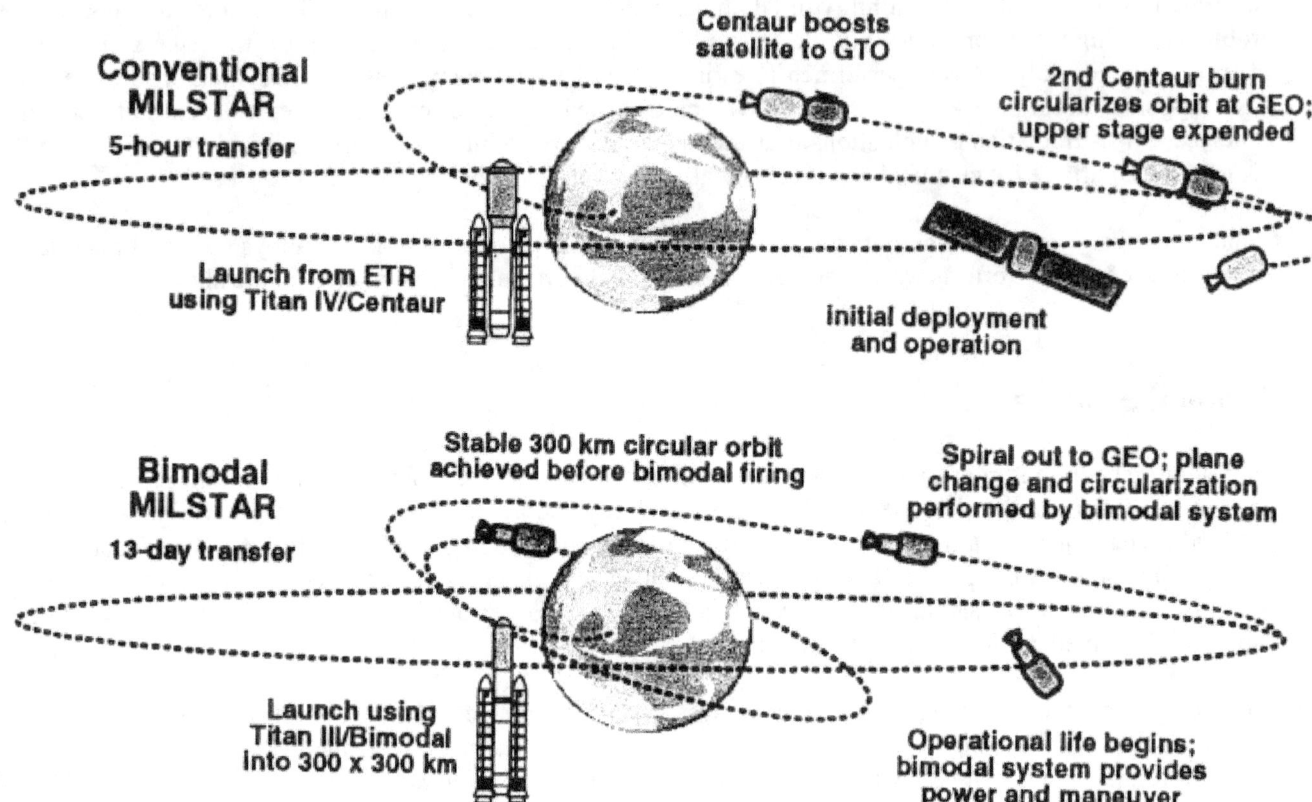

Fig. 29. Conventional and bimodal military communications satellite mission profiles (bimodal transfer assumes 50 N continuous thrust 770 s specific impulse). *From Fred G. Kennedy and Vlad Vanek, 1994*

It was found that Titan IV/Centaur payloads could be stepped down to either Titan III/NUS or Atlas IIAS, although the latter would require a high I_{sp} system (940 s) in addition to a very lightweight reactor system (600 kg or less). Use of Atlas payloads on Delta vehicles may be possible, but only within stringent mass requirements ($\leq$ 1,000 kg at 770 s, $\leq$ 1,250 kg at 940 s). Fig 30. illustrates the potential advantages of a bimodal nuclear system using a Titan III launch vehicle.

Planetary orbiter missions to Jupiter, Saturn or Neptune are also greatly enhanced by bimodal nuclear reactor systems. Using an Atlas 2AS launcher, times to reach the planets can be reduced to 1 to 3 years and provide two orders of magnitude higher data transmission rate than a conventional RTG powered outer solar system spacecraft.[25]

Several studies have been performed to modifying space power systems to a bimodal arrangement. SP-100 could add a propellant heat exchanger to heat hydrogen from a propellant tank and discharge the heated hydrogen through a nozzle (see Fig. 31). The specific impulse could be on the order of 550 - 650 s. A comparison with the Bimodal SP-100, electric propulsion, and chemical orbit transfer is shown in Table.22.[26] Bimodal nuclear is significantly better than chemical, but takes a transfer time of 10 - 40 days. Electric propulsion provides the best payload, but the transfer times are considered unacceptably long.

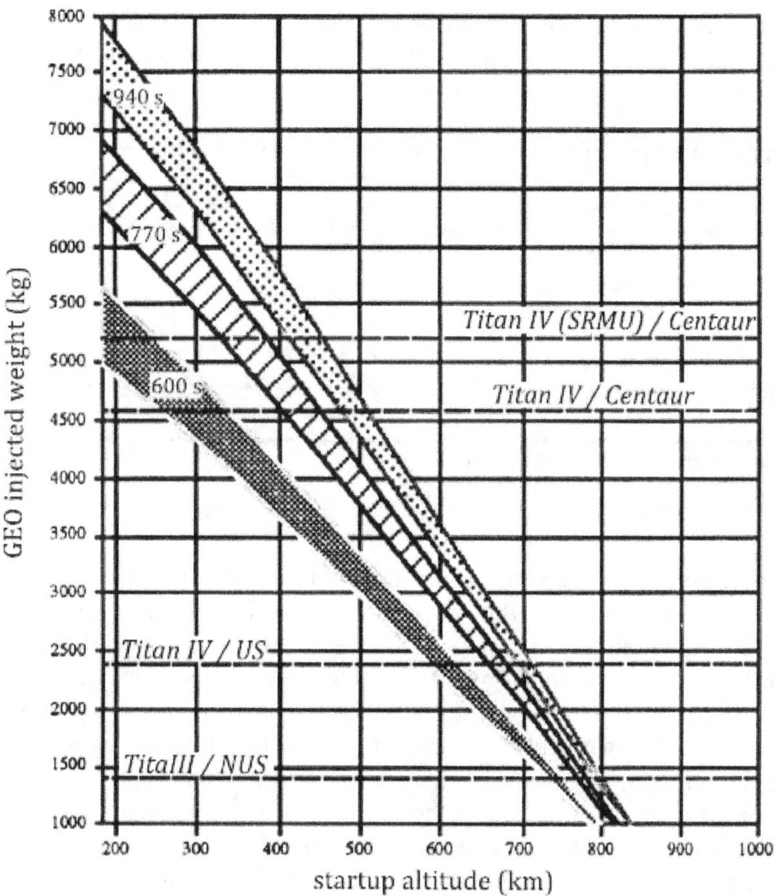

Fig. 30. Titan III inject weight to GEO as a function of bimodal startup altitude and specific impulse. *From Fred G. Kennedy and Vlad Vanek, 1994*

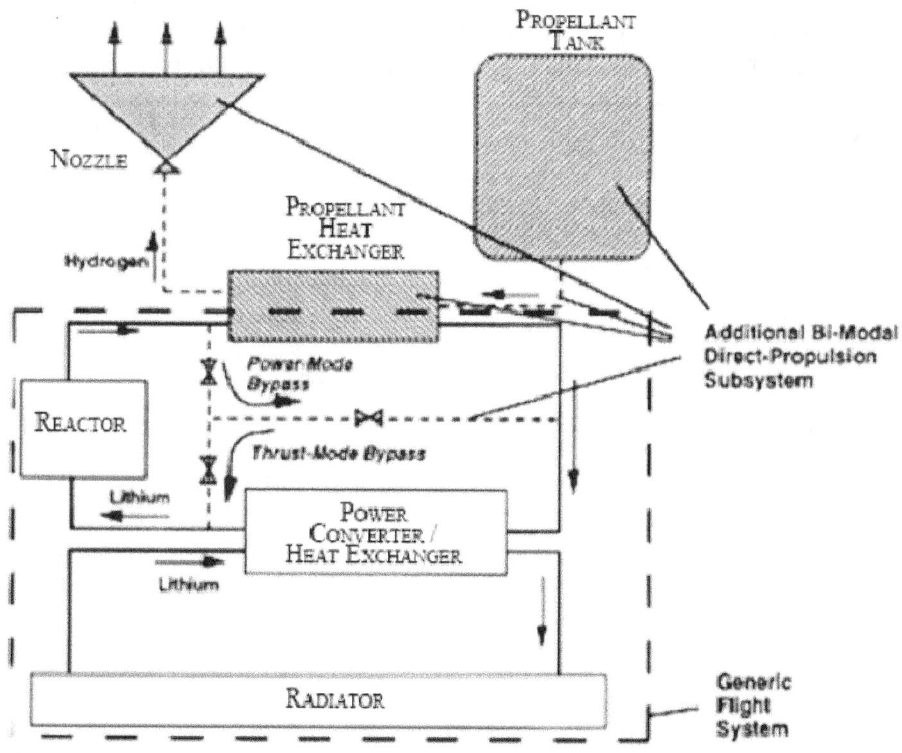

Fig. 31. SP-100 bimodal schematic. *From John J. Buksa, Scott DeMuth, Todd Huber, 1994.*

Table 22. Comparison of SP-100 Bimodal, electric Propulsion, and Chemical Propulsion Systems for a LEO-to-GEO orbit transfer. *From John J. Buksa, Scott DeMuth, Todd Huber, 1994.*

System	Power Level (kWe)	Mass Savings (metric tons)	Payload[a] (metric tons)	Transfer Time (days)
Bi-Modal	20 @ 650 sec Isp	8.3	18.3	40
	100 @ 640 sec Isp	6.3	16.3	10
MPD-OTV[b]	25 @ 1000 sec Isp	8	18	500
	25 @ 2000 sec Isp	12	22	575
	100 @ 1000 sec Isp	4	14	150
	100 @ 2000 sec Isp	9	19	200
LO_2/LH_2 Centaur	N/A	0	10	minutes

[a] For IMLEO=35 metric tons (Titan IV SRMU)

- From Choueiri 1993[27] and Interavia 1992[28]

Several other reactor power systems were studies in a bimodal configuration. For instance, the single cell, in-core thermionic system can be adapted to the bimodal application by having the hydrogen flow through the center fuel zone of the TFE. The thermal power can be increased during the propulsion mode since the power is removed directly by the hydrogen propellant flow and is relatively independent of the thermionic system.[29, 30] Also, STAR-C out-of-core could be modified into a bimodal configuration by providing hydrogen passages through the center of the core.[31] Heat pipe reactor designs can be modified by including propellant heating channels, the change of the refractory metal heat conduction discs and core can from molybdenum to tungsten, and the introduction of an extra core can brazed to the tungsten can.[32] A pin and cermet hybrid bimodal design uses SP-100 UN-Nb1Zr fuel in a central power section surrounded by 710 program fuel UO_2-W cermet fin the propulsion section.[33] Gas-cooled bimodal configurations are also possible using the same reactor core configuration for both power and propulsion modes.[34]

Restricting the bimodal system to operate above some specified minimum altitude could seriously handicap its advantages. Safety studies indicate that a minimum altitude for nuclear operations can be 200 km with sufficient propellant to move the reactor to a sufficiently high orbit after operation by bleeding cold gas from the tank.[35] Another consideration in developing bimodal systems is the facilities needed to test and qualify the power plant. It must now accommodate testing for both a short   duration propulsion mode using hydrogen as well as the long duration power testing.

Chapter 10

Where We Are and Visions of the Future

In this chapter we will summarize the characteristics of various types of fission power systems and the state of the technologies necessary for space nuclear reactor power systems development. The most significant short coming is the need for a more extensive materials data base necessitated by the effects of radioactive interactions with other materials.

This leads into visions of the future. With mankind's continued exploration and expansion into space, space nuclear reactor power systems are a necessity to fulfill these dreams. Adequate power is fundamental to achieving many future goals and aspirations. We have embarked on a journey of going back to the Moon and on to Mars. To make the most of this journey, we will need power plants raising to the megawatt level for manufacturing and resource utilization. Also, the unmanned exploration of our solar system will require systems with much expanded capabilities. So, as we go forward, power requirements will raise and the need for nuclear power to fulfill these requirements is enabling.

Power Plant Design Characteristics

The selection of power plant configurations for past developments were based on the design requirements, status of technology at that time, cost, schedule and risk constraints. Thus, as the mission lifetime needs changed, the constraints imposed by the available launch vehicles planned for given missions differed, and consequently the power plant designs selected for development were different. When one looks back over the power plant selection process, it was found that the selected power plant concepts were logical for the design requirements that were defined at that time.

We have summarize some of the key characteristics of the more significant types of reactor power plant designs in Table 1. Most of the design activity went into liquid-metal cooled reactors, first the SNAP reactors with uranium-zirconium-hydride fuel, than the SP-100 with uranium nitride fuel, and now Fission Surface Power using uranium oxide. A U-ZrH reactor was the only U.S. reactor power plant that actually was flown in space. However, as design requirements became more demanding, other technologies were determined to be needed. Much of the basic U-ZrH reactor design ideas have been incorporated in later designs. Also, the thermionic reactors use many of the U-ZrH components outside of the thermionic fuel elements.

The SP-100 design, with thermoelectric conversion, offered a high degree of redundancy in the power plant outside of the reactor. It was well on the way to have demonstrated all of the components for a seven year lifetime system. To reduce mass and size, the reactor needed to use refractory metals. This involved a considerable amount of work to characterize these materials in a radiation environment associated with a fast spectrum reactor. Also, the working fluid in the primary loop was selected to be lithium because of the operating temperature levels. High temperatures were driven by the need to keep the radiator a reasonable size and weight. Interactions between lithium and the fuel pins were resolved with a rhenium barrier. The SP-100 was ready to build and test an engineering prototype at program termination.

The SP-100 reactor could also be used with other power conversion systems that are more efficient than thermoelectric converters. In particular, a Stirling engine was being developed, with its increased efficiency; this provides a higher electrical output from the same SP-100 reactor or allows the reactor size to be scaled down for the same power output. Stirling engine development is continuing for use with radioisotope generators and the new generation of Fission Surface Power systems.

The Fission Surface Power program has selected a temperature range below 900 K in order to allow use of non-refractory materials and current power conversion technologies. This brings the design closer to terrestrial nuclear power plant experience. The working fluid is NaK with experience on all reactors flown by the U.S. and Russia in space to date. With the reactor operating at reduced temperatures, the materials in the power conversion subsystem and radiator are also in a range where there is greater experience. Redundancy to avoid single failure points is designed into the non-nuclear subsystems by a modular approach. The program so far has developed a reference design and is now working on developing and validating prototype components.

Gas-cooled power plant configurations offer the advantage of using an inert gas as coolant in the reactor--eliminating much of the material interactions between coolant and fuel elements. With a Brayton cycle conversion system operating at temperatures of 1,150 K, operating temperatures are significantly less than with a lithium cooled liquid-metal reactor system. Refractory metals are still needed in the reactor, but, the conditions are much less stringent and demanding. A major concern is the lack of redundancy inherent in the power conversion subsystem. Being debated is whether to have a single, very high efficient Brayton engine or multiple Brayton engines with the added complexity of isolation valves, position sensors, control logic, etc. that could possibly reduce overall system reliability and add mass to the overall power plant.

Because of the characteristics of Brayton cycles, the radiator temperature is quite low--about 500 K. This means that the radiator becomes very large. The radiator area required is somewhat reduced by the high efficiency of the Brayton converter--about 30% compared to thermoelectric converter efficiency of 6 - 8%.
To further minimize radiator size, a regenerator is included in the power plant cycle. However, the radiator still must be large because of rejecting heat at an area proportional to the fourth power of the temperature.

Table 1. Key characteristics of past reactor power plant development programs.

Parameters	Liquid-Metal Cooled Reactors			Gas-Cooled Reactors	Thermionic Reactors	Heat Pipe Reactors
Representative Systems	SNAP-2, 10A, 8	SP-100	FSP	Prometheus	TFE Topaz I Topaz II	SPAR/SP-100
Fuel						
Form	Pin			Ceramic	Pellets	Wafers
Fuel	U-ZrH	UN	UO_2	UO_2	UO_2	UO_2
Cladding	Hastelloy N tube	Refractory metals with rhenium barrier	SS-316	Refractory metals	None	None
Temperatures, K						
Reactor Outlet	810-975	1450 EOL	900	1,150	1,000	1,500
Peak Fuel			1095	1775 EOL	1,800-2,000	1,730
Radiator	575-595	790	250	500	1,000	800
Primary Coolant Fluid	NaK	Li	Na or NaK	He-Xe	Li NaK	Li
Heat Transport	EM Pumps			Turpopumps	DC conduction pump	Heat pipes
Power Conversion (Primary)	Thermo-electrics Rankine	Thermo-electrics Stirling	Stirling Brayton	Brayton Stirling	Thermionic multi-cell fuel element Thermionic single cell element	Thermoelectrics Brayton Stirling

Reliability	Redundant thermoelectric elements. Lacks redundancy in lines and distribution elements. Redundant dynamic conversion units need isolation valves. Significant experience with pin fuels.			Lacks redundancy in lines and distribution elements. Redundant dynamic conversion units need isolation valves. Significant experience with ceramic fuels.	Lacks redundancy in pumped loops to radiator, Coolant temperatures are lower than for liquid-metal or gas-cooled reactors and can possibly use stainless steel outside the TFE elements	Inherent redundancy with no single failure points. Heat pipes have proven to be highly reliable at lower temperatures but lack experience at higher temperatures of interest.
Mass	Relatively low temperatures increased mass	Between Gas-cooled and Heat Pipe because of lighter flow distribution system and lower pressure containment vessel.	Very heavy because of low fuel temperatures.	Heavy NSS because of higher pressures, larger core diameter from poorer heat transfer, longer flow distribution, and heavier shield from larger core diameter.	Reactor mass not comparable because NSS includes electric converters. Needs pressure vessel and shield; tends to be bigger and heavier because reactor diameter larger.	Lighter mass because no pressure vessel and fluid pumps.
Safety						
Subcritical On Liquid Immersion	Intrinsically subcritical using highly spectrum-dependent thermal-neutron absorbers in the core	Safety rods in core, intact reentry using heat shield.	Built-in safety with low core void volume.	Could follow SP-100 design approach.	Could follow SP-100 design approach.	Safety rod in core. Designed to disassemble on reactor reentry.
Subcritical On Land Impact		Incorporated reentry heat shield to assure intact reentry and reactor integrity. Safety rods in core.	Could be fueled on surface.	Could incorporated reentry heat shield to assure intact reentry and reactor integrity Safety rods in core.	Could incorporated reentry heat shield to assure intact reentry and reactor integrity. Safety rods in core.	Banded reflector and core should result in core splitting apart on major impact. Curved structure of reflector also helps avoid core compression.
Significant Negative Temperature Coefficient of Reactivity	All US configurations designed to possess a negative temperature coefficient of reactivity. (Russian designs sometimes designed with a positive temperature coefficient.)					

Independent Reactivity Subcriticality	All designs meet this requirement by using separate scram and reactivity actuators.					
Decay Heat Removal After Unplanned Shutdown	Demonstrated in SNAPSHOT shutdown	Auxiliary coolant loop	Inherent afterheat removal	May need to add an auxiliary coolant loop	May need to add an auxiliary coolant loop	Multiple, independent parallel heat removal devices inherent in design.
Launch Environment	SNAPSHOT demonstrated launch solutions	Designed to meet launch environments	Designed to meet launch environments	Designed for launch environments. No barriers anticipated for successful design.	Topaz I has been successfully launched by the Russians. No barriers anticipated for successful design.	Use ductile materials. Preliminary studies indicate NSS can be designed to meet launch loads.

Thermionic reactors are attractive because all of the components outside of the thermionic fuel element are at low temperatures. These are similar to those components developed in the SNAP program. The Russians have developed and flown thermionic fuel element reactors--though the lifetimes are quite limited compared to ones now needed in projected U.S. applications. The major concerns are: (1) the high temperatures in the fuel element of 1,800 - 2,000 K leads to a quite limited choice of materials and (2) the close tolerances that must be maintained throughout the life of the converter. Much work is needed to characterized material behavior in a radiation environment. Another big advantage of thermionic power plant is the radiator temperatures are high--about 1,000 K. This leads to significant smaller radiators. The integrated configuration of fuel and converter leads to a very challenging development process.

The thermionic single-cell element has an advantage that complete system test can be performed using electric heaters. However, this does not provide nuclear interactions testing.

Heat pipe reactors do not need any independent coolant pumping systems, have redundant heat transport systems, a simple fuel configuration, and no pressure vessel. This makes the system mass less than other configurations. Also, the lack of a pressure vessel makes designing the system to meet safety requirements easier in that the core can be banded and designed to either break apart or remain intact on reentry. The major problem with heat pipe reactors is demonstrating heat pipes that operate at 1,500 K for 7 - 20 years in a radiation environment. With the lithium in the heat pipe, any damage to the refractory materials that could allow the lithium to escape even very slowly will result in the lost of the heat pipe and if the mechanism is common to all of the heat pipes, the redundancy would not really be there.

Technology Status

Reactor Fuels[1]

Uranium oxides, carbides, and nitrides have been developed and tested for space power reactors. Table 2 summarizes the characteristics of these fuels. Desirable fuel characteristics are high density, high thermal conductivity, high melting point, high temperature stability, chemical compatibility, predictable irradiation performance and ease of fabrication. The oxides have predictable irradiation performance and are the easiest to fabricate; but they have relatively low thermal conductivity and density, and they react with liquid metals. Irradiation causes oxides to release more fission gas, but fuel swelling is less than carbides or nitrides. Carbides swell more than nitrides, while nitrides are less stable and more difficult to fabricate. SNAP reactors and SP-100 used uranium nitride fuel. However, the thermionic programs, Prometheus and now Fission Surface Power use uranium oxide fuels. Rover/NERVA nuclear rocket used uranium carbide fuels. Because the uranium carbide fuels are mainly of interest for nuclear rockets, the state of technology of uranium carbide fuels is summarized in Book 2 of this series.

Table 2. Space reactor fuel characteristics.

Characteristic	UO_2	UN	UC	UC_2	$(U_{0.2}Zr_{0.8})C_{0.99}$
U Density, g/cc	9.66	13.52	12.97	10.60	2.88
Melt Point, K	3100	3035	2775	2710	3350
Thermal Cond. W/mK	3.5	25	23	18	30
Relative Stability	Moderate	Low	High	High	High
Relative Swelling	Low	Mid	Mid	Low	High
Fission Gas Release	High	Low	Mid	Low	Mid
Fabricability	Easy	Moderate	Easy	Difficult	Difficult

Uranium Nitride Fuel Development Status

During the SP-100 program, approximately 90 UN fuel pins were tested in the Experimental Breeder Reactor (EBR-II) and the Fast Flux Test Facility (FFTF) fast spectrum reactors with the following variables:

Design Features:
        Cladding: 5.84 - 7.62 mm-o.d. Nb-1%Zr or PWC-11 (Nb-1%Zr +0.1%C)
        Liner: Tungsten, rhenium, and bonded-rhenium
        UN density: 87% - 96% theoretical density
Irradiation Conditions:
        Temperature: 1,200, 1,400, and 1,500 K cladding
        Burnup: 1 - 6 at. %.

Centerline temperatures in the fuel were as high as 1,900 K without fuel pin failures. It was concluded that the testing demonstrated robust design behavior. Metallic fission products, especially ruthenium, were found to migrate down the thermal gradient to the fuel pellet surface and into the cladding. No breakaway swelling or fission gas release was observed, and high density fuel released less fission gas than low density fuel. Large density gas pores formed in the center of the low-density UN fuel pellets, while the high density fuel showed no evidence of restructuring or cracking. The chemical-vapor-deposited tungsten liners cracked during irradiation, but a thin liner of wrought rhenium metallurgically bonded to the inside diameter of Nb-1Zr was found to be an effective barrier to fuel/cladding/chemical and mechanical interactions. Mechanistic performance models, predicting the performance of UN fuels for space reactors have been developed.

Fission gas release is plotted as a function of burnup and temperature in Fig. 1. The high density UN released considerably less fission gas than the low density fuel by a factor of about three because of the greater open porosity in low density fuel allows better access to the plenum gap. Temperature trends suggest that low density UN begins to release fission gas at temperatures greater than 1,600 K, while high density UN appears to retain most of the generated fission gas up to 1,800 K.

Uranium nitride burnup was measured on 45 fuel pins irradiated during the SP-100 program. Dimensional changes of the fuel pellets were measure by digitizing densitometry traces of neutron radiographs of the irradiated pins.

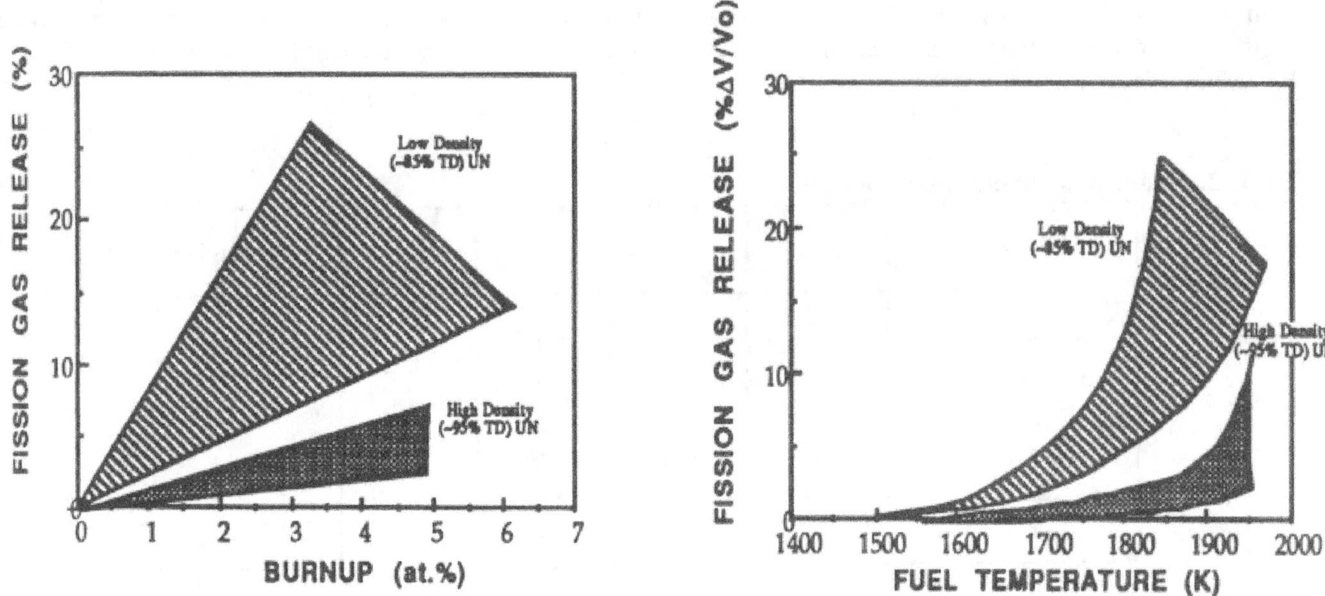

Fig. 1. UN fission gas release as a function of burnup and temperature. *From R. Bruce Mathews, et al., 1994.*

Swelling data includes cladding temperatures from 1,250 K to 1,550 K and high and low density UN (Fig. 2). The decrease in swelling rate at higher burnup is probably caused by resistance of the cladding, and the initial high swelling represents unrestrained UN swelling.

Fuel pellet testing showed that high density UN (> 95 % TD) operating at moderate temperatures (fuel center line < 1,650 K, cladding <1,400K) has low swelling, low fission gas release, no fission product interaction and the potential of operating to high burnups. Low density, hypostoichiometric UN, and high operating temperatures are undesirable characteristics that limit the useful lifetime of UN fuel.

The conclusions on uranium nitride fuel are:

- Fuel density greater than 95% TD to reduce restructuring and fission gas release.
- Peak fuel operating temperature less than 1,800 K to reduce fission gas release, swelling, and fission product migration.
- Hyperstoichiometric UN_{1+x} to eliminate metallic fission product formation.
- Pin diameter, cladding thickness, fuel/cladding gap thickness, plenum volume based on performance requirements, but constrained by Storms[2] fission gas release (FGR) (eq. 1) correlation and swelling at 1.5% $\Delta V / V_0$ per at. % burnup equation. The Storms fission gas release correlation is an empirical fit to the available fission gas release data where TD is the percent of theoretical fuel density, BU is burnup in atomic percent of uranium, and T is the average fuel temperature in Kelvin. The correlation suggests that release declines with increasing burnup, accelerates rapidly with increasing temperature, and is sensitive to fuel

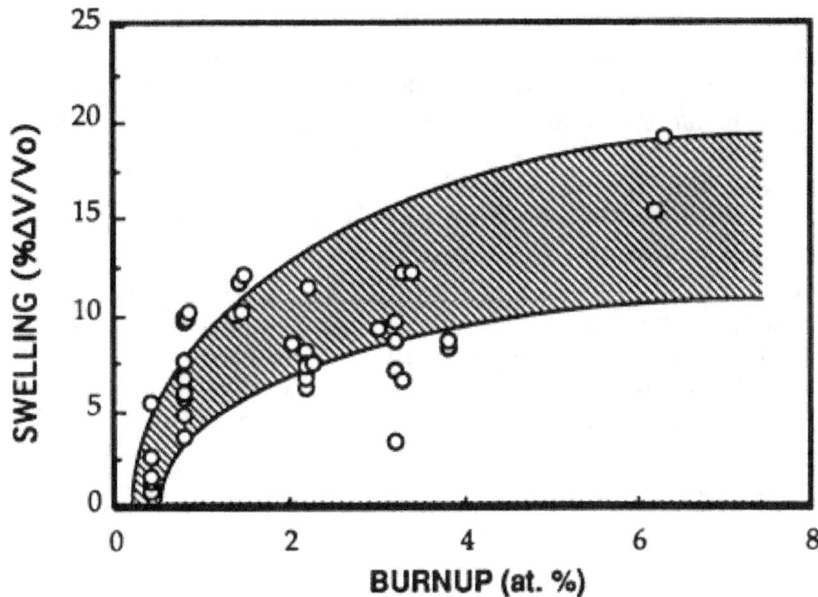

Fig. 2. Swelling rate of SP-100 UN fuel irradiated in FFTF and EBR-II. Differences in operating temperature and fuel density were not discriminated. *From R. Bruce Mathews, et al., 1994.*

$$FGR = 100/[exp(0.0025\{90TD^{0.77}/BU^{0.09} - T\}) +1] \tag{1}$$

From the irradiation testing, the SP-100 fuel specifications were established as:

Density:	$96.5 \pm 1.5\%$ TD
Stoichiometry:	1.00 - 1.05 (N+C+O)/U
Grain size:	$> 30\ \mu m$
Carbon:	$< 3,000$ ppm
Oxygen:	$< 1,000$ ppm

Compatibility of the fuel pin with lithium required a fuel cladding lined with rhenium to protect the Nb-1Zr from reacting with UN fuel. Out-of-pile tests shows that UN and rhenium will react at operating temperatures according to:

$$UN_{1+x} + Re = URe_2 + N_2 \tag{2}$$

if x is sufficiently small. Design conditions for SP-100 were selected to avoid reaction of URe_2.

Post-irradiation examinations of test pins show the formation of molten fission product uranium/ruthenium alloys near the pellet/cladding interface. Though the molten fission product alloys will react with cladding components and present a threat to the fuel pin, calculations predict that the reaction will not occur during the seven-year lifetime of the SP-100 reactor if stoichiometry, temperature, and liner integrity are controlled.

The SP-100 program demonstrated for UN fuel operation to 6 at. % burnup at 1,400 K with operation to 10 at % burnup plausible. Unresolved issues concern UN dissociation at high temperatures and burnup. Two-year lifetime is verified, seven-year lifetime is achievable, processes to fabricate rhenium-bonded, niobium-1% zirconium cladding have been developed, and a full core of UN fuel pellets has been fabricated to tight specifications.

Uranium Oxide Fuels[3]

The thermionic fuel elements operate at much higher temperatures than other reactors using UO_2 fuel. Several fuel elements were fabricated and irradiated tested in the Experimental Breeder Reactor and the Training, Research, and Production General Atomic (TRIGA) reactor. Two year lifetime was demonstrated at 1,800 K with potential for five-year lifetimes at 2,000 K. Issues include fuel swelling, emitter distortion, and UO_2 vaporization.

With the high operating temperatures, fission gas release from thermionic fuel appears to be very high. The release seems to be higher than predicted by the LIFE[4] code--the LIFE code was developed to predict behavior of liquid-metal, fast-breeder reactors. The release appears to be immediate, though not complete. Appreciable emitter deformation is observed, implying fuel swelling beyond that caused by solid fission products. Some small amount of gas must be retained to account for the swelling; retention of gas in the central void has been proposed as a mechanism. Cladding (emitter) strain in thermionic fuel elements is not caused by released fission gas pressure because fission gas is vented. Fig. 3 shows predictions of fuel swelling after adjusting the LIFE code.

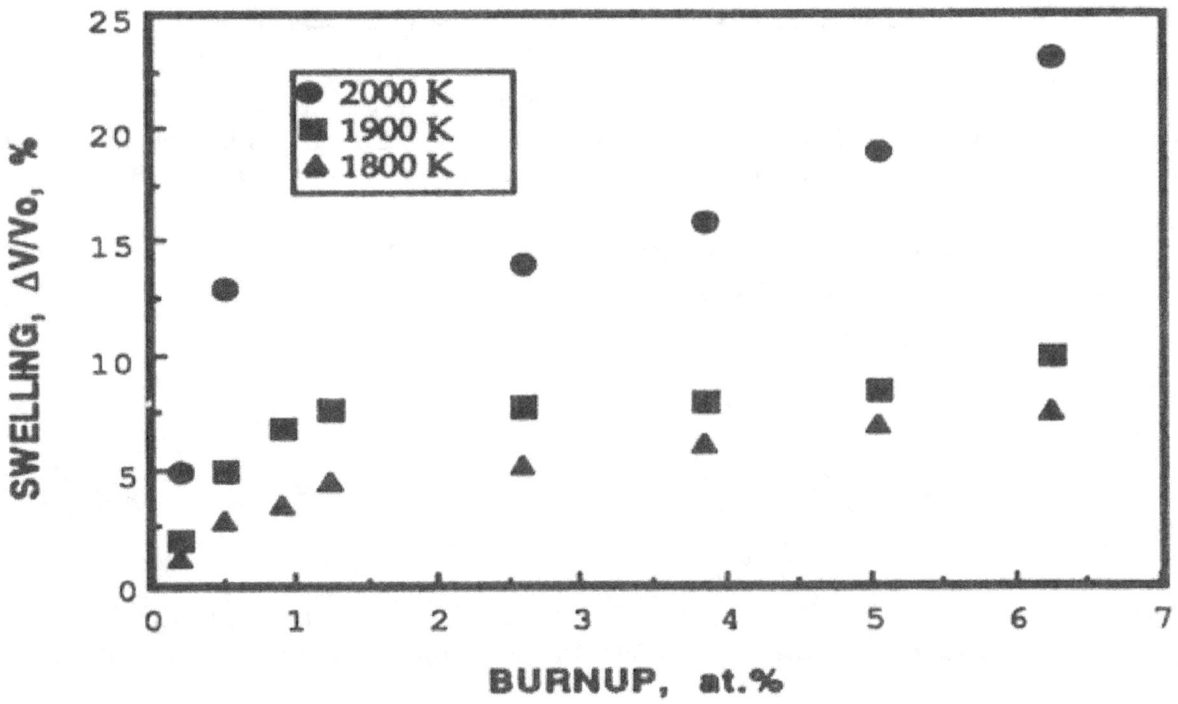

Fig. 3. Unrestrained UO_2 swelling. *From R. Bruce Mathews, et al., 1994.*

Prometheus operates at a much lower fuel temperatures than the thermionic fuel element program. Its reactor outlet temperature is ~1,150 K with the fuel at 1,450 K and a design lifetime of 15 years. Data from terrestrial power reactors provides some information on the fuel behavior, but these have lifetimes of 3 - 6 years, operate at relatively low temperatures, and operate in an environment where periodic inspection and replacement of defective fuel elements are possible. As a result, the following additional data is needed:[5]

    · Irradiation data at low fission rates, high temperatures, and low burnups to develop design relations;
    · Experimental verification of UO_2 restructuring behavior;
    · Determination of the best relationship for fission gas release from UO_2;
    · Alleviating uncertainty in UO_2 swelling calculations; and
    · Long term chemical compatibility of fuel, fission products, and fuel system materials.

For Prometheus, several materials were candidates for cladding and fuel element materials (see Table 3).[6] As far as reaction with UO_2, in general few incompatibilities are expected between UO_2 and the candidate structural materials. The only potential reaction with zirconium is marginal and this has not been a problem in nuclear reactors currently in service.

Fission Surface Power uses SS-316 for cladding and reactor structural materials. An extensive terrestrial reactor data base exist for the combination of UO_2 and SS-316 materials, including experience from the Experimental Breeder Reactor and the Fast Flux Test Facility.

Incompatibilities due to fission products is considered a much more serious threat. Most of the oxygen released from uranium atoms fissioning combines with fission products to form stable binary oxides or multicomponent compounds. Thus, potentially problematic fission products are tied up as stable oxides and the liberated oxygen is effectively immobilized. Cesium molybdate is formed and thus the relative thermodynamic stability of these oxides compared to cesium molybdate is important. The materials molybdenum, tungsten, and rhenium should not be oxidized and thus are good candidates for fuel pin liners. Conversely, alloys and ceramics with a substantial quantities of tantalum, zirconium, silicon, carbide, graphite, or niobium may require a liner to ensure adequate mechanical performance after long-term exposure to irradiated fuel.

Table 3. Candidate core materials for use with UO_2.

Refractory Metal Systems	Mo-41.5Re
	Mo-47.5Re
	Ta-10W
	ASTAR 811C (Ta-8W-1Re-0.7Hf-0.025C)
	FS-85 (Nb-27Ta-10W-1Zr)
Silicon Carbide Systems	SiC (with Re liner)
	SiC (with C liner)

Embrittlement from fission products is another potential problem. Unfortunately, this effect is difficult to assess. The observed dependence of embrittlement is not only between interactions in fission products and cladding materials, but also between the fission products themselves.

Reactor Components

Other reactor components, such as moderator, reflector, support structure and periphery control elements, controls and instrumentation and shielding are reasonably similar in the various reactor designs.

Reflectors have been designed in the form of stationary and moveable elements surrounding the reactor core. Reflector material candidates include beryllium and BeO. Beryllium is the lowest density material (~ 1.85 g / cm^3) and is sufficiently strong to be its own structure. The material properties for Be are largely know. BeO is another potential material with a density of ~ 3.01 g / cm^3. However, it will require structural and canning materials and additional testing to qualify it for long lifetimes. This includes the need for irradiation induced swelling of BeO, which affects mechanical stability and to some extend must be done in the design configuration.

In drum systems, B_4C, a neutron absorber, is usually used in the control sector. An extensive data base exist for this material.

Pressure boundary materials, including the reactor vessel and piping, concerns include thermal creep and irradiation embrittlement. Stainless steel is well characterized. Nickel-based superalloys and refractory metal alloys require a substantial amount of additional irradiation testing for long lifetimes. Mass transport between refractory metal core components and the pressure boundary components could limit power plant lifetimes.

Instrumentation and controls architectures usually provide for autonomous operation and redundant elements with fault management. Technologies are available for all key plant technologies including temperature, neutron flux, pressure (for gas cooled systems), control drive mechanism position, and coolant flow. Table 4 summarizes the sensor technology status for the gas-cooled Prometheus system. Neutron flux sensors are well developed. The most challenging sensors are hot leg temperature sensors due to lifetime degradation and the need to integrate them with internal components. A promising sensor technology for this is the sapphire Fiber Bragg Gratings.

Computing electronics located within the spacecraft are considered low risk. Analog electronics associated with the sensor conditioning circuits require development because of the high radiation hardness and reliability necessary.

For shielding, material options for neutron or combined gamma/neutron shielding include lithium hydride, water, beryllium, boron carbide and tungsten. The shield thickness is driven by reactor power, lifetime, core to payload distance, payload neutron dose limits, and shield materials and design. Designs consider thermal growth, swelling, thermal management, and piping streaming paths. Shielding materials have been well characterized and design capabilities exist for future power plant development.

Converter Systems

The extensive U.S. radioisotope generator flight experience have used thermoelectric converters. Thermoelectric converters for reactor power plants were included in the SP-100 design. Prototype TE cells were tested to up to 4,100 hours before having problems with the graphite to SiGe 'N' leg bonds. The SiGe thermoelectric material is very mature having also been used in MHW-RTG and GPHS-RTG's. Spacecraft using SiGe have been flown in a series of missions since 1976 with over thirty years of continuous, successful operation. The SP-100 demonstrated an improved version of thermoelectrics, SiGe-GaP with a figure-of-merit increase from 0.67×10^{-3} / K to 0.72×10^{-3} / K for 5 - 7 years mission lifetimes. Currently, as part of the advanced converter programs for RTGs, development of higher figure-of-merit materials might be made available for future space reactor power plants.

Table 4. Sensor technology. *From "Prometheus Project Reactor Module Final Report, for Naval Reactors Information," Knolls Atomic Power Laboratory Report No. SPP-67110-0008.*

Parameter Technology	Key Advantages	Key Challenges
Temperature Optical Fiber Bragg Grating	-Potential for many sensors on a single fiber -Demonstrated sensor robustness, accuracy, and low drift at high temperature -Non-amplitude based signal measurement, tolerant to fiber optic transmission line degradation -Small sensor mass and size	-Radiation tolerance of specialized components, including electro-optical components (diodes, amplifiers, lasers) -Signal processing complexity -Technology largely limited to laboratory applications

Table 4. (continued)

Parameter / Technology	Key Advantages	Key Challenges
Flux *Fission Chamber*	-- Mature technology, widely used in nuclear plant applications - High energy neutron output pulses eases discrimination from gamma induced pulses and allows large separation (up to ~100m) between sensor and electronics. Local preamplification unnecessary. - Capable of full range operation. - Moderate detector bias voltage compared to other ion chamber technologies. - High temperature performance (~800K)	- Handling, shipment, and storage issues associated sensor with highly enriched Uranium coating - Providing startup flux sensitivity and a neutron-moderating environment within a reasonable volume - Complex signal processing may be required to provide pulse counting during startup and current measurement during power operation - Non-traditional materials required for fore-of-shield service at ~1000K. Performance must be demonstrated. - Large size and mass required to achieve necessary sensitivity.
Position *Ultrasonics*	- Magnetostrictive ultrasonic position technology commercially available. - Magnetic and insulation materials exist for the magnetostrictive components (magnets, waveguides, coils) that are robust in sensor radiation and thermal environment. - Flexible sensor range of measurement - Relatively moderate sensor mass.	- Performance and lifetime in thermal and radiation environment must be verified. - Significant susceptibility to electromagnetic interference from control element motor. - Large sensor space envelope. Mass of sensor unknown.
Flow, Compressor Outlet, 550K *Ultrasonics*	- Non-invasive flow measurement - Widely available piezoelectric materials can be used in transducer. - UT flow systems in wide usage in natural gas industry at this service temperature. - Current materials and design methods may be used to design acoustic couplers and standoffs.	- Radiation and aging effects of piezoelectric transducers for long duration service must be verified.
Pressure *Ultrasonics Thin-Walled Bellows*	- Non-invasive flow measurement - Accurate, high resolution measurements demonstrated for low, moderate, and high system pressures. - Ultrasonic transducers and mounting accessories exist for moderate temperature service less than 550K. UT sensors in wide usage in natural gas industry at this service temperature.	- Ultrasonic transducers for high temperature service above 850K in significant radiation fields do not exist. High risk research and development required. - Radiation and aging effects of piezoelectric transducers for long duration service must be verified.

Thermionic power plants operate at temperatures of 1,800 - 2,000 K. Though significant progress has been made in developing components capable of withstanding the neutron fluence (4×10^{22} n / cm^2, E > 0.1 MeV) and the burnup (5.3%) that would be experienced in a 2-MWe, 7 year lifetime thermionic space nuclear reactor, much more needs to be done. TFE testing has demonstrated 18 months operation with no mechanism identified that would limit the TFE lifetime to less than 7 years. Also, higher burn up for fuel emitters to 3% at 1,800 K and 4% at 1,700 K is equivalent to over 5 y operation   Sheath insulators were tested under 10 - 12 V for 8 mo. (the test were stopped by a water leak with the test samples still in good condition)   Insulator seals in converters were still functional after testing for 39,000 h. Graphite reservoirs were radiation tested to equivalent of over 10 y operation. However, a significant amount of research and development still needs to be performed on such items as: (1) seals that reduce intercell spacing to increase the effective fuel density and ease criticality limits; (2) tungsten-to niobium joints that don't use tantalum to reduce parasitic neutron-absorption and ease reactor criticality limits; and (3) on life testing.

Brayton cycles were most extensively studied in the Prometheus Program. The Brayton cycle subsystem including the recuperator, gas cooler, Brayton turboalternator, valves, bearings, and piping are within the bounds of current technology assuming non-refractory materials can be used. The major concern is demonstrating long-life operation under space conditions and that the components remaining leak tight.

Mercury Rankine cycle converters were being developed as part of the SNAP-2 and SNAP-8 programs and potassium Rankine cycle converter for SNAP 50/SPUR program. In SNAP-8, all of the major conversion subsystem components were operated at least 10,000 hours without maintenance and a "breadboard" configuration run successfully for 8,700 hours. In the SNAP 50/SPUR program converter components were tested for up to 10,000 hours, but no complete system tests performed. No development has been done for longer life systems. Various molybdenum alloys, such as TZM (Mo-0.6 Ti-0.1 Zr-0.035), and tantalum alloys, such as T-111 (Ta-8W-2Hf), are used in the subsystem. To extend the lifetime, extensive material properties data is needed followed by component and subsystem tests.

Free-Piston Stirling engines were being developed as part of the SP-100 program and continue to be developed in support of advanced radioisotope generators and Fission Surface Power. Endurance testing on a 2 kW$_e$ free piston Stirling engine, called EM-2, operated at a heater temperature of 1,033 K for 5,385 h with only minor scratches from the 262 dry starts/stops. Current Stirling generators under development use an inconel displacer at an operating temperature of 925 K. By the end of 2009, it is expected to have sufficient testing data from an engineering unit that mission planners can consider Stirling engines with a high degree of confidence. The heater head of Stirling power conversion systems is the major design challenge because heater head creep is predicted to be the life-limiting mechanism for Stirling engines. Additional data is needed at elevated temperatures on material behavior and for creep analysis and to improve inelastic analysis techniques. High operating temperatures and long operating periods are taxing the ultimate capabilities of even the strongest superalloys. Fission Surface Power, limiting the converter temperature to 830 K, uses IN718 for the heater head.

Conversion systems such as AMTEC and thermophotvoltaic are considered at a technology level where analytical and experimental critical functions have been demonstrated, but component or laboratory breadboards still need to be done.

Heat Rejection Radiators

Several configurations of radiators can be used including pumped loops with fins and heat pipes with fins. Lifetime performance can be effected by degradation of effective emissivity and in heat pipe radiators by heat pipe failures. The Prometheus Program considered using high temperature water heat pipes in the radiator; however, the performance and life capability has not been verified. The relatively high temperature (500K) and evaporator surface heat flux (10 W / cm^2) and lifetime are beyond established water heat pipe performance.

SP-100 used an integral carbon-carbon tube and fin radiator design with Nb-1Zr potassium heat pipes. Nb-1Zr potassium screen-wick heat pipes have been tested for over 10,000 hours without failures or degradation. Heat pipes with carbon-carbon armor and fins were fabricated, but not tested.

System Designs

The U.S. has only demonstrated in space or even in full systems in a simulated ground environment uranium-zirconium-hydride reactor power plants. These power plants were designed for a limited lifetime of one year and the mass of scaled up power plants would probably be unacceptable to meet future mission needs. Extensive development was performed on the liquid-metal cooled SP-100 power systems and components were well on their way to have been tested in a relevant environment. A generic flight system design was completed for a seven year operating lifetime power plant, but not built or tested.

Prometheus requirements called for 20 years or more lifetimes at full operating power. This resulted in a thorough review of all past programs and what it would take to demonstrate a design to meet these expanded mission needs. The selection was a gas-cooled reactor power system. The design had progressed to the conceptual level at the termination of the program.

Thermionic reactor power plants have been flown by the Russians, but have relatively short lifetimes (about 1 y). Multi-cell and single cell designs are possibilities. The U.S. has concentrated on the multi-cell designs in the development efforts to date. However, the Russian's provided the U.S. a Topaz II unfueled reactor for testing that establishes a data base for single cell thermionic reactor development. The single cell reactor configuration makes it possible to electrically heat a prototype system; however, this does not test the affects of induced radiation damage.

Preliminary heat pipe reactor designs have been completed to define component requirements. Some questions have been raised about fuel behavior during operation reactor. The main problem appears to be demonstrating long life, high temperature heat pipes in a reactor environment, not designing the power plant.

Fission Surface Power, with less restrictions on size and mass, utilizes lower temperatures and more conventional materials. The non-refractory materials are much better characterized. So far, a reference design to guide component development has been formulated. Activities are concentrating on demonstrating component readiness.

Visions of the Future

Power requirements continue to evolve as the planning for returning to the Moon and going on to Mars goes on. The Moon is a natural space station, orbiting Earth every 27 days. It is by far the closest extra terrestrial body to Earth, a quarter million miles away. In spaceflight terms only three days away and by radio only 1.5 seconds. In contrast, Mars is 35 to 230 million miles from Earth and radio messages take seven to 40 minutes.

Lunar habitats and bases will eventually need to use in-situ resources--transportation cost are too great to continuously maintain bases supplied from Earth. The Moon soil is 40% by weight chemically bound oxygen. This could provide the raw materials to produce oxygen for life support, rovers, and propellant. Producing oxygen for propellant could save up to 80% of launch weight from earth for exploration of the Moon and beyond.[7] Also, the lunar soil is rich in iron, aluminum and other metals. This can provide the raw materials for manufacturing some products on the Moon. Fig. 4 shows an artist concept of the Fission Surface Power deployed on the Moon.[8]

The major near-term reason for establishing lunar settlements, in addition to learning to live and work on other solar system entities, is science. The Moon has undergone a distinct and complex geological history, and much of the understanding of the terrestrial planets, especially Mars and Mercury, derives from or is constrained by our knowledge of the Moon. The Moon has been called a Rosetta stone that allows us to read the otherwise indecipherable text of planetary evolution written in the crust of the terrestrial planets.[9]

In addition, the Moon is an ideal location for astronomical observatories. It has no obscuring atmosphere and with its 28 days rotation a telescope has 28 times as much time to gather light from a distant object as would be possible on Earth.[10] The resolution on the Moon could be a million times better than Hubble. Such observatories would allow the mapping of Earth-sized planets around nearby stars out to a distance of around 100 light years. Much of what we know of physical laws comes from astronomical observations. How much more can we learn?

Fig. 4. Illustration of Fission Surface Power on the moon. *Courtesy of NASA.*

Environmentally non-polluting power from space to satisfy the Earth's needs is currently a dream. The cost of transporting the mass of materials to build large power satellites in Earth orbit makes the concept uneconomical at the present time. However, with resources from the Moon, this dream could become a reality (see Fig. 5). Or, mining the Moon for helium-3 for fueling terrestrial fusion power plants might become feasible.

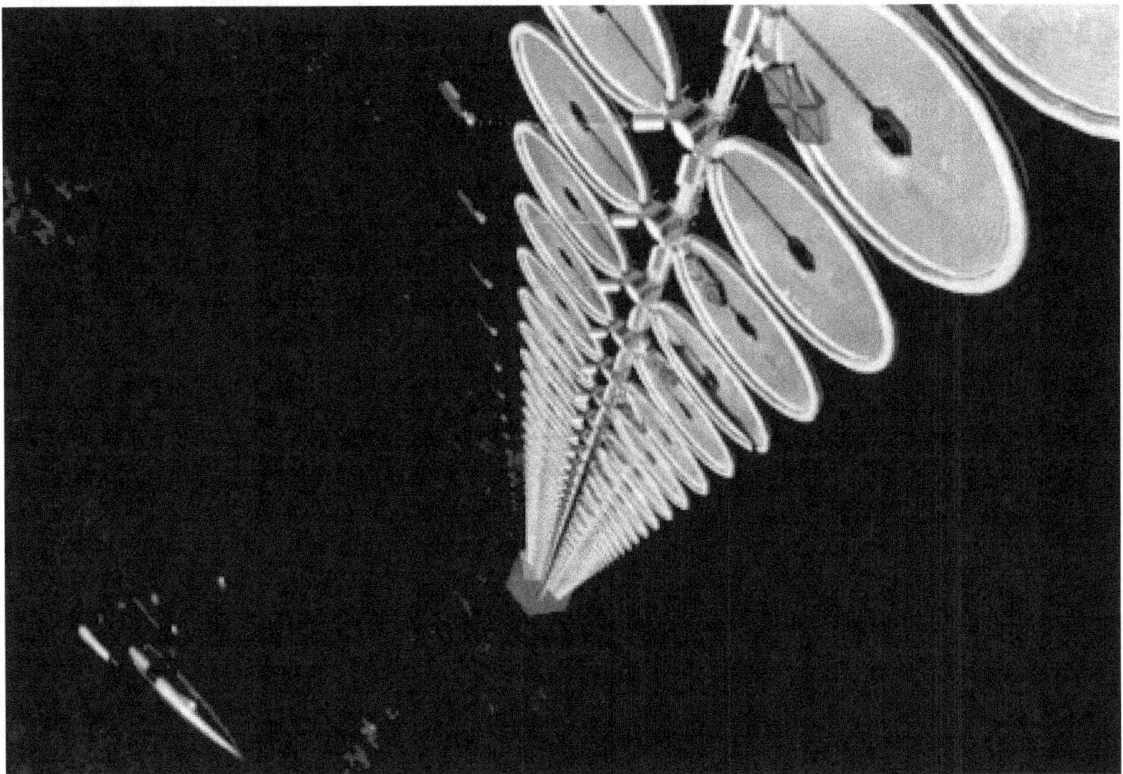

Fig. 5. Solar-powered satellites could produce electricity and beam it down to Earth via microwaves. *Artwork by Pat Rawlings, courtesy of NASA*

Mars is the Solar System planet most like Earth. It has a thin atmosphere, weather, seasons, water and a 25-hour day. Mars was once warmer, wetter and had a much denser atmosphere in its history. Scientist hope by understanding Mars that we can gain a better understanding of Earth's environment and the effects of pollution and greenhouse gases. Initially, the emphasize will be on exploring the Mars landscaping and understanding the geology. However, for many the eventual goal of going to Mars is to terraform the planet and become the second home for mankind.

Manned missions to Mars can not only utilize nuclear thermal propulsion, but require significant surface power for habitats and in-situ resource processing. In the Mars Direct approach for manned Mars missions, propellant is processed on the Mars surface for the return to Earth leg of the mission (see Fig. 6). The Marian atmosphere is 95 percent carbon dioxide and this can be processed to produce methane and water. The methane is liquefied while the water is electrolyzed to produce oxygen. The propellants can be used for rovers and rocket propulsion to return to Earth.[11]

Estimate have been made of what power levels are needed for the initial lunar and Mars manned missions. Table 5 provides estimates in the Synthesis Group Report.[12] Habitats and laboratories power requirements are up to 100 kW_e and nuclear power becomes a necessary option. For the Mars Direct approach to Mars, a 100 kW_e nuclear power system was proposed.

Fig. 6. Manned mission to Mars using Mars resources for return to Earth. Artwork courtesy of Robert Murray, Pioneer Astronautics.

Turning to outer solar system exploration, nuclear fission power is becoming a necessity to perform the more demanding missions now being contemplated. This is to provide electric power for propulsion and electric power for much higher data transmission rates (Table 6).[13] The higher data rates would make possible very high resolution multi-spectral images (with high resolutions both spatially and spectrally), something not possible currently. Instead of seeing things the size of cars, you will be able to see things the size of cats. Also, movies will be possible to study atmospheric phenomena and surface activities such as avalanches, floods, and volcanic activity. In addition, electromagnetic waves in the form of radar and radio occultation that require higher power levels than currently available become feasible. In summary, higher powers makes it possible to perform much more detailed surface and underground planetary studies in solar system exploration.

Table 5. Estimates of Lunar and Mars near term power needs.

Functions	Mars Power	Moon Power	Suggested Technology
Transportation			
Spacecraft			
Piloted	to 20 kWe	to 30 kWe	Fuel cells (Moon) Nuclear/photovoltaics (Mars)(1)
Cargo	5 kWe	5 kWe	Fuel cells (Moon) Photovoltaics (Mars)
Lander	20 kWe	20 kWe	Fuel cells (w/wo photovoltaics)
Electric propulsion	to 5 MWe	to 5 MWe	Nuclear
Surface Activities			
Day only	20 kWe		Photovoltaics
Habitat/lab			
Initial Operational Capability	to 30 kWe	to 50 kWe	Photovoltaics or nuclear (1)
Next Operational Capability	50 kWe	100 kWe	Nuclear
Base Power			
Initial Operational Capability	to 100 kWe	to 100 kWe	Nuclear
Next Operational Capability	to 800 kWe	to 1 MWe	Nuclear
Rovers			
Unloader/Construction	240 kWe-h	240 kWe-h	Fuel cells (2)
Pressurized			
Initial Operational Capability (per trip)	1900 kWe-h	1900 kWe-h	Fuel cells (2)
Next Operational Capability (per trip)	4800 kWe-h	4800 kWe-h	Fuel cells (2)
Unpressurized	100 kWe-h	100 kWe-h	Fuel cells (2)

(1) Depends on final power level

(2) In situ methane and oxygen produced on Mars may substitute for fuel cells.

The initial application in the class of missions using nuclear power for electric propulsion was to be the Jupiter Icy Moons Orbiter (JIMO) to explore Callisto, Ganymede and Europa. Other possible missions included: Saturn and its moons; Neptune and its moons; comet and multi-asteroid sample return; and Kuiper Belt rendezvous. These missions were to use nuclear electric propulsion with a power system of about 200 kW_e and operation times of up to 20 years. Interest remains high to perform these missions.

To summarize, nuclear fission power is an absolute necessity to perform the challenging scientific and exploration missions projected as mankind goes forward in his search for knowledge and desire to conquer the Solar System! The technology is proven! It is only a matter of time when constrained budgets will permit nuclear power plants to be incorporated into the space infrastructure!

NOTES

OVERVIEW

[1] Dennis Miotla, "Assessment of Plutonium-238 Production Alternatives," Briefing for Nuclear Energy Advisory Committee, April 21, 2008.

[2] H, M. Dieckamp, *Nuclear Space Power Systems,* Atomics International, Canoga Park, CA, September 1967

[3] Atomic Industrial Forum, *Guidebook for the Application of Space Nuclear Power Systems,* New York, January 1969.

[4] D. Buden et al., "Selection of Power Plant Elements for Future Reactor Space Electric Power Systems," LA-7858, Los Alamos National Laboratory, NM, September 1979.

[5] Ibid, D. Buden, et al.

[6] U.S. Department of Energy, "Environmental Development Plan (EDP)--Space Applications," DOE/EDP-0026, April 1978.

[7] "Nuclear Energy in Space," DOE Office of Nuclear Energy, Science & Technology, DOE/NE0071.

[8] Bennett, Gary, et al, "Mission of Daring: The General-Purpose Heat Source Radioisotope Thermoelectric Generator," 4th International Energy Conversion Engineering Conference, AIAA 2006-4096, June 2006.

[9] Jet Propulsion Laboratory Internet Site, www.jpl.nasa.gov.

[10] U. of Wisconsin lectures, fti.neep.wisc.edu/neep602/SPRING00/lecture34.pdf

[11] U.S. Department of Energy, ebid.

CHAPTER ONE

[1] Yen, C. L. and C. G. Sauer, "Nuclear Electric Propulsion for Future NASA Space Science Missions," *in AIDAA, AIAA/DGLR/SASS 22nd International Electric Propulsion Conference, Paper No. IEPC-91-035*, Viareggio, Italy, October 1991.

[2] Gary L. Bennett, "Radar Men on the Moon: A Brief Survey of Fission Surface Power Studies," *CP969. Space Technology and Applications Forum--STAIF-2008*, edited by M. S. El-Genk, 2008 American Institute of Physics, 978-0-7354-0486-1-08, 2008, pp. 372-379.

[3] George P. Dix and Susan S. Voss, "The Pied Piper--A Historical Overview of the U.S. Space Power Reactor Program," *Space Nuclear Power Systems 1984*, Edited by M. S. El-Genk and M. D. Hoover, Orbit Book co., Malabar, FL., 1985, pp. 23-30.

[4] R. B. Harty, M. P. Moriarty, and T. A. Moss, "Capabilities of Brayton Cycle Space Power Systems," in *Proceedings of the Second Symposium on Space Nuclear Power Systems*, held in Albuquerque, NM, Jan. 1985.

[5] E. C. Hylin and M. P. Moriarty, "Impact of Stirling Engine Characteristics Upon Space Nuclear Power Systems Design," in *Proceedings of the Second Symposium on Space Nuclear Power Systems*, held in Albuquerque, NM, Jan. 1985.

[6] W. S. Chiu, "Impact of Stirling Engines to a Space Nuclear Power System," in *Proceedings of the Second Symposium on Space Nuclear Power Systems*, held in Albuquerque, NM, Jan. 1985.

[7] G. L. Yoder and R. R. Graves, "Analysis of Alkali Liquid Metal Rankine Space Power Systems," in *Proceedings of the Second Symposium on Space Nuclear Power Systems*, held in Albuquerque, NM, Jan. 1985.

[8] W. Terrill and L. R. Putman, "SP-100 Thermocouple Assembly, Design and Technology Development," in *Proceedings of the Second Symposium on Space Nuclear Power Systems*, held in Albuquerque, NM, Jan. 1985.

[9] NASA, "Fission Surface Power System Technology for NASA Exploration Missions," January 2008.

[10] Joseph A. Angelo, Jr. and David Buden, *Space Nuclear Power,* Orbit Book Company, 1985.

[11] John R. Lamarsh, *Introduction to Nuclear Engineering* (2nd Edition), Addison-Wesley Publishing Co., Reading, Mass., 1983.

[12] S. Glasstone and M. C. Edlund, *The Elements of Nuclear Reactor Theory*, VanNostrand Co. Inc., Princeton, NJ, 1952.

[13] John R. Lamarsh, *Introduction to Nuclear Reactor Theory*, Addison-Wesley Publishing Co., Reading, Mass., 1972.

[14] D. Buden et al., "Selection of Power Plant Elements for Future Reactor Space Electric Power Supplies," LA-7858, Los Alamos Scientific Laboratory, September 1979.

[15] Atomic Industrial Forum, "Guidebook for the Application of Space Nuclear Power Systems," January 1969.

[16] H. M. Dieckamp, "Nuclear Space Power Systems," Atomics International, September 1967.

[17] J. H. Van Osdol et al., "Design and Integration of Zirconium Hydride Reactor Power Systems," AI-AEC-13072, Atomics International, June 1973.

[18] P. N. Stevens and H. C. Claiborne, *Weapons Radiation Shielding Handbook*, Chapter 2, DASA-1892-5, Defense Atomic Support Agency, June 1970.

[19] A. M. Weinberg and E. P. Wigner, *The Physical Theory of Nuclear Chain Reactors*, University of Chicago Press, 1958.

[20] Argonne National Laboratory, "Reactor Physics Constants", ANL- 5800 (2nd edition), July 1963.

[21] D. Buden, et al, "Selection of Power Plant Elements for Future Reactor Space Electric Power Systems," LA-7858, Los Alamos National Laboratory, September 1979.

[22] R. V. Anderson, et al, "Space Reactor Electric Systems," ESG-DOE-13398, Rockwell International, 29 March 1983.

[23] Ibid D. Buden, et al.

[24] C. Wood, "Optimization of Thermoelectric Materials," *Proceedings of 1982 Working Group on Thermoelectrics*, 7-9 Dec 1982, Jet Propulsion Laboratory, JPL-D-497, January 1983.

[25] Joseph A. Angelo, Jr. and David Buden, *Space Nuclear Power,* Orbit Book Company, Malabar, Florida 1985.

[26] Rasor Associates Inc., "Direct Conversion Nuclear Reactor Space Power Systems, Volume I," NSR-17-I, Sunnyvale, CA, December 1981.

[27] Joseph A. Angelo, Jr. and David Buden, *Space Nuclear Power*, Orbit Book Company, Malabar, Florida, 1985.

[28] J. P. Holman, *Thermodynamics* (2nd Edition), McGraw-Hill Book Company, New York, 1974.

[29] W. Reynolds and H. Perkins, *Engineering Thermodynamics,* McGraw-Hill Book Company, 1974.

[30] K. Wark, *Thermodynamics* (3rd Edition), McGraw-Hill Book Company, New York, 1977.

[31] I. Asimov, *Biographical Encyclopedia of Science and Technology* (Revised Edition), Doubleday and Company, New York, 1972.

[32] Ibid Joseph A. Angelo, Jr. and David Buden

[33] B. Goldwater, "Study of a Free Piston Stirling Engine-Linear Power Conversion System for a 10.5 kWe and 51.0 kWe Output Power," Mechanical Technology Inc., November 1975

[34] G. Walker, *Stirling Cycle Machines*, Oxford University Press, Ely House, 1973.

[35] W. Beale et al., "Free Piston Stirling Engines- A Progress Report," SAE Paper 730647, June 1973.

[36] G. M. Grover, T. P. Cotter and G. F. Erickson, "Structures of Very High Thermal Conductance," *J. Applied Physics*, Vol. 35, No.6, p. 1990, June 1964.

[37] T. P. Cotter, "Theory of Heat Pipes," LA-3246-MS, Los Alamos Scientific Laboratory, 1965.

[38] NASA, "Space Shuttle Program-Space Shuttle System Payload Accommodations, Level II Program Definition and Requirements, Vol. XIV," JSC-07700, Rev H, 16 May 1983.

[39] NASA, *Space and Planetary Environment Criteria Guidelines for Use in Space Vehicle Development*, 1982 Revision (Volume I), NASA TM-82478, January 1983.

[40] F. Kreith, *Radiation Heat Transfer for Spacecraft and Solar Power Plant Design*, International Textbook Co., Scranton, 1962.

[41] Joseph A. Angelo, Jr. and David Buden, *Space Nuclear Power,* Orbit Book Company, Malabar, FL, 1985.

[42] J. Angelo, Jr. and D. Buden, "Shielding Considerations for Advanced Space Nuclear Reactor Systems," 1982 IEEE Conference on Nuclear and Space Radiation Effects, Las Vegas, Nev., 20-21 July 1982.

[43] D. Buden et al., "Selection of Power Plant Elements for Future Reactor Space Electric Power Systems," LA-7858, Los Alamos National Laboratory, September 1979.

[44] Ibid D. Buden, et al.

[45] Atomic Industrial Forum, *Guidebook for the Application of Space Nuclear Power Systems*, January 1969.

CHAPTER TWO

[1] Joseph A. Angelo, Jr. and David Buden, "Shielding Considerations For Advanced Space Nuclear Reactor Systems," *1982 IEEE Conference on Nuclear and Radiation Effects*, July 1982, Las Vegas, Nevada, Los Alamos National Laboratory Report LA-UR-82-2002.

[2] *Space Transportation System Payload Safety Guidelines Handbook, NASA Lyndon B. Johnson Space Center, Houston, TX, July 1976, Table 3.14-III.*

[3] Calina C. Seybold, "Characteristics of the Lunar Environment," *seybold@utcsr.ae.utexas.edu*, August 1995.

[4] Wikipedia, "Geology of Mars," August 2008.

[5] NASA facts, "Constellation Program: America's Fleet of Next-Generation Launch Vehicles, The Ares I Crew Launch Vehicle." *FS-2007-08-110-MSFC.*

[6] Hardy, E. B., and J. A Hartung, "SNAP-10A Aerospace Nuclear Safety: A Good Foundation for the SP- 100 Program:, *UNM Symposium*, January 1984.

[7] Ibid Harty, January 1984.

[8] Buden, D., The Acceptability of Reactors in Space," *16th Intersociety Energy Conversion Engineering Conference Proceedings*, Atlanta, GA, August 9-14, 1981, pp. 278-287.

[9] Ibid Harty, January 1984.

[10] Voss, Susan S. and G. P. Dix, "A Review of the Rover Safety Program, 1959-1972", Presented at the *Space Nuclear Propulsion Workshop*, Los Alamos, NM, 11-13 December 1984.

[11] Lee, J., et al, 1990, "SIREN Search Intercept Retrieve Expulsion Nuclear," *AIAA/NASA/DOD Orbital Debris Conference: Technical Issues and Future Directions*, Baltimore, MD, 18 April 1990.

[12] "Nuclear Safety Criteria and Specifications for Space Nuclear Reactors", *Dept. of Energy Document OSNP-1, Revision-O*, Washington, D.C., August 1982.

[13] Buden, D. and P. W. Garrison, "Space Nuclear Power System and the Design of the Nuclear Electric Propulsion OTV", *AIAA/SAI/ASME 20th Joint Propulsion Conference*, Cincinnati, OH, June 11-13,1984.

[14] United Nations Working Paper by the Union of Soviet Socialist Republics, Use of Nuclear Power Sources in Outer Space," *Committee on the Peaceful Uses of Outer Space, Scientific and Technical Subcommittee, NAC.105/C.1IWG.5/L.24/Add.2*, 26 Feb. 1990.

[15] Atkinson, Dale, et al, "Collision Damage To Nuclear Satellites," *American Nuclear Society Meeting Nuclear Technologies For Space Exploration-92*, Jackson Hole, WY, Aug. 1992, pp.843-849.

[16] "Prometheus Project Reactor Module Final Report, Revision 1," *Knolls Atomic Power Laboratory, SPP-67110-0008*, April 24, 2006.2

CHAPTER THREE

[1] James H. Van Osdol, W. B. Thompson, and O. S. Merrill, "Uranium Zirconium Hydride Reactor Space Power Systems," International Astronautical Federation XXVIIth Congress, Anaheim, Calif., 10-16 October 1976, Paper No. 76-256, 250 Rue Saint-Jacques, 75005 Paris, France, p. 6.

[2] A. A. Jarrett, "SNAP-2 Summary Report," Atomics International Report No. AI-AEC-13068, 27 July 1973.

[3] Ibid D. W. Staub

[4] H. M. Dieckamp, "Nuclear Space Power Systems," Atomics International report issued September 1967, Canoga Park, Calif.

[5] Ibid H. M. Dieckamp

[6] Gary Bennett, "Overview of the U.S. Flight Safety Process for Space Nuclear Power," *Nuclear Safety*, Vol. 22, No. 4, July-August 1981, pp. 423-434.

[7] D. Buden, "The Acceptability of Reactors in Space," Proceedings of the 16th Intersociety Energy Engineering Conference, Aug. 9-14, Atlanta, Ga.

[8] D. Buden and G. Bennett, "On the Use of Nuclear Reactors in Space," *Physics Bulletin*, Vol. 33, 1982, pp. 432-434.

[9] "Studies on Technical Aspects and Safety Measures of Nuclear Power Sources in Outer Space," U.S.A. working paper to Working Group on the Use of Nuclear Power Sources in Outer space, U.N. Committee on Peaceful Uses of Outer Space, 23 Jan. 1980.

[10] Ibid D. W. Staub

BOOK 3
SPACE NUCLEAR FISSION ELECTRIC POWER SYSTEMS

[11] J. N. Hodgson, R. G. Germer, A. H. Kreeger, "SNAP-8-A Technical Assessment," Proceedings of Intersociety Energy Conversion Engineering Conference, 1967, pp. 135-143.

[12] W. Corliss, "SNAP Nuclear Space Reactors," U.S. Atomic Energy Commission, Library of Congress Catalog No. 66-62772 (1966).

[13] J. N. Hodgson and R. P. Macosko, "A SNAP-8 Breadboard System-Operating Experience," Intersociety Energy Conversion Engineering Conference (1968), pp. 338-351.

[14] J. K. Asquith and D. G. Mason, "Zirconium Hydride Space Power Reactor Design," Proceedings of Intersociety Energy Conversion Engineering Conference, 1972, pp. 572-580.

[15] J. R. Pope and A. Stromquist, "SNAP-8 Rotating Machinery Components-I0,000 Hour Endurance Achievement," Proceedings of Intersociety Energy Conversion Engineering Conference, 1970, pp. 11-67 to 11-70.

[16] H. Derow, "SNAP-8 Power Conversion System Breadboard Assembly-Materials Evaluation After 8,700 Hours Operation," Proceedings of Intersociety Energy Conversion Engineering Conference, 1970, pp. 11-24 to 11-27.

[17] Ibid J. N. Hodgson and R. P. Macosko

[18] R. L. Detterman, L. M. Maki, R. F. Wilson, and D. E. Reardon, "Technology Status-ZrH Power Reactors," Intersociety Energy Conversion Engineering Conference (1970), p. 12-26 to 12-32.

[19] "A Management Summary of the ZrH Reactor-Thermoelectric Space Power System Program," Atomic International Report AI-BD-AEC-73-23, Canoga Park, Calif.

[20] "10 to 75 kWe Reactor Power Systems for Space Applications," Rockwell International Report No. N652T1140013, 24 March 1976, Canoga Park, Calif.

[21] D. W. Staub (Ed.), "SNAP Programs Summary Report," Atomics International Report No. AI-AEC-13067, 30 July 1973.

[22] "ZrH Reactor Thermoelectric Space Power System Program Review with the DOD/AEC Ad Hoc Space Power Working Group," Rockwell Report AI-BD-AEC 73-27, 11 September 1973.

CHAPTER FOUR

[1] Redd, F. J. and E. V. Fomoles (1985) "Emerging Space Nuclear Power Needs," in *Space Nuclear Power Systems 1984, Proceedings of the First Symposium on Space Nuclear Power Systems*, held in Albuquerque, NM, Jan. 1984, Chapter 9, 41-42.

[2] Ambrus, J. H. and R. G. Beatty (1985) "Potential Civil Mission Applications for Space Nuclear Power Systems," in *Space Nuclear Power Systems 1984, Proceedings of the First Symposium on Space Nuclear Power Systems*, held in Albuquerque, NM, Jan. 1984, Chapter 10, 43-51.

[3] Alan T. Marriott and Toshio Fujita, "Evolution of SP-100 system Designs," *CONF 940101, 1994 American Institute of Physics*, pp. 157-172.

[4] Truscello, V. C. and L. L. Rutger (1992) "The SP-100 Power System," in *Proceedings Ninth Symposium on Space Nuclear Power Systems*, held in Albuquerque, NM, Jan. 1992, 1:1-23.

[5] Mondt, J. F. (1989) "Development Status Of The SP-I00 Power System," in *Proceedings AIAAIASMEISAEIASEE 25th Joint Propulsion Conference, Paper No. AIAA-89-259l*, held in Monterey, CA, July 1989.

[6] Josloff, A. T. (1988) "SP-I00 Space Reactor Power System Scaleability," in *Proceedings 23th Intersociety Energy Conversion Engineering Conference*, held in Denver, CO, Aug. 1988, 1:263-265.

[7] Mondt, J. (1992) "SP-I00 Accomplishments", in SP-I00 Project Review held June 15, 1992, *JPL D-9650*.

[8] Disney, R. K., J. E. Sharbaugh, and J. C. Reese (1990) "Shielding Approach And Design For SP-100 Space Power Applications," in *Proceedings Seventh Symposium On Space Nuclear Power Systems*, held in Albuquerque, NM, 1:121-124.

[9] Deane, N. A. et al. (1989) "SP-I00 Reactor Design and Performance," in *Proceedings 24th Intersociety Energy Conversion Engineering Conference*, held in Washington, DC, Aug. 1989, 2:1225-1226.

[10] Atwell, J. C., et al. (1989) "Thermoelectric Electromagnetic Pump Design For The SP-100 Reference Flight System," in *Transactions Sixth Symposium On Space Nuclear Power Systems*, held in Albuquerque, NM, Jan. 1989,280-283.

[11] Bond, J. A., D. N. Matteo, and R. J. Rosko (1993) "Evolution of the SP-100 Conductivity Coupled Thermoelectric Cell," in *Proceedings Tenth Symposium Space Nuclear Power and Propulsion*, held in Albuquerque, NM, January 1993,2:753- 758.

[12] Hwang, Choe et al. (1993) "SP-l00 Lithium Thaw Design, Analysis, and Testing," in *Proceedings Tenth Symposium on Space Nuclear Power and Propulsion Systems*, held in Albuquerque, NM, Jan. 1993, 2:697-702.

[13] Schmitz, P., H. Bloomfield, and J. Winter (1992) "Near Term Options For Space Reactor Power," in *Proceedings of Nuclear Technologies For Space Exploration, American Nuclear Society Meeting* held in Jackson, WY, August 1992, 2:313-321.

[14] Josloff, A. T., H. S. Bailey, and D. N. Matteo (1992a) "SP-l00 System Design and Technology Progress," in *Proceedings Ninth Symposium on Space Nuclear Power Systems*, held in Albuquerque, NM, Jan. 1992, 1:363-371.

[15] Ibid Truscello and Rutger (1992).

[16] Josloff, A. T., D. N. Matteo, and H. S. Bailey (1992b) "SP-l00 Technology Development Status," in *Proceedings Nuclear Technologies For Space Exploration, NTSE-92*, American Nuclear Society Meeting held in Jackson, WY, August 1992, 1:238-246.

[17] Mondt, J. (1992) "SP-l00 Accomplishments, in SP-100 Project Review held June 15, 1992, *JPL D-9650*.

[18] Ibid Mondt 1992.

[19] Matteo, D. N. (1993) Personal Communication, Martin Marietta, Valley Forge, PA, 25 May, 1993.

[20] Bond, J. A., D. N. Matteo, and R. J. Rosko (1993) "Evolution of the SP-100 Conductivity Coupled Thermoelectric Cell," in *Proceedings Tenth Symposium Space Nuclear Power and Propulsion*, held in Albuquerque, NM, January 1993, 2:753-758.

[21] Mondt, J. F. (1991) "Overview of the SP-100 Program," in *AIAA/NASA/OAI Conference on Advanced SEI Technologies, AIAA 91-3585*, held in Cleveland, OH, September 1991.

[22] General Electric Co. (1989b) "SP-l00 Safety Feature," presented at the *Sixth Symposium On Space Nuclear Power Systems*, held in Albuquerque, NM, Jan. 1989.

[23] General Electric Co. (1989a) "SP-l00 Mission Risk Analysis," in *Report No. GESR· 00849*, San Jose, CA, Aug. 1989.

[24] Mondt, J. F. (1991) "Overview of the SP-I00 Program," in *AIAA/NASA/OAI Conference on Advanced SEI Technologies, AIAA 91-3585*, held in Cleveland, OH, September 1991.

[25] Mondt, J. F. (1993) "SP-100 Project Memorandum No. 8501-93-004," 25 March 1993.
Phillips Laboratory (1992) *Thermionic Space Nuclear Power Technical Program Review*, held in Albuquerque, NM, 9-10 November 1992.

[26] Jack F. Mondt, Vincent C. Truscello, Alan T. Marriott, "SP-100 Power Program," *CONF 940101, 1994 American Institute of Physics*, pp. 143-155.

[27] Scott F. DeMuth, "SP-100 Reactor Subsystem Development," *CONF 940101, 1994 American Institute of Physics*, pp. 171-172.

[28] Scott F. DeMuth, "SP-100 control Drive Assembly Development Plan," *CONF 940101, 1994 American Institute of Physics*, pp. 173-174.

[29] John J. Buksa, "SP-100 Heat Transport Technology Development," *CONF 940101, 1994 American Institute of Physics*, pp. 175-181.

[30] Christopher England and Richard C. Ewell, "Progress On The SP-100 Power Conversion Subsystem," *CONF 940101, 1994 American Institute of Physics*, pp.183-193.

[31] Slaby, J. G. (1987) "Overview of NASA Lewis Research Center Free-Piston Stirling Engine Technology Activities Applicable to Space Power Systems," in *Space Nuclear Power Systems 1987*, Chapter 5,35-39.

[32] Dudenhoefer, J. E. (1990) "Programmatic Status of NASA's CSTI High Capacity Power Stirling Space Power Converter Program," in *Proceedings 25th Intersociety Energy Conversion Engineering Conference*, held in Reno, Nevada, Aug. 1990, 1:40.

[33] Dudenhoefer, 1. E. and J. M. Winter (1991) "Status of NASA's Stirling Space Power Converter Program," in *Proceedings 26th Intersociety Energy Conversion Engineering Conference*, held in Boston, MA, Aug. 1991, 2:38-43.

[34] Dudenhoefer, J. E., D. Alger, and J. M. Winter (1992) "Progress Update of NASA's Free-Piston Stirling Space Power Converter Technology Project," in *Proceedings Nuclear Technologies For Space Exploration,NTSE-92, American Nuclear Society Meeting* held in Jackson, WY, August 1992, 1:294-303.

[35] Jack F. Mondt, Vincent C. Truscello, and Alan T. Marriott, "SP-100 Power Program," *CONF 940101, 1994 American Institute of Physics*, pp. 143-155.

CHAPTER FIVE

[1] J. W. Holland et al., "Thermionic Reactor Development," *Intersociety Energy Conversion Engineering Conference, 1971 Proceedings*, pp. 939-947.

[2] A. J. Gietzen and W. G. Homeyer, "Thermionic Reactor Power Systems," *Intersociety Energy Conversion Engineering Conference., 1972 Proceedings*, pp. 556-558.

[3] D. S. Beard, "Thermionic Reactor Technology-An Overview," *Intersociety Energy Conversion Engineering Conference., 1971 Proceedings*, pp. 933-938.

[4] L. Yang and J. Chin, "Developmental Status of Thermionic Materials," *Intersociety Energy Conversion Engineering Conference., 1972 Proceedings*, pp. 1041-1049.

[5] D. S. Beard, "Thermionic Fuel Development," *Intersociety Energy Conversion Engineering Conference., 1969 Proceedings*, pp. 131-136.

[6] "Development of a Thermionic Reactor Space Power System," *Gulf General Atomic Report No. Gulf-GA-AI208*, San Diego, Calif., 30 June 1973.

[7] L. K. Price, "Thermionic Reactor Concepts," *Intersociety Energy Conversion Engineering Conference., 1969 Proceedings*, pp. 122-130.

[8] "Direct Conversion Nuclear Reactor Power Systems," *Razor Associates*, Sunnyvale, Calif. 1982.

[9] D. R. Koenig and W. A. Ranken, "Reactor Design for Nuclear Electric Propulsion," *14th Intersociety Energy Conversion Engineering Conference*, Boston, Mass., 1979 Proceedings, pp. 1418- 1424.

[10] W. M. Phillips, "Nuclear Electric Power System for Solar System Exploration," *J. Spacecraft, Vol. 17, No.4, (Article No. 79-1337R)*, pp. 348-353.

[11] W. M. Phillips and E. V. Pawlik, "Design Considerations for a Nuclear Electric Propulsion System," *13th Intersociety Energy Conversion Engineering Conference*, San Diego, Calif., 1978 Proceedings, pp. 218-223.

[12] E. J. Britt and G. O. Fitzpatrick, "Direct Conversion Nuclear Reactor Power Systems," *17th Intersociety Energy Conversion Engineering Conference*, 1982 Proceedings, pp. 1894-1901.

[13] J. W. Holland et al., "Thermionic Reactor Development," *Intersociety Energy Conversion Engineering Conference, 1971 Proceedings*, pp. 939-947.

[14] L. Yang and J. Chin, "Developmental Status of Thermionic Materials," *Intersociety Energy Conversion Engineering Conference., 1972* Proceedings, pp. 1041-1049

[15] "Development of a Thermionic Reactor Space Power System," *Gulf General Atomic Report No. Gulf-GA-AI208*, San Diego, Calif., 30 June 1973.

[16] L. K. Price, "Thermionic Reactor Concepts," Intersociety Energy Conversion Engineering Conference, 1969 Proceedings, pp. 122-130.

[17] D. S. Beard, "Thermionic Fuel Development," *Intersociety Energy Conversion Engineering Conference., 1969 Proceedings*, pp. 131-136.

[18] "Direct Conversion Nuclear Reactor Power Systems," *Razor Associates*, Sunnyvale, Calif. 1982.

[19] W. M. Phillips and E. V. Pawlik, "Design Considerations for a Nuclear Electric Propulsion System," *13th Intersociety Energy Conversion Engineering Conference*, San Diego, Calif., 1978 Proceedings, pp. 218-223.

[20] "Direct Conversion Nuclear Reactor Power Systems," *Razor Associates*, Sunnyvale, Calif. 1982.

[21] E. J. Britt and G. O. Fitzpatrick, "Direct Conversion Nuclear Reactor Power Systems," *17th Intersociety Energy Conversion Engineering Conference*, 1982 Proceedings, pp. 1894-1901.

[22] Brown, C. E. and D. Mulder (1992) "An Overview of the Thermionic Space Nuclear Power Program," presented at the *1992 Space Nuclear Power Systems Symposium, held in Albuquerque*, NM, Jan. 1992.

[23] Homeyer, W. G., et al. (1984) "Thermionic Reactors for Space Nuclear Power," in *Space Nuclear Power Systems 1984, Chapter 25*, Orbit Book Co., Melabar, FL, 197-205.

[24] Snyder, H., J. and J. H. Mason (1985) "Thermionic Space Power System Concept," in *Space Nuclear Power Systems 1985, Chapter 29*, Orbit Book Co., Melabar, FL, 223-229.

[25] Holland, J.W. et al. (1985) "SP-100 Thermionic Technology Program Status," in *Space Nuclear Power Systems 1985, Chapter 24*, Orbit Book Co., Melabar, FL, 179-188.

[26] Strohmayer, W.H. and T. H. Van Hagan (1985) "Parametric Analysis of a Thermionic Space Nuclear Power System," in *Space Nuclear Power Systems 1985*, Orbit Book Co., Melabar, FL, Chapter 20, 141-149.

[27] Michael G. Houts, et al., "Technical Accomplishments of the Thermionic Fuel Element Verification Program," *CONF 940101, 1994 American Institute of Physics*, pp.997-1007

[28] Samulelson, S., L. and R. C. Dahlberg (1990) "Thermionic Fluid Element Verification Program - Summary of Results," in *Proceedings Seventh Symposium on Space Nuclear Power Systems*, held in Albuquerque, NM, Jan. 1990, 1:77-84.

[29] General Atomics (1988) TFE Verification Program Semiannual Report for the Period Ending September 30, 1987, *General Atomics Documents GA-A19115*, La Jolla, CA, March 1988.

[30] Richard C. Dahlberg, et al., "Review of Thermionic Technology: 1983-1992," *A Critical Review of Space Nuclear Power And Propulsion, 1984-1993*, Editor Mohamed S. El-Genk, AIP Press, L.C. catalog Card No. 94-70780, pp. 121-165.

[31] Michael G. Houts, et al., "Technical Accomplishments of the Thermionic Fuel Element Verification Program," *CONF 940101, 1994 American Institute of Physics*, pp. 997-1007.

[32] William R. Determan and Tom H. Van Hagan, "Space Power Reactor In-Core Thermionic Multicell Evolutionary (S-PRIME) Design." *CONF 930103, 1993 American Institute of Physics*, pp. 1235-1239.

[33] Joseph C. Mills, William R. Detterman, and Tom H. Van Hagan, "S-PRIME/TI-SNPS Conceptual Design Summary," *CONF 940101, 1994 American Institute of Physics*, pp. 695-700.

[34] L. L. Begg, T. L. Wuchte, and W. D. Otting, "STAR-C Thermionic Space Nuclear Power System," in *Proceedings Ninth Symposium on Space Nuclear Power Systems*, held in Albuquerque, NM, Jan. 1992, 1:114-119.

[35] Hyop S. Rhee, et al., "SPACE-R Thermionic Space Nuclear Power System With Single Cell In-Core Thermionic Fuel Elements," *CONF 920104, 1992 American Institute of Physics*, pp. 120-129.

[36] Begg, L. L., T. J. Wuchte, and W. D. Otting (1992) "STAR-C Thermionic Space Nuclear Power System," in *Proceedings Ninth Symposium on Space Nuclear Power Systems*, held in Albuquerque, NM, Jan. 1992,1:114-119.

[37] Allen, D. T., G. O. Fitzpatrick, and D. M. Ernst (1991), "The STAR-C Nuclear Power With High-Efficiency Transparent Thermionic Converters," in *Proceedings Eighth Symposium on Space Nuclear Power Systems, Conf-910116*, M. S. El-Genk and M. D. Hoover, eds., American institute of Physics, New York, AlP Conf. Proc. No. 217,3:1093-1099.

CHAPTER SIX

[1] U. of Wisconsin lectures, fti.neep.wisc.edu/neep602/SPRING00/lecture34.pdf

[2] Nicholas L. Johnson, "A New Look At Nuclear Power Sources And Space Debris," *Proceedings of the Fourth European Conference on Space Debris*, Darmstadt, Germany, ESA SP587, August 2005, pp.551.

[3] Gary L. Bennett, "Soviet Space Nuclear Reactor Incidents: Perception Versus Reality," *Space Nuclear Power Systems 1989*, Edited by M. S. El-Genk and M. D. Hoover, Orbit Book co., Malabar, FL, 1992, pp. 273-278.

[4] Topaz CoDR, "Flight Topaz Conceptual Design Review," *U.S. Air Force Phillips Laboratory*, review hold in Albuquerque, NM, Sept. 16-17, 1992.

[5] D. Brent Morris, "The Thermionic System Evaluation Test (TSET): Descriptions, Limitations, And The Involvement Of The Space Nuclear Power Community," *CONF 930103, 1993 American Institute of Physics*, pp. 1251-1256.

[6] Gary L. Bennett, "Soviet Space Nuclear Reactor Incidents: Perception Versus Reality," *Space Nuclear Power Systems 1989*, Edited by M. S. El-Genk and M. D. Hoover, Orbit Book co., Malabar, FL., 1992, pp.273-278.

[7] G. Bennett, "A Look At The Soviet Space Nuclear Power Program," in *Proceedings 24th Intersociety Energy Conversion Engineering Conference*, held in Washington, D.C., August 1989, 2:1187-1194.

[8] Sven Grahn, "The USSR program (Radar Ocean Reconnaissance Satellites – RORSAT) and Radio Observations Thereof," www.ssvengrahn.pp.se/trackind/RORSAT/RORSAT.htm

[9] N. N. Ponomarev-Stepnoi, N. E. Kulkharkin, and V. A. Usov, "Romashka Reactor Converter," Atomic Energy, Vol 89, Number 6, Dec. 2000, pp. 1057 - 1068.

[10] "Questions Relating to the Use of Nuclear Power in Outer Space," *U.S.S.R.: Working Paper Studies on Technical Aspects and Safety Measures of Nuclear Power Sources in Outer Space, U.N. General Assembly A/AC.105/C.1/WG.V/L.10*, dated 25 January 1980.

[11] U.S. Department of Energy, "Operation Morning Light," *Nevada Operations Office, NV-198*, Las Vegas, Nev., September 1978.

[12] W. R. Gummer, et al., "COSMOS 954, the Occurrence and Nature of Recovered Debris," *Canadian Government Publishing Center, Catalogue No. CC 172/1980E*, May 1980.

[13] U.S.S.R Information to Committee on the Peaceful Uses of Outer Space, *U.N. Document ST/SG/SER.E/72*, dated 29 Dec. 1982.

[14] U.S.S.R. Information to Committee on the Peaceful Uses of Outer Space, *U.N. Document ST/SG/SER,E/72/Add. 1*, dated 20 Jan 1983.

[15] U.S.S.R Information to Committee on the Peaceful Uses of Outer Space, *U.N. Document ST/SG/SERE/72/Add. 2*, dated 21 Jan. 1983.

[16] U.S.S.R Information to Committee on the Peaceful Uses of Outer space, *U.N. Document ST/SG/SER.E/72/Add. 3*, dated 9 Feb. 1983.

[17] "Cosmos 1402's Uranium Remains", *Science News, Vol. 132,1 No. 18, Oct. 31, 1987*, pp. 132-133.

[18] G. M. Gryaznov, et al., "Radiation Safety Of The Space Nuclear Power Systems And Its Realization On The Satellite 'Cosmos 1900'," presented at the *Sixth Symposium on Space Nuclear Power Systems*, Albuquerque, NM, Jan. 1989.

[19] G. L. Bennett, Gary L. Bennett, "Soviet Space Nuclear Reactor Incidents: Perception Versus Reality," *Space Nuclear Power Systems 1989*, Edited by M. S. El-Genk and M. D. Hoover, Orbit Book Co., Malabar, FL., 1992, pp.273-278.

[20] Hyder, A. K., et al, *2000 Spacecraft Power Technologies*, Imperial College Press, London.

[21] G. Bennett, "A Look At The Soviet Space Nuclear Power Program," in *Proceedings 24th Intersociety Energy Conversion Engineering Conference*, held in Washington, D.C., August 1989, 2:1187-1194.

[22] Igor P. Bogusth, et al., "The Main Purposes and Results of Flight Test of SNPSs within the "TOPAZ" Program," *Seventh Symposium on Space Nuclear Power Systems,* Albuquerque, NM, 1990.

[23] Nicitin, V. P., et al. (1992) "Special Features And Results Of The "Topaz II" Nuclear Power System Tests With Electric Heating," in *Proceedings Ninth Symposium on Space Nuclear Power Systems, Conference 920104, American Institute of Physics*, 1992, 1:41-46.

[24] Bennett, G. (1989) "A Look At The Soviet Space Nuclear Power Program," in *Proceedings 24th Intersociety Energy Conversion Engineering Conference*, held in Washington, D.C., August 1989,2:1187-1194.

[25] TSET/NEPSTP Workshop (1993) held in the Albuquerque Convention Center, sponsored by Phillips Laboratory (AFSC), 15 January 1993.

[26] Topaz CoDR (1992) Flight Topaz Conceptual Design Review, *U.S. Air Force Phillips Laboratory, Review* hold in Albuquerque, NM, Sept. 16-17, 1992.

[27] Space Power Inc. (1990) "Characteristics Of The Soviet Topaz II Space Power System," *Report SPI-52-1*, San Jose, CA, October 1990.

[28] Carl W. Hoth, et al., "Design And Performance Of The UO_2 Fuel For The Topaz-II Reactor," *CONF 940101, 1994 American Institute of Physics*, pp. 55-61.

[29] U.S. Topaz II Flight Safety Team (1992), Topaz II Flight Program Safety Assessment, *Draft Report* Aug. 24, 1992.

[30] Carl W. Hoth, et al., "Design And Performance Of The UO_2 Fuel For The Topaz-II Reactor," *CONF 940101, 1994 American Institute of Physics*, pp. 55-61.

[31] Gary F. Polansky, et al., "Utilizing A Russian Space Nuclear Reactor For A United States Space Mission: Flight Qualification Issues," *CONF 940101, 1994 American Institute of Physics*, pp. 1171-1177.

[32] *TSET/NEPSTP Workshop (1993)* held in the Albuquerque Convention Center, sponsored by Phillips Laboratory (AFSC), 15 January 1993.

[33] Ray A Haaman, Susan S. Voss, and Veniamin Usov, "Topaz II Reactor Modifications Overview," *CONF 940101 American Institute of Physics*, pp. 1179-1183.

[34] Gafarov, Albert A., Boris I. Bakhtin, Alexey V. Kosov, "Principles Of Radiation Safety For Reactor Space Nuclear Power Sources And Methods Of Their Realization," *The Scientific-Research Institute of Thermal Processes 8, Onexhskaya*, Moscow, 125438, USSR, pp. 19-24.

[35] *Voss, Susan, "Topaz II: Safety Program Overview," presented at the* American Nuclear Society Nuclear Technologies For Space Exploration- 92, *Jackson, Wyoming, Aug. 1992.*

[36] Topaz CoDR (1992) Flight Topaz Conceptual Design Review, U.S. Air Force Phillips Laboratory, review hold in Albuquerque, NM, Sept. 16-17, 1992.

CHAPTER SEVEN

[1] "Project Prometheus Reactor Module Final Report," prepared by Knolls Atomic Power Laboratory and Bettis Atomic Power Laboratory, SPP-67110-0008, April 24, 2006.

[2] John Ashcroft, et al., "Key Factors Influencing the Decision on the Number of Brayton Units for the Prometheus Space Reactor," CP880, Space Technology and Applications International Form--STAIF 2007, edited by M. S. El-Genk, 2007 American Institute of Physics 978-0-7354-0386-4.07, pp. 522-540.

CHAPTER EIGHT

[1] NASA facts, "Fission Surface Power System Technology for NASA Exploration Missions," Jan. 2008.

[2] Donald T. Palac, et al, "Fission Surface Power Technology development Status," NASA/TM 2010-216249 and AIAA-2009-6535, April 2010.

[3] Donald T. Palac and Lee S. Mason, "Fission Surface Power Technology Development Status," NASA/TM--2009-215602, AIAA-2008-7812, March 2009.

[4] *Ibid* Donald T. Palac and Lee S. Mason

[5] Lee Mason, David Poston, and Louis Qualls, "System Concepts for Affordable Fission Surface Power," NASA/TM--2008-215166, January 2008.

[6] David I. Poston, et al, "Reference Reactor Module for the Affordable Fission Surface Power System," *CP989, Space Technology and Applications Forum--STAIF 2008*, edited by M. S. El-Genk, American Institute of Physics 078-0-7354-0486, 2008, pp. 277-284.

[7] *Ibid* Lee Mason, David Poston, and Louis Qualls

[8] Steven M. Geng, et al, "Overview of Multi-Kilowatt Free-Piston Stirling Power Conversion Research at GRC," *CP989, Space Technology and Applications Forum--STAIF 2008*, edited by M. S. El-Genk, American Institute of Physics 078-0-7354-0486, 2008, pp. 617-624.

[9] Krause, David L. and Pete T. Kantzos, "Accelerated Life Structural Benchmark Testing for a Stirling Convertor Heater Head," *CP813, Space Technology and Applications International Forum--STAIF 2006*, edited by M. S. El-Genk, 2006 American Institute of Physics 0-7354-0305-8/06, pp. 623-630.

[10] Wood, J. Gary, et al, "Advanced Stirling Convertor Update," *CP813, Space Technology and Applications International Forum--STAIF* 2006, edited by M. S. El-Genk, 2006 American Institute of Physics 0-7354-0305-8/06, pp. 640-652.

[11] Wood, J. Gary, et al, "Continued Development of the Advanced Stirling Convertor (ASC)", *Sunpower Report Doc. 0104* issued by AIAA.

[12] Ibid Wood, J. Gary, et al.

[13] Ibid Wood, J. Gary, et al.

[14] *Ibid* Lee Mason, David Poston, and Louis Qualls

[15] *Ibid* Donald T. Palac and Lee S. Mason

CHAPTER NINE

[1] John E. Allen, "SNAP-50/SPUR Power Plant Requirements Report," Pratt and Whitney Aircraft, Middletown, Conn., *Report No. CNLN-6291*, 28 May 1965.

[2] "Design No.5 of SNAP-50/SPUR Reactor Test Systems," Pratt and Whitney Aircraft, Middletown, Conn., *Report No. CNLM- 5734*,9 July 1964.

[3] L. M. Raring, "Fuel and Materials Highlights: SNAP-50 and Advanced Programs," Pratt and Whitney Aircraft, Middletown, Conn., *Report No. CNLM-6155*, February 1965.

[4] J. A. Heller, T. A. Moss, and G. J. Barna, "Study of a 300- Kilowatt Rankine-Cycle Advanced Nuclear-Electric SpacePower System," *NASA Technical Memorandum, NASA TM X-1919*, November 1969.

[5] Ibid J. A. Heller, et al.

[6] H. J. Branch, "Preliminary Design of the 2 MWth Reactor and Shield (PWAR-20) for the SNAP-50/SPUR Power Plant," Pratt and Whitney Aircraft, Middletown, Conn., *PW AC-445*, 30 December 1964.

[7] Ibid L. M. Raring

[8] J. R. Peterson et al., "Status of Advanced Rankine Power Conversion Technology," *17th Annual Meeting of American Nuclear Society*, Boston, Mass., 13-17 June 1971.

[9] A. P. Fraas, "Summary of the MPRE Design and Development Program," *Oak Ridge National Laboratory Report, ORNL- 4048*, June 1967.

[10] General Electric Co., "710 High-Temperature Gas Reactor Program Summary Report," Cincinnati, Ohio, 1968.

[11] John H. Pitts and C. E. Walters, "Conceptual Design of a 2 MWt (375 kWe) Nuclear-Electric Space Power System," *Journal of Spacecraft and Rockets, Vol. 7, No. II*, November 1970, pp. 1282-1286.

[12] Lawrence Radiation Laboratory, "Advanced Space Nuclear Power Programs Progress Report No. 5 (October through December 1966)," *LRL Report C-92b, M-3679*, 25 April 1967.

[13] John H. Pitts and C. E. Walters, "Conceptual Design of a 10-MWe Nuclear Rankine System for Space Power," *J. Spacecraft, Vol. 7, No.3*, March 1970, pp. 259-265.

[14] Lawrence Radiation Laboratory, "Quarterly Report, Advanced Space Nuclear Power Program, January Through March 1968," *LRL Report UCRL-50004-68-1*.

[15] "SP-100 Conceptual Design Description, No. 1" *Jet Propulsion Laboratory*, Pasadena, Calif., June 1982.

[16] Michael A. Merrigan, "Space Nuclear Power Heat Pipe Technology Issues Revisited," *A Critical Review of Pace Nuclear Power and Propulsion 1984-1993*, Editor Mohamed S. El-Genk, 1994 American Institute of Physics, L.C. Catalog Card No. 94-70780, 1994, pp. 167-178.

[17] Michael G. Houts, David I. Poston, William J. Emrich, Jr., " Heat Pipe Power System and Heat Pipe Bimodal System Development Status," *CP420, Space Technology and applications International Forum-1998*, edited by Mohamed S. El-Genk, 1998 The American Institute of Physics 1-56396-747-1/98, pp. 1189-1196.

[18] Daniel R. Koenig, "Heat Pipe Reactor Designs for Space Power," *Space Nuclear Power Systems 1984*, Edited by M. S. El-Genk and M. D. Hoover, Orbit Book Co., Malabar, FL., 1985,, pp. 217-228.

[19] J. A. Angelo, Jr. and D. Buden (1985) *Space Nuclear Power*, Orbit Book Company, Malabar, FL.

[20] Ibid J. A. Angelo, Jr. and D. Buden

[21] Schmidt, J. E., J. F. Wett, and J. W. H. Chi (1988) "The NERVA Derivative Reactor and a Systematic Approach to Multiple Space Power Requirements," in *Space Nuclear Power Systems 1988*, Chapter 19, Orbit Book Co., Melabar, FL, 95-100.

[22] Chi, 1. W. H. and W. L. Pierce (1990) "NERVA Derivative Reactors and Space Electric Propulsion Systems," in *Proceedings Seventh Symposium on Space Nuclear Power Systems*, held in Albuquerque, NM, Jan. 1990, 1:208-213

[23] Powell, J. R. and F. L. Horn (1985) "High Power Density Reactors Based on Direct Cooled Particle Beds," in *Space Nuclear Power Systems 1985*, Chapter 39, Orbit Book Co., Melabar, FL, 319-329.

[24] Fred G. Kennedy and Vlad Vanek, "Summary of the Bimodal Satellite Mission Study," *CONF 940101 American Institute of Physics*, 1994, pp. 1385-1390.

[25] Robert Zubrin and Jack Mondt, " An Examination of Bimodal Nuclear Power and Propulsion Benefits for Outer Solar System Missions," *CONF 960109, 1996 American Institute of Physics*, 1996, pp. 1495-1504.

[26] John J. Buksa, Scott DeMuth, Todd Huber, "Assessment of an SP-100 Bi-Modal Propulsion and Power System," *CONF 940101, 1994 American Institute of Physics*, 1994, pp. 1391-1400.

[27] Edgar Y. Choueiri, Arnold J. Kelly and Robert G. Jahn, "Mass Savings Domain of Plasma propulsion for LEO to GEO Transfer," in *Proc. 10th Symposium on Space Nuclear Power Systems, CONF 930103, American Institute of Physics*, 1993, pp. 1187-1196.

[28] Interavia Space Directory, edited by Andrew Wilson, Biddles Ltd. Guildford, UK, 1991-92.

[29] William D. Otting, et al, "Bimodal Power and Propulsion Mission Benefits and Candidate Designs," *CONF 940101, 1994 American Institute of Physics*, 1994, pp. 1401-1405.

[30] Nikolai N. Ponomarev-Stepnoi, "Conceptual Design of the Bimodal Nuclear Power and Propulsion System Based on the "TOPAZ-2" Type Thermionic Reactor Converter with the Modernized Single-Cell Thermionic Fuel Elements," *CONF 950110, 1995 American Institute of Physics*, 1995, pp. 755-760.

[31] Ibid William D. Otting.

[32] Michael G. Houts, et al, "ALERT-Derivative Bimodal Space Power and Propulsion Systems," *CONF 950110, 1995 American Institute of Physics*, pp. 733-740.

[33] Abraham Weitzberg and John W. Warren, "Pin and Cermet Hybrid Bimodal Reactor," *CONF 950110, 1995 American Institute of Physics*, 1995, pp. 741-746.

[34] Donald W. Culver and Melvin J. Bulman, "Multimodal Space Nuclear Thermal Propulsion and Power System for Space Industrialization," *CONF 950110, 1995 American Institute of Physics*, 1995 pp. 747-754.

[35] David Buden, et al, "Bimodal Technology Readiness Assessment," *CONF 940101, 1994 American Institute of Physics*, 1994, pp. 1379-1383.

CHAPTER TEN

[1] R. Bruce Mathews, et al., "Fuels for Space Nuclear Power and Propulsion: 1983-1993," in *A Critical Review of Space Nuclear Power and Propulsion 1984-1993*, Editor Mohamed S. El-Genk, 1194 American Institute of Physics, L.C. Catalog Card No. 94-70780, 1994, pp. 179-220.

[2] E. K. Storms, "An Equation Which Describes Fission Gas Release from UN Reactor Fuel." *J. Nuclear Materials*, 1988, p. 158.

[3] Ibid, R. Bruce Mathews.

[4] T. Roth, "LIFE-4 Programmer's Manual," *Ward-94000-12, Vol. 2*, Rev. 0, 1982

[5] "Prometheus Project Reactor Module Final Report, for Naval Reactors Information," *Knolls Atomic Power Laboratory Report No. SPP-67110-0008*.

[6] D. C. Noe, K. B. Gibbard, M. H. Krohn, "Fuel System Compatibility Issues for Prometheus-1," *Bechtel Bettis, Inc. Report No. B-MT(SRME)-55*, dated Jan. 20, 2006.

[7] Dora M. Stefanescu, Peter a. Curreri, and Subhayu Sen, "Molten Materials Transfer and Handling on the Lunar Surface," *CP969, Space Technology and Applications International Forum--STAIF 2008*, edited by M. S. El-Genk, 2008 American Institute of Physics, 978-0-7354-0486-1 08, 2008, pp. 162-169.

[8] NASA facts, "Fission Surface Power System Technology for NASA Exploration Missions," Jan. 2008.

[9] Paul D. Spudis, "The Once and Future Moon," Smithsonian Institution Press, Washington and London, *ISBB 1-56098-634-4*, 1996.

[10] Robert Zubrin, "Entering Space: Creating a Spacefaring Civilization," Jeremy P. Tarcher/Putnam, New York, *ISBN 0-87477-975-8*, 1999.

[11] Ibid Zubrin

[12] Report of the Synthesis Group on America's Space Exploration Initiative, "America At The Threshold," May 6, 1991.

[13] Ibid Zubrin